Methods of Detection and Identification of Bacteria

Author:

Brij M. Mitruka, M.S., Ph.D.

Director, BHP Clinical Laboratory
Westchester, Pennsylvania
and
Associate Professor of Microbiology
Temple University School of Dentistry
Philadelphia, Pennsylvania

With Special Assistance From:

Mary J. Bonner, M.S., Ph.D.

Department of Research Medicine
University of Pennsylvania,
School of Medicine
Philadelphia, Pennsylvania

CRC Press
Taylor & Francis Group
Boca Raton London New York

CRC Press is an imprint of the
Taylor & Francis Group, an **informa** business

First published 1976 by CRC Press
Taylor & Francis Group
6000 Broken Sound Parkway NW, Suite 300
Boca Raton, FL 33487-2742

Reissued 2018 by CRC Press

© 1976 by Taylor & Francis
CRC Press is an imprint of Taylor & Francis Group, an Informa business

No claim to original U.S. Government works

A Library of Congress record exists under LC control number: 76028809

Publisher's Note
The publisher has gone to great lengths to ensure the quality of this reprint but points out that some imperfections in the original copies may be apparent.

Disclaimer
The publisher has made every effort to trace copyright holders and welcomes correspondence from those they have been unable to contact.

ISBN 13: 978-1-138-55056-8 (hbk)
ISBN 13: 978-1-138-56085-7 (pbk)
ISBN 13: 978-0-203-71134-7 (ebk)

Visit the Taylor & Francis Web site at http://www.taylorandfrancis.com and the CRC Press Web site at http://www.crcpress.com

PREFACE

During the past decade, many new methods have been developed for the detection and identification of bacteria in clinical specimens and biological samples. The rapidly increasing work load in the laboratory (particularly in clinical microbiology), recent advances in electronic technology, and successful application of technological know-how in other clinical laboratories (such as clinical chemistry) have inspired many bacteriologists to evaluate the usefulness of newer methods for the identification of bacteria. Explorations of outer space coupled with a curiosity to determine the existence of life beyond the boundaries of the planet earth have also necessitated the invention of rapid, ultrasensitive, and specific methods for the determination of bacteria. The purpose of this book is to summarize the methods applicable to bacteriology, including conventional methods that are currently used as well as new approaches and automated methods in diagnostic, public health, and industrial bacteriology laboratories.

Most of the newer methods of detection and identification of bacteria utilize the basic properties of bacterial cells, such as their size, staining characteristics, colonial morphology, chemical constituents, biochemical reactions, and immunologic properties. Therefore, a comprehensive review of current methods of detection, enumeration, characterization, and identification of bacteria is presented. This is followed by a discussion of the newer techniques applied to bacterial identification, utilizing principles of morphologic, biochemical, and serological methods. The next section deals with automated methods, with a view towards their possible applications in the bacteriology laboratory. The last chapter includes a discussion on the use of computers for the identification of bacteria, storage and retrieval of taxonomic information, and data processing in the bacteriology laboratory. The application of computers in bacteriology may be useful in coping with the increasing work load and achieving greater precision and accuracy in methodology.

I hope this book will be useful for everyone involved in the bacteriology laboratory.

Brij M. Mitruka, M.S., Ph.D.
Philadelphia, Pennsylvania
July 12, 1976

THE AUTHOR

Brij M. Mitruka, M.S., Ph.D., is Director of the BHP Clinical Laboratory, Westchester, Pennsylvania and Associate Professor of Microbiology, Temple University School of Dentistry, Philadelphia, Pennsylvania. He was formerly Associate Professor, Department of Research Medicine, University of Pennsylvania, Philadelphia and Professor and Head, Department of Microbiology, Punjab Agricultural University, Ludhiana, India.

Dr. Mitruka graduated in 1955 from Dungar College, Bikaner, Rajasthan, India with an I.Sc. degree and obtained his B.V.Sc. and A.H. degrees in 1959 from Rajasthan Veterinary College, Bikaner. He received his M.S. and Ph.D. degrees in 1962 and 1965, respectively, from Michigan State University, East Lansing.

Dr. Mitruka is a member of the American Society of Microbiology, American Association of Clinical Chemists, American Association for the Advancement of Science, American Veterinary Medical Association, and the New York Academy of Sciences.

Dr. Mitruka has published more than 40 research papers and he is the author of *Gas Chromatographic Applications in Microbiology, Medicine and Animals for Medical Research: Models for the Study of Human Disease,* and *Clinical, Biochemical, and Hematological Reference Values in Normal Experimental Animals and Humans.* He was the principal investigator of a grant from the National Institute of General Medical Sciences entitled *Gas Chromatography as an Aid in Clinical Pathology.* He is currently working on GC-MS applications in the study of oral microorganisms and development of specific diagnostic methods including immuno globulins, antibiotic sensitivity testing, and isozymes.

for Devyani

ACKNOWLEDGMENT

I would like to thank Dr. Howard M. Rawnsley and his associates for reviewing the entire manuscript. Thanks are due to Dr. Mary Jo Bonner for helping at various stages of the preparation of the manuscript. I thank Joann Doven and Hazel Harber for excellent typing and proofreading and the Art Department of CRC Press, Inc. for preparing the illustrations.

The courtesy of many authors and publishers who promptly granted their permission to utilize previously published material is gratefully acknowledged. Sources are acknowledged in the legends of figures and tables.

The cooperation of the staff of CRC Press, Inc., especially Gerald A. Becker, Gayle Tavens, Peg Latham, and Sandy Pearlman, is also acknowledged gratefully.

The work included in this book was supported in part by Research Grants 5 R01 and GM 21443 from the National Institute of General Medical Sciences, U.S. Public Health Service and in part by funds from the Connecticut Lung Association.

B. M. M.

TABLE OF CONTENTS

Chapter 1

INTRODUCTION

Mary J. Bonner and Brij M. Mitruka

DIFFERENTIATION OF PROCARYOTIC AND EUCARYOTIC CELLS

Concepts proffered in Darwin's *Origin of the Species*[1] provided the framework for the description of a third kingdom to serve as the transition form between the plant and animal kingdoms. Haeckel[2] termed this kingdom the Protista. Chatton[3] recognized two general patterns of cellular organization in Protista, the eucaryotes (Greek, true nucleus), which include protozoa, fungi, and most algae, and the procaryotes, which include all bacteria and the small group of blue-green algae (Cyanophyceae). While the original division of Protista was based on the simple organization of the procaryotic bacteria, the electron microscope has revealed a fundamental division based on complexity of organization.

Bacteria form a heterogeneous group of procaryotic cells anatomically, physiologically, and biochemically. In general, the main distinguishing features of the procaryotic cells (bacteria and related organisms such as rickettsiae, chlamydiae, and mycoplasmas) are as follows.

1. Their nucleus is a simple homogeneous body without a nuclear membrane separating it from the cytoplasm. They lack nucleolus, a spindle, and a number of separate nonidentical chromosomes.

2. Procaryotes lack the internal membranes isolating the respiratory and photosynthetic enzyme systems in specific organelles comparable to the membrane-bound mitochondria and the chloroplasts of eucaryotic cells. Thus, the respiratory enzymes in bacteria are located mainly in the peripheral cytoplasmic membrane, and their effective functioning is dependent on the integrity of the cell protoplast as a whole.

3. Procaryotes have a rigid cell wall structure containing a specific mucopeptide not found in eucaryotic cells.

4. Morphologically, bacteria are characterized by their small cell size (usually between 0.4 and 1.5 μm in short diameter); characteristic shape, which may be spherical (coccus), rod shaped (bacillus), comma shaped (vibrio), spiral (spirillum and spirochete), or filamentous (actinomyces); and arrangement, such as clusters, chains, rods, filaments, or mycelia.

5. Although unicellular bacteria may grow attached to one another so as to appear multicellular (in clusters, chains, rods, etc.), each cell is physiologically self-sufficient and, if isolated artificially, able to nourish itself, grow, and reproduce the species by binary fission.

6. Another distinctive feature of the procaryotic cell is that its ribosomes are small[4,5] (10 to 20 nm) with a sedimentation constant of 70S as compared to the larger 80S ribosome of eucaryotes. The 70S ribosome is composed of 30S and 50S subunits, whereas the 80S ribosome of the eucaryotic cell is composed of 40S and 60S subunits.

7. The basic chemical composition of all microorganisms is essentially similar, i.e., made up of compounds of lower molecular weight which are about 10% of dry weight in procaryotes and 15% in eucaryotes. Protein, nucleic acid, and lipid contents are also slightly lower in procaryotic cells as compared to eucaryotic cells.

The procaryotic bacteria show a considerably narrow range of structural and biochemical variations as compared to those in eucaryotic cells. Thus, evolutionary specialization among bacteria is expressed in metabolic rather than structural terms. However, there is great metabolic variation among bacteria; for example, representatives of all four primary nutritional categories (photoautotrophs, photoheterotrophs, chemoautotrophs, and chemoheterotrophs) occur in bacteria. Almost every type of metabolic activity present in eucaryotic cells can be found in bacteria. The physiological diversity of bacteria, which explains both their numerical abundance and their ubiquity throughout the biosphere, is reflected in the role they play as agents in the cycles of carbon, nitrogen, sulfur, oxygen, and phosphorus.

Comparative cytology using the light microscope and classical staining methods, extension of comparative cytology to the ultrastructural level with the electron microscope (EM), and advances in biochemical techniques have established the uniqueness of bacteria.[6,7] By using these three basic experimental approaches, it was recognized that bacteria have a wide range of anaerobic energy-yielding reactions, synthesize unique cell wall polymers (except for the "L"-form *Mycoplasma* and those without walls and exceptional bacteria such as *Halobacterium*), fix nitrogen, and accumulate poly-β-hydroxybutyrate as a reserve material. Such properties are virtually or completely absent from eucaryotes.

The cytological recognition of procaryotic organization supported by biochemical and physiological data provided the foundation for a coherent, systematic view of the nature of bacteria such as that described in *Bergey's Manual of Determinative Bacteriology.*[14]

TAXONOMY OF BACTERIA

Bacteria are usually classified according to the Linnean binomial scheme of genus and species, although many kinds of systematic compilations are possible. Detailed classification of bacteria utilizes results of cytological, biochemical, and physiological tests as well as specialized techniques of strain identification (such as serotyping, biotyping, bacteriophage typing, and genetic analyses including measurement of guanine and cytosine (GC) contents and DNA-DNA homology analysis).[7-13]

Phylogenetic Scheme

The most widely used system of classification and nomenclature in the United States is *Bergey's Manual of Determinative Bacteriology.*[14] The first edition appeared in 1923 and the eighth edition in 1974. Bacteria are placed into the following 19 parts:

Kingdom — Procaryotae
    Division I — Cyanobacteria
    Division II — Bacteria
        Part 1 — Phototrophic bacteria
        Part 2 — Gliding bacteria
        Part 3 — Sheathed bacteria
        Part 4 — Budding and/or appendaged bacteria
        Part 5 — Spirochetes
        Part 6 — Spiral and curved bacteria

Part 7 — Gram-negative aerobic rods and cocci
        Part 8 — Gram-negative facultatively anaerobic rods
        Part 9 — Gram-negative anaerobic bacteria
        Part 10 — Gram-negative cocci and coccobacilli
        Part 11 — Gram-negative anaerobic cocci
        Part 12 — Gram-negative chemolithotrophic bacteria
        Part 13 — Methane-producing bacteria
        Part 14 — Gram-positive cocci
        Part 15 — Endospore-forming rods and cocci
        Part 16 — Gram-positive, asporogenous, rod-shaped bacteria
        Part 17 — Actinomycetes and related organisms
        Part 18 — Rickettsias
        Part 19 — Mycoplasmas

Each part is also divided, where appropriate, into classes, orders, families, tribes, genera, and species.

In the older phylogenetic scheme of classification, the class Schizomycetaceae included all true bacteria, filamentous bacteria (including mycobacteria), spirochetes, and mycoplasmas. For example, in this scheme *Escherichia coli* was placed as follows:

Kingdom — Monera
    Class (-aceae) — Schizomycetaceae
        Order (-ales) — Eubacteriales
            Family (-aceae) — Enterobacteriaceae
                Genus — *Escherichia*
                Species — *coli*

However, since the publication of *Bergey's* eighth edition, *Escherichia coli* is placed as follows:

Kingdom — Procaryotae
    Division II — Bacteria
        Part 8 — Gram-negative facultatively anaerobic rods
            Family I — Enterobacteriaceae
                Genus I — *Escherichia*
                    Species — *coli*

As specified in the International Code of

Nomenclature of Bacteria,[15] scientific names of all taxa, or those words used to refer to any taxonomic group, are either to be taken from Latin or are Latinized if taken from other languages. The major categories of taxa are

Category	Name ending
Individual	
Species	
Series	
Section	
Genus	
Tribe	-eae
Family	-aceae
Order	-ales
Class	
Division	

A nomenclatural type is recognized as the constituent element of a taxon to which the name of the taxon is permanently attached, for example, the type species of a genus. Provisions are made for changes in rules and for rule interpretation through the International Committee of Nomenclature of Bacteria and its Judicial Commission. These bodies are organized by the International Association of Microbiological Societies.

Well-known, trivial, or commonly used names (such as "tubercle bacillus" for *Mycobacterium tuberculosis* and "typhoid bacillus" for *Salmonella typhi*) appear frequently in medical literature. In the phylogenetic classification, the higher groupings of order and family are distinguished by characteristics such as cell shape, Gram reaction, spore formation, and flagellation, whereas genus and species are distinguished by characteristics such as fermentation reactions, nutritional requirements, and pathogenicity. The fundamental weakness of such a system is the arbitrary importance attached to the distinguishing characteristics. Other schemes for classification are currently being developed, usually to meet specific requirements, and some of these are described in the following sections.

Genetic Basis of Classification

There are two approaches to classification on a genetic basis. One approach is based on the nucleic acid homology of bacteria; the other approach groups bacteria together according to the biochemical manifestation of gene-controlled stable metabolic patterns, cell polymers, and organelle structures. Surface polymers (including capsules, teichoic acids, and O antigens) can be used for comparison of bacterial relatedness.

In the genetic approaches to bacterial taxonomy, certain techniques give insight into the genotypic properties of bacteria, thereby complementing the hitherto exclusively phenotypic characterization of these organisms. Two different kinds of analyses performed on isolated nucleic acids (the analysis of the base composition of DNA[16-18] and the study of chemical hybridization between DNA and DNA or DNA and RNA) furnish information about the genotype.[19-21] The genetic information encoded in the DNA base sequence changes in different environments by processes such as mutation, recombination, transduction, and selection. The genes of the organisms change in size, nucleotide base composition, and nucleotide base sequence. Base composition has been shown by both chemical and physiochemical methods to be constant and characteristic in each organism (Table 1.1). Base composition is generally expressed in terms of the mole fraction of guanine plus cytosine (G + C/G + C + A + T) expressed as a percentage. The G + C ratio or DNA base composition may be determined by several methods, one of which is denaturation or melting temperature of DNA. An equation relating G + C content and melting temperature has been derived by Marmur and Doty.[16] Values of GC content for different organisms vary from 30 to 75%; this reflects the differences in the amino acid composition of the cellular proteins of different organisms, some having a preponderance coded by A + T-rich triplets and others by G + C-rich triplets. A substantial divergence between two organisms with respect to mean DNA base composition reflects a large number of individual differences between the specific base sequences of their respective DNAs. This in turn indicates a major genetic divergence and, hence, a wide evolutionary separation. However, two organisms with identical mean DNA base composition may differ greatly in genetic constitution. Although the values for the species in a particular genus differ, the total for most bacterial genera is fairly narrow and can be considered a useful character for definition of a bacterial genus. Organisms that are clearly related will have close G + C ratios, but the converse does not necessarily hold true. Data concerning G + C ratios tend to confirm existing classifications. Base ratio data also point out the organisms that are not adequately differentiated by present laboratory tests.

TABLE 1.1

Nucleotide Base Composition of the DNA of Various Bacteria

Organism	% GC
Spirillum linum	28–30
Clostridium perfringens, C. tetani, Fusobacterium fusiforme	30–32
Staphylococcus aureus	32–34
Bacillus alvei, B. cereus, B. anthracis, B. thuringiensis, Clostridium kluyverii, Mycoplasma gallisepticum, Streptococcus faecalis, Treponema pallidum, Veillonella parvula	34–36
Diplococcus pneumoniae, Streptococcus salivarius, Streptococcus pyogenes, S. bovis, S. viridans, Leuconostoc mesenteroides, Lactobacillus acidophilus, Listeria monocytogenes, Proteus vulgaris, P. mirabilis, P. rettgeri, Haemophilus suis, H. influenzae, H. parainfluenzae, H. aegypti, Bacillus megaterium	38–40
Moraxella bovis, Neisseria catarrhalis, Leptospira biflexa	40–42
Bacillus subtilis, B. polymyxa, B. stearothermophilus, Coxiella burnetii, Vibrio metschnikovii	42–44
Vibrio comma, Pasteurella pestis, Corynebacterium acnes	46–48
Neisseria perflava, N. gonorrhoeae, N. flava	48–50
Neisseria meningitidis, N. sicca, N. subflava, Proteus morganii, Escherichia coli, Citrobacter freundii, Shigella dysenteriae, Salmonella typhosa, S. typhimurium, S. enteritidis, S. arizona, S. ballerup	50–52
Klebsiella aerogenes, Corynebacterium diphtheriae	52–54
Klebsiella pneumoniae, K. rhinoscleromatis, Azotobacter agilis, Alcaligenes faecalis, Brucella abortus, Nitrosomonas sp.	54–56
Azotobacter vinelandii, Lactobacillus bifidis	56–58
Agrobacterium tumefaciens, Serratia marcescens	58–60
Pseudomonas fluorescens, Rhizobium japonicum, Rhodospirillum rubrum	60–62
Xanthomonas phaseoli	62–64
Desulfovibrio desulfuricans	64–66
Pseudomonas aeruginosa, Mycobacterium phlei, M. tuberculosis	66–68
Mycobacterium smegmatis, Micrococcus lysodeikticus, Nocardia sp.	70–80

Marmur et al.[10] arranged individual organisms into groups on the basis of the homology of their DNA base sequences. The basis of DNA homology classification is the fact that double strands re-form from separated strands during controlled cooling of a heated preparation of DNA. This "annealing" process can be readily demonstrated with suitably heated homologous DNA extracted from a single species, but it can also occur when a mixture of DNA from two related species is used; in the latter case, hybrid pairs of DNA strands are produced. These hybrid pairings frequently occur between complementary regions of two bits of DNA, and the degree of hybridization can be assessed if labeled DNA preparations are used. mRNA binding studies can also give information to complement these observations providing genetic evidence of relatedness among bacteria. Measurement of DNA homology has recently been quantified by several procedures which determine the extent of formation of molecular hybrids from two DNA strands of different origin. The techniques of nucleic acid hybridization show relatedness between different bacterial strains (Table 1.2).

Biochemical Basis of Classification

Metabolic reactions and the chemical composition of the cell wall are characteristic of various bacterial species. Although the mucopeptide framework (repeating units of N-acetylglucosamine and N-acetylmuramic acid cross-linked by peptide structures) in the cell wall of bacteria is common to all, there are structural differences between Gram-positive and Gram-negative bacteria. Individual differences of taxonomic value in chemical constituents of bacteria are also present; for example, Gram-positive bacteria contain considerable amounts of glycerol teichoic

TABLE 1.2

DNA Homologies among Certain Bacteria

DNA source	% relatedness
	to *E. coli*
Escherichia coli	100
Shigella dysenteriae	71
Aerobacter aerogenes	45
Salmonella typhimurium	35
Proteus vulgaris	14
Serratia marcescens	7
Pseudomonas aeruginosa	1
Bacillus subtilis	1
Brucella neotomae	0
	to *N. meningitidis*
Neisseria meningitidis	100
Neisseria gonorrhoeae	80
Neisseria perflava	55
Neisseria subflava	48
Neisseria catarrhalis	15
Neisseria caviae	10
Mima sp.	5
Herellea	5
Escherichia coli	3
Monkey kidney	0.1

Data from Brenner, D. J., Martin, M. A., and Hoyer, B. H., *J. Bacteriol.*, 94, 486, 1967; McCarthy, B. J. and Bolton, E. T., *Proc. Natl. Acad. Sci. U.S.A.*, 50, 156, 1963; Kingsbury, D. T., *J. Bacteriol.*, 94, 870, 1967.

acid and ribitol teichoic acid. Also found are a wide variety of monosaccharides that are, by and large, group specific, for example, rhamnose for *Streptococcus, Lactobacillus,* and *Clostridium* and arabinose for *Corynebacterium, Mycobacterium,* and *Nocardia.*

Gram-negative bacteria, on the other hand, contain a group-specific polysaccharide composed of glucose, galactose, N-acetylglucosamine, 2-keto-3-deoxyoctonate, and a haptose moiety together with a highly specific side chain bearing a type-specific arrangement of sugar. Various chromatographic and spectroscopic methods are employed to analyze chemical constituents of bacteria in order to distinguish genera, species, and strains based on their biochemical properties (see Chapter 3).

Numerical Taxonomy or Adansonian System

Because of the difficulties in constructing phylogenetic classifications which are based on a few arbitrarily weighted characteristics, descriptive taxonomy has been revived for many strains in the form of computerized numerical comparisons of large numbers of diagnostic features. Computer taxonomy has been developed for groups of bacteria in which large numbers of strains exist. They are described in terms of 100 or more clear-cut taxonomic properties such as presence or absence of certain enzymes and presence or absence of certain morphologic structures.[22,23] Punch cards are prepared for each strain. The computer compares the cards and prints out a list of the strains so that each strain is followed by the strain with which it shares the most characteristics. When this is done, the list often reveals several broad subgroups or strains, with each subgroup characterized by a large number of shared characteristics (see Chapter 6). The median strain within each subgroup can then be considered a type species.

Based on two main criteria, the mechanism of motility and the character of the cell wall, bacteria can be classified into four major groups.

1. Eubacteria, which have thick rigid cell walls. Motility, when present, is by means of flagella.

2. Spirochetes, which have thick rigid cell walls. Motility is by means of contraction of an axial filament wound helically about the cell and anchored at both ends.

3. Gliding bacteria, some of which have thick rigid walls and others with thin flexible walls. Although the organelles of motility have not been detected in these organisms, they move smoothly along solid surfaces by means of an unknown mechanism.

4. Mycoplasma, which do not have cell walls but do have a triple-layered unit membrane and are highly pleomorphic.

Artificial Classification or Simplified Working System

In this scheme, descriptive properties are arranged so that an organism may be readily identified. Organisms are grouped together in a "key;" they are not necessarily related phylogenetically, but are listed together because they share certain easily recognizable characteristics. For example, it is reasonable to include a group of organisms which form red pigments in the "key" type of classification even though this would necessitate the inclusion of such unrelated forms as *Serratia marcescens* and purple sulfur bacteria. Another approach to the classification of bacteria is to avoid the conflicting arguments of the taxonomists and use a simplified working system such as that used by medical bacteriologists (Table 1.3).

TABLE 1.3

Bacteria of Medical Importance

Part 5
  Spirochetes
    Order I. Spirochaetales
      Family I. Spriochaetaceae
        Genus III. *Treponema*
        Genus IV. *Borrelia*
        Genus V. *Leptospira*

Slender, flexuous, helically coiled, unicellular, 3–500 μm in length; multiplication by transverse fission; motility may be rapid rotation about long axis of helix, flexion, or serpentine; no endospores; larger cells Gram negative; may have inclusions; aerobic, facultatively anaerobic, or anaerobic; chemoheterotrophic; free living, commensal, or parasitic

Part 7[a]
  Gram-negative aerobic rods and cocci
    Family I. Pseudomonodaceae
      Genus I. *Pseudomonas*

Family I – Straight or curved rods, motile by polar flagella, Gram negative, metabolism respiratory, never fermentative, strict aerobes, catalase +; oxidase usually +; growth 4 to 43°C

Part 8[b]
  Gram-negative facultatively anaerobic rods
    Family I. Enterobacteriaceae
      Genus I. *Escherichia*
      Genus II. *Edwardsiella*
      Genus III. *Citrobacter*
      Genus IV. *Salmonella*
      Genus V. *Shigella*
      Genus VI. *Klebsiella*
      Genus VII. *Enterobacter*
      Genus VIII. *Hafnia*
      Genus IX. *Serratia*
      Genus X. *Proteus*
      Genus XI. *Yersinia*
      Genus XII. *Erwinia*
    Family II. Vibrionaceae
      Genus I. *Vibrio*
      Genus II. *Aeromonas*

Family I – Small Gram-negative rods; motile by peritrichate flagella or nonmotile; capsulated or noncapsulated; no spore formation; not acid fast; aerobic and facultatively anaerobic; chemoorganotrophic; metabolism respiratory and fermentative; acid produced from carbohydrates and alcohols; catalase + except one serotype of *Shigella*; nitrates reduced to nitrites except some strains of *Erwinia*; type genus *Escherichia*, Castellani and Chalmers, 1919, 941

Family II – Rigid Gram-negative rods, straight or curved; motile by polar flagella and some lateral flagella; chemoorganotrophs, metabolism both fermentative and respiratory; oxidase positive; facultative anaerobes; type genus *Vibrio*, Pacini 1854, 411

[a]Genera of uncertain affiliation include *Alcaligenes, Acetobacter, Brucella, Bordetella,* and *Francisella.*
[b]Genera of uncertain affiliation include *Haemophilus* and *Pasteurella.*

TABLE 1.3 (continued)

Bacteria of Medical Importance

Part 9
 Gram-negative anaerobic bacteria
  Family I. Bacteroidaceae
   Genus I. *Bacteroides*
   Genus II. *Fusobacterium*
   Genus III. *Leptotrichia*

Gram negative; uniform or pleomorphic rods; nonmotile or motile with peritrichous flagella; no spore formation; chemoorganotrophs; obligate anaerobes

Part 10
 Gram-negative cocci and coccobacilli
  Family I. Neisseriaceae
   Genus I. *Neisseria*
   Genus II. *Moraxella*

Spherical in pairs or masses with adjacent sides flattened, may also be rod shaped in pairs or short chains; no flagellus; some twitching motility; Gram negative; most species catalase + cytochrome oxidase +; aerobic; some species have complex growth requirements following isolation and may later grow in simple defined media; type genus *Neisseria*, Trevisan 1885, 105

Part 11
 Gram-negative anaerobic cocci
  Family I. Veillonellaceae
   Genus I. *Veillonella*

Cocci with varying diameter (0.3–2.5 μm); pairs, single cells, chains, or masses; no endospores; no flagellum, nonmotile; Gram negative but resist decolorization; Chemoorganotrophic; complex nutritional requirements; anaerobic; cytochrome oxidase –; catalase –; some strains have pseudocatalase; type genus *Veillonella*, Privot 1933, 118

Part 14
 Gram-positive cocci
  a. Aerobic and/or facultatively anaerobic
  Family I. Micrococcaceae
   Genus I. *Micrococcus*
   Genus II. *Staphylococcus*
   Genus III. *Planococcus*

Family I – Cells spherical; 0.5 μm in diameter; divide in more than one plane, yielding packets; motile or nonmotile; Gram positive; no endospores, chemoorganotrophs; metabolism respiratory or fermentative; acid but no gas produced from glucose when utilized; catalase +; type genus *Micrococcus*, Cohn 1972, 151

  Family II. Streptococcaceae
   Genus I. *Streptococcus*
   Genus II. *Leuconostoc*

Family II – Cells spherical or ovoid, chains, or tetrads; rarely motile; no endospores; Gram positive; chemoorganotrophs; metabolism fermentative; lactic, acetic, formic acids, ethanol, CO_2 produced from carbohydrates; catalase variable, facultatively anaerobic; type genus *Streptococcus*, Rosenback 1884, 22

  b. Anaerobic
  Family III. Peptococcoceae
   Genus I. *Peptococcus*
   Genus II. *Peptostreptococcus*
   Genus IV. *Sarcina*

Family III – Cocci (0.5–2.5 μm) singly, pairs, tetrads, masses, cubic packets; no flagellum; no endospores; nonmotile; Gram positive; chemoorganotrophic; carbohydrate fermentation ±

Part 15
 Endospore-forming rods and cocci
  Family I. Bacillaceae
   Genus I. *Bacillus*
   Genus III. *Clostridium*

Rod-shaped cells except in one genus, spherical endospores; mostly Gram positive; motile by lateral or peritrichous flagella or nonmotile; aerobic, facultative, or anaerobic

TABLE 1.3 (continued)

Bacteria of Medical Importance

Part 16[c]
Gram-positive, asporogenous rod-shaped bacteria
  Family I. Lactobacillaceae
    Genus I. *Lactobacillus*
	Straight or curved rods, single or in chains;
	rarely motile; Gram positive; anaerobic or
	facultative; marked lactate production;
	catalase –

Part 17
Actinomycetes and related organisms
  Coryneform group of bacteria
    Genus I. *Corynebacterium*
	Straight to slightly curved rods;
	division yields palisade arrangements of
	cells; usually nonmotile; Gram positive; not
	acid fast; chemoorganotrophs; carbohydrate
	metabolism; mixed fermentative and respira-
	tory; aerobic and facultatively anaerobic;
	catalase +; type species
	Corynebacterium diphtheriae
	(Kruse), Lehmann and Neumann 1896, 350

Part 18
Rickettsias
  Order I. Rickettsiales
    Family I. Rickettsiaceae
      Tribe I. Rickettsiae
        Genus I. *Rickettsia*
        Genus II. *Coxiella*
	Usually rod shaped, coccoid, pleomorphic;
	no flagellum; typical bacterial cell walls;
	Gram negative; multiply only inside
	host cells

Part 19
Mycoplasmas
  Class I. Mollicutes
    Order I. Mycoplasmatales
      Family I. Mycoplasmataceae
        Genus I. *Mycoplasma*
	Small cells sometimes 200 nm, pleomorphic;
	coccoid to filamentous; bounded single
	triple-layered nonmotile; Gram negative;
	minute colonies; grow into media; most
	species completely resistant to penicillin

[c]Genera of uncertain affiliation include *Listeria* and *Erysipelothrix.*

Adapted from Stonier et al.,[30] Joklik and Smith,[31] Cruickshank et al.,[32] Davis et al.,[33] and Buchanan and Gibbons.[14]

DETECTION, ENUMERATION, AND IDENTIFICATION OF BACTERIA

Many methods of detection, enumeration, and identification of bacteria have been established and are routinely used in diagnostic bacteriology in hospitals; public health laboratories; and dairy, food, and soil microbiology laboratories. If the concentration of bacteria in a given sample is high enough and if sufficient time is available for a detailed examination, the detection, specific identification, and quantitative assessment of the bacteria does not present much of a problem to the bacteriologist. However, immediate or rapid detection or specific identification of a small number of bacteria in a sample is much more difficult and requires specially developed methods.[24] Recent trends in the development of newer and better methods of bacterial detection

and identification exploit many of the known chemical, physical, and physiological charac-teristics of bacteria. Some of these methods are listed below.

1. Methods of detection and enumeration of bacteria

    A. Selective media, preformed media, and reagents (for the determination of growth and metabolism)

    B. Staining methods

    C. Coulter Counter® (for particle counting)

    D. Partichrome analyzer (for the detec-tion and counting of microscopic particles)

    E. πMC system (for the detection of particles in a microscopic field)

    F. Membrane filtration

    G. Velocimeter (a sound velocity probe)

H. Luminol® chemiluminescence and firefly bioluminescence (ATP determination)

I. Colorimetric or spectrophotometric method (turbidity measurement)

J. Radiometric method (for the detection of radioactive CO_2)

K. Gas-chromatographic method (for the detection of chemical components or metabolic activity of bacteria)

2. Methods of identification and characterization of bacteria

A. Gas chromatography

B. Mass spectrometry and mass fragmentography

C. Autoanalyzer®

D. Electrophoresis (density gradient, zone, gel or disk method for the separation of specific macromolecules of bacteria)

E. Continuous particle electrophoresis (CCPE)

F. Countercurrent immunoelectrophoresis (CIE)

G. Densitometric scanning of antigen-antibody reaction

H. Radioimmunoassay (RIA)

I. Fluorescent antibody staining method (FAST)

J. Fluorescent enzyme staining technique (FEST)

These methods and other special techniques for the detection and identification of bacteria are described in detail in subsequent chapters.

Microorganisms are ubiquitous in the human environment and abound in an individual's intimate biosphere. Bacteria isolated directly (whether from humans, animals, plants, or the environment) are rarely obtained in pure culture. Specimens of feces, sputum, and throat secretions contain a variety of resident bacteria in addition to any pathogens that may be present. Blood, pus, cerebrospinal fluid, and urine are free from resident bacteria and are likely to be infected with only a single pathogen, unless there is a secondary infection present or contamination has occurred during the course of collection. Contamination of blood or cerebrospinal fluid is rare. However, significant bacteriuria requires 100,000 colonies or more per milliliter in a clean-voided midstream specimen. Any count in a suprapubic aspirate is significant, but recovery of skin contaminants such as diphtheroids and staphylococci may not be significant and may require repetition of the suprapubic aspirate to confirm or rule out their presence in urine. Significant bacteriuria is established with 80 to 90% accuracy by microscopic examination of a Gram-stained smear of a well-mixed, unspun urine or a wet preparation of unstained urinary sediment.[25-28]

The first step in bacterial identification is, of necessity, to obtain the organism in pure culture. Pure cultures of strains may be stored for later reference by freeze-drying and may be purchased from several culture-type collections. The *World Directory of Collections of Cultures of Microorganisms* provides information on 329 collections in 52 countries.[29] Knowledge of the source of the specimen and the conditions of isolation, together with morphologic observations and Gram-stain reaction, will narrow the identification to a small group of genera. Obvious separations (such as division into one of the four primary nutritional categories of chemolithotrophic, autotrophic, photosynthetic, and chemoheterotrophic bacteria) become apparent during the initial isolation procedures. The simple or true bacteria are routinely isolated in most bacteriological laboratories and are classified primarily by a combination of morphology and physiology with emphasis on oxygen requirement; this includes the manner of utilization of carbohydrates (particularly glucose), whether or not they possess certain enzymic activities (such as oxidase, catalase, and urease), whether or not certain end products (such as nitrites and H_2S) are produced, tests of amino acid metabolism, temperature range of growth, response to NaCl, and results of antibiotic susceptibility tests. In the case of bacteria suspected of being pathogenic, inoculation into plants or animals may be a necessary step in identification. Antibiotic sensitivity studies are of limited use at the species level. Notable exceptions are Optochin® sensitivity of the pneumococcus and bacitracin sensitivity of *Streptococcus pyogenes*. Determining the type of a strain within a species may be of epidemiological value in tracing the source of spread of infection. Typing of strains involves special biochemical or serological tests to distinguish biotypes or serotypes, respectively. Serotyping is particularly well developed for certain genera, for example, *Salmonella*. Phage typing, which depends on susceptibility to lysis by each of a set of type-specific lytic bacteriophages, is used in epidemiological studies of staphylococci.

Detection and identification methods for bacterial species of eubacteria are primarily dealt with in the following chapters, with occasional reference to the mycoplasma. Only a few bacterial species among the major groups of eubacteria are described in detail since it is beyond the scope of this book to give the methods of characterization and enumeration for all of the organisms in this group. Particular emphasis is given to medically important bacteria (such as pneumococci, streptococci, and mycobacteria) in order to demonstrate the extensive methodology involved in the detection, enumeration, and identification of bacteria in vitro and in vivo.

The objective of this book is to present a critical review and evaluation of the so-called conventional methods currently being used for bacterial identification, as well as to discuss the new approaches for the detection and identification of bacteria. Morphological, biochemical, and serological methods of detection and identification of bacteria in clinical specimens are emphasized, and current methods of characterization and enumeration of bacteria in air, water, milk, and other food materials are also described.

REFERENCES

1. Darwin, C. R., *Origin of the Species*, Atheneum Press, New York, 1964.
2. Haeckel, E. H., *Systematische Phylogenie der Protisten und Pflanzen*, I. G. Reimer, Berlin, 1894.
3. Chatton, E., *Titres et Travauz Scientifiques*, Séte, Sottano, 1937.
4. Taylor, M. M. and Storck, R., Uniqueness of bacterial ribosomes, *Proc. Natl. Acad. Sci. U.S.A.*, 52, 958, 1964.
5. Taylor, M. M., Eubacterium fissicatena sp. nov. isolated from the alimentary tract of the goat, *J. Gen. Microbiol.*, 71, 457, 1972.
6. Ris, H., Ultrastructure and molecular organization of genetic systems, *Can. J. Genet. Cytol.*, 3, 95, 1961.
7. Ainsworth, G. C. and Sneath, P. H. A., *Microbial Classification*, Cambridge University Press, London, 1962.
8. Bousfield, I. J., A taxonomic study of some coryneform bacteria, *J. Gen. Microbiol.*, 71, 441, 1972.
9. Mandel, M., New approaches to bacterial taxonomy, perspective and prospects, *Annu. Rev. Microbiol.*, 23, 329, 1969.
10. Marmur, J., Falkow, S., and Mandel, M., New approaches to bacterial taxonomy, *Annu. Rev. Microbiol.*, 17, 329, 1963.
11. Skerman, B. V. D., *A Guide to the Identification of the Genera of Bacteria*, 2nd ed., Williams & Wilkins, Baltimore, 1967.
12. Carmichael, J. W. and Sneath, P. H. A., Taxometric maps, *Syst. Zool.*, 18, 402, 1969.
13. Wheat, R. W., The classification and identification of bacteria, in *Zinssers Microbiology*, 15th ed., Joklik, W. K. and Smith, D. T., Eds., Meredith, New York, 1972.
14. Buchanan, R. E. and Gibbons, N. E., Eds., *Bergey's Manual of Determinative Bacteriology*, 8th ed., Williams & Wilkins, Baltimore, 1974.
15. International Code of Nomenclature of Bacteria, *Int. J. Bacteriol.*, 16, 459, 1966.
16. Marmur, J. and Doty, P., Determination of the base composition of deoxyribonucleic acid from its thermal denaturation temperature, *J. Mol. Biol.*, 5, 109, 1962.
17. Skyring, G. W. and Quadling, C., Soil bacteria: A principal component analysis and guanine-cytosine contents of some arthrobacter-coryneform soil isolates and of some named cultures, *Can. J. Microbiol.*, 16, 95, 1970.
18. Yamada, K. and Komagata, K., Taxonomic studies on coryneform bacteria. III. DNA base composition of coryneform bacteria, *J. Gen. Appl. Microbiol.*, 16, 215, 1970.
19. McCarthy, B. J. and Bolton, E. T., An approach to the measurement of genetic relatedness among organisms, *Proc. Natl. Acad. Sci. U.S.A.*, 50, 156, 1963.
20. Brenner, D. J., Martin, M. A., and Hoyer, B. H., Deoxyribonucleic acid homologies among some bacteria, *J. Bacteriol.*, 94, 486, 1967.
21. Kingsbury, D. T., Deoxyribonucleic acid homologies among species of the genus *Neisseria*, *J. Bacteriol.*, 94, 870, 1967.
22. Sokal, R. R. and Sneath, P. H. A., *Principles of Numerical Taxonomy* Freeman Press, New York, 1963.
23. Chatelain, R. and Second, L., Taxonomie numérique de quelques *Brevibacterium*, *Ann. Inst. Pasteur* (Paris), 111, 630, 1966.
24. Strange, R. E., Rapid detection and assessment of sparse microbial populations, *Adv. Microbiol. Physiol.*, 8, 105, 1972.
25. Loudon, I. S. L. and Greenhalgh, G. P., Urinary tract infections in general practice, *Lancet*, 2, 1246, 1962.
26. Mond, N. C., Percival, A., Williams, J. D., and Brumfitt, W., Presentation, diagnosis, and treatment of urinary-tract infections in general practice, *Lancet*, 1, 514, 1965.
27. Pryles, C. V. and Lustik, B., Laboratory diagnosis of urinary tract infection, *Pediatr. Clin. North Am.*, 18, 233, 1971.
28. Kunin, C. M., *Detection, Prevention, and Management of Urinary Tract Infections: A Manual for the Physician, Nurse, and Allied Health Worker*, Lea & Febiger, Philadelphia, 1972.
29. Martin, S. M. and Skerman, V. B. B., *World Directory of Collections of Cultures of Microorganisms*, John Wiley & Sons, New York, 1972.
30. Stanier, R. Y., Doudoroff, M., and Adelberg, E. A., *The Microbial World*, Prentice-Hall, Englewood Cliffs, N.J., 1970.
31. Joklik, W. K. and Smith, D. T., *Zinsser Microbiology*, 15th ed., Appleton-Century-Crofts, New York, 1972.
32. Cruickshank, R., Duguid, J. P., Marmion, B. P., and Swain, R. H. A., *Medical Microbiology*, 12th ed., Churchill Livingstone, Edinburgh, 1973.
33. Davis, B. D., Dulbecco, R., Eisen, H. N., Ginsberg, H. S., and Wood, W. B., *Microbiology*, 2nd ed., Harper and Row, 1973.
34. Lennette, E. H., Spaulding, E. H., and Truant, J. P., Eds., *Manual of Clinical Microbiology*, 2nd ed., American Society for Microbiology, Washington, D.C., 1974.

Chapter 2

REVIEW OF CURRENT METHODS USED IN BACTERIOLOGY

INTRODUCTION

Bacteria in clinical specimens or biological mixtures are usually identified after the organisms have been isolated on the surface of a solidified agar medium. A 12- to 24-hr period of incubation is required to promote the growth of most of the organisms in the isolation procedures. In those instances where the isolation methods fail to isolate the organism, the direct analysis of samples may occasionally provide presumptive but not definitive identification. After the bacteria are isolated in discrete colonies, the organism can be identified by morphological, biochemical, serological, or other special tests. Although, in many instances, the bacteria can be presumptively identified by microscopic examination of a stained smear preparation or by other morphological examinations (such as colonial morphology, cultural characteristics, and requirements of certain growth and nutritional conditions), further specific tests are often performed to establish specific identity of the organisms. For example, in the diagnostic bacteriology laboratory, tests are routinely performed for the identification of pathogenic bacteria; these include specific biochemical, serological, and antibiotic sensitivity tests. Other methods, such as pathogenicity of the unknown organism to laboratory animal species and bacteriophage typing, are sometimes used to aid in the identification of bacteria. A number of biochemical tests (including fermentation of sugars, utilization of a specific substrate, production of typical biochemical products, and measurement of enzyme activities), which provide clues toward the identification of bacteria, have been established. Serological tests are performed in which serum is examined for the presence of specific antibodies in response to the bacterial antigens. Many types of antigen-antibody reactions (such as agglutination, precipitation, flocculation, or complement fixation, etc.) are used routinely in bacterial identification procedures. However, isolated organisms are needed in order to perform these specific serological tests for definitive identification.

This chapter briefly reviews the methods commonly used to identify bacteria in clinical specimens and food and water samples. Although many methods employed in a diagnostic bacteriology laboratory can also be used in a public health laboratory, certain screening tests using biochemical or immunologic techniques, phage typing procedures, and enumeration of viable bacteria are frequently employed in epidemiological research and public health laboratories. These methods of detection and identification of bacteria are also briefly described in this chapter.

METHODS USED IN DIAGNOSTIC BACTERIOLOGY

General Procedures
Morphology and Staining Methods

Bacteria have been classified by microscopic examination on the basis of size, shape, and staining properties. Although unstained preparations may be used, especially with the phase contrast microscope, medical bacteriologists generally use heat-fixed, stained preparations. Stains and dyes permit morphological examination of bacteria. Methylene blue and dilute carbolfuchsin are examples of simple stains which stain the organism itself. In relief staining the background is darkened, and the cells are light objects in contrast to the dark field; India ink and nigrosin are examples of such stains.

The Gram stain (Hans Christian Gram, Denmark, 1884) — A Gram-stained smear is sufficient to show the Gram reaction, size, shape, and grouping of the bacteria; whether they possess endospores; and the shape, size, and intracellular position of such spores. The bacterial cells are fixed to the slide by heat and stained with a basic dye (e.g., crystal violet) which is taken up by all bacteria in similar amounts. The specimen is treated with an I_2-KI mixture (Gram's iodine) to fix (mordant) the stain. It is then washed with a 1:1 mixture of acetone and ethanol and finally is counterstained with a paler dye of a different color (e.g., safranin). The slide is washed gently with water, blotted dry, and examined microscopically under the oil immersion lens. Gram-positive organisms retain the initial violet stain, while Gram-negative organisms are decolorized by the organic solvent and hence show the counter-stain. Young growing cultures showing a Gram-positive reaction may tend to lose this characteristic with age.

Acid-fast stains — Acid-fast organisms retain stain when treated with dilute acid followed by alcohol wash. The Ziehl-Neelsen carbolfuchsin stain is usually used. Fluorescent dyes such as auramine O and rhodamine B are also used for acid-fast organisms, particularly *Mycobacterium tuberculosis* and Runyon's Groups I, II, and III of the atypical mycobacteria which retain the stain and fluoresce when the stain is excited by light of suitable wavelength.

Granule stains — Granule stains such as methylene blue are used for throat and nose smears. These stains are used to accentuate granules in bodies of organisms such as *Corynebacterium diphtheriae*.

Capsule stains — These are used by the Anthony and Muir staining methods. With the Anthony method, the capsule is unstained against deep purple-stained cells and background. The Muir method uses Ziehl-Neelsen carbolfuchsin followed by mordant, tannic acid, mercuric chloride, and potassium alum. Ethanol is used to decolorize the stained specimen to a faint pink color. The specimen is counterstained with 0.3% methylene blue. Capsules stain blue and cells stain red. Flagella are visualized with the Leifson or Gray method. The former utilizes the Leifson stain (5% potassium alum solution, 2% tannic acid solution, plus 1% solution of basic fuchsin) and, if necessary, a methylene blue counterstain. The latter utilizes the Gray mordant (saturated solution of potassium alum, 20% solution of tannic acid, and saturated solution of mercuric chloride) followed by the carbolfuchsin stain. Care must be exercised to use scrupulously clean slides and coverslips and young vigorous cultures.

Spore stains — With sample stains such as methylene blue or the Gram stain, spores appear as unstained holes in the bacterial cell body. Special methods, such as the Dorner method, may be used to visualize the spores more clearly. This method utilizes Ziehl-Neelsen carbolfuchsin and nigrosin Dorner solutions. The spores are stained red, vegetative cells remain unstained, and the background appears gray. For a discussion of the various stains, see Lillie[1] and Conn et al.[2]

Size and Shape

In medical bacteriology the size and shape of

certain organisms in pathological specimens may be sufficient for presumptive diagnostic identification. Bacteria are generally unicellular organisms averaging from 0.2 to 10 μm in size. The spherical bacteria (cocci) (for example, staphylococci and streptococci) average 0.5 to 1.0 μm in size. They occur as irregular clusters (e.g., *Staphylococcus*), chains (e.g., *Streptococcus*), and pairs (e.g., *Diplococcus*). These organisms may also be seen in a cluster, chain, pair, or singly, depending on specimen preparation. Triads and tetrads may also exist, depending on the plane of multiplication. Cuboidal packets of eight cells are typical of the genus *Sarcina*. Bacilli are rod-shaped organisms that vary considerably in size. The shape of the rod may vary, e.g., club shaped, rounded end, or polar staining. Rods of *Mycobacterium tuberculosis* range in size from 0.3 to 0.6 × 0.5 to 4 μm, whereas rods of *M. leprae* and the saprophytic mycobacteria (e.g., *M. phlei* and *M. smegmatis*) range in size from 0.2 to 0.5 × 1 to 8 μm.

The spatial arrangement of bacilli is not as characteristic as that of the cocci, although some regular arrangements occur as with the cuneiform figures or picket fence of *Corynebacterium diphtheriae*. Occasionally, pairs (diplobacilli) and chains (streptobacilli) may be seen.

The family Spirillaceae has curved cells. *Spirillum* cells are spiral shaped, 5 to 10 μm in length, with polar flagella. *Vibrio cholerae* are curved rods 1 to 5 μm long. They may arrange singly, in pairs, as commas, or in C, S, or spiral forms. Morphology of the organism may be examined after staining with dilute carbolfuchsin.

Spirochetes are characterized by slender, flexous bodies, spiral in shape, with distinctive types of motility. They vary in size from 0.2 to 0.3 × 8 to 20 μm (for example, *Borrelia*) to 0.1 to 0.2 × 5 to 20 μm (e.g., *Treponema*). The former are larger and longer with shallow, coarse, irregular coils; the latter have tightly wound coils. *Leptospira*, on the other hand, have fine, regular coils 0.25 × 6 to 40 μm in size. The larger size of *Leptospira* is indicative of degeneration. The spirals are so fine and tight that when viewed under dark-field examination, they appear only as small dots. One or both ends bend to form a hook.

Some bacteria are pigmented. If stable and distinctive, the pigmentation is an aid in the identification of the organism. This is particularly true with the atypical mycobacteria, since *Mycobacterium tuberculosis* is not pigmented.

Pigmentation may be cream, lemon yellow, orange, white, pink, or red. The Runyon Group I photochromogen colonies are nonpigmented in the dark and yellow after light exposure. Group II scotochromogen colonies are yellow to orange in the light or dark. Group III nonphotochromogen colonies are usually nonpigmented. Group IV contains rapid growers, which obviates the need for pigment study.

Although spores are formed by some species of *Vibrio* and *Sarcina*, the majority of spore formers belong to the genus *Bacillus*; all are primarily aerobic. Spore formers which develop under strictly anaerobic conditions belong to the genus *Clostridium*. Spores are usually clear and highly refractile and can be stained only with strong dyes such as warm fuchsin in phenol.

Some bacteria produce a capsular material which gives a smooth, glistening appearance to colonies on solid media as opposed to uncapsulated organisms whose colonies appear rough and wrinkled on solid media. Capsules may be composed of polysaccharide (as in the pneumococcus); polysaccharide, phospholipid, and protein (as in *Bacillus dysenteriae, Shigella shiga*); or almost totally polypeptide (as in some of the spore-forming *Bacillus* groups such as *B. anthracis*). With organisms such as *Streptococcus* and *Pneumococcus*, the polysaccharide capsules are serotype-specific substances. This material is the basis for the Neufeld quellung or capsular precipitin reaction.[3]

Flagella are organs for motility that are found most often on rod-shaped organisms. A few cocci are motile with peritrichous (along the sides) flagella. Rods with peritrichous flagella include the Enterobacteriaceae. Some rod-shaped organisms such as lactobacilli have no flagella and are therefore nonmotile. Rod-shaped organisms with polar flagella are members of the genus *Pseudomonas*, family Pseudomonadaceae. Spirilla have flagella at both poles, usually in bunches, which act as a unit organ. Vibrios are monotrichous. Flagella 0.12 μm thick may be observed microscopically by means of dark-field illumination and special stains.

Cultural Characters and Growth Requirements

Organic matter from dead bodies and excreta is the natural media for saprophytes, which include fungi and most bacteria. Parasites utilize living matter for growth. Often they adapt to host conditions so well that cultivation outside of the host is difficult, and special substances must be supplied. Autotrophic organisms can provide for growth by the reduction of carbon dioxide. The range of conditions that support growth of bacteria is characteristic of particular organisms. The ability or inability of the organism to grow on medium containing a selective inhibitory factor (e.g., bile salt, Optochin®, bacitracin, low pH, high pH, or tellurite) may also be a differentiating characteristic of the bacteria.

Culture media may be divided into collection, transport, plating, and diagnostic types and are further subdivided into fluid and solid media according to consistency. Colonies may be isolated only on solid media. Choice of medium depends on the organism. The medium should be readily obtainable and support ease of growth and recognition of the organism. Ideally, it should support growth selectively. Most media contain peptone and frequently contain meat extract and/or yeast extract. Media may be further supplemented with carbohydrate, antibiotics, vitamins, salts, etc. Heart infusion, tryptone soy (or Trypticase® soy agar), and broth media containing 5 to 10% blood are frequently used for culturing and isolating organisms. For example, staphylococci are isolated from a primary culture, where contamination with other bacteria is expected, on a high salt medium (such as mannitol salt agar)[4] in addition to a blood agar medium. Because of the 7.5% NaCl and the presence of mannitol and phenol, *Staphylococcus aureus* appears as yellow colonies surrounded by a yellow zone. Other staphylococci usually form small colonies in red to purple zones; other organisms, including those which are Gram negative, are usually inhibited. Other differential media have also been used to separate staphylococci.[5,6] Blood agar base medium supplemented with 0.01% phenolphthalein phosphate serves to screen for coagulase-positive organisms.[7] All coagulase-positive staphylococci contain phosphatase and split off phenolphthalein when grown in this medium. When exposed to strong ammonia, coagulase-positive colonies become deep pink. However, one must keep in mind that staphylococci can produce phosphatase and can be coagulase negative. For mycobacteria, egg medium[8] or the Lowenstein-Jensen medium[9] is favored; Middlebrook 7H-10 agar[10] is also widely used. For pneumococci, the media of choice are beef-heart infusion broth, Trypticase® soy, or thioglycollate broths. Cultivation of catalase-negative organisms is facilitated by the addition of thioglycollate to the medium. Addition of glucose to the thioglycollate medium suppresses growth of Gram-negative organisms in a mixed culture from which anaerobes are to be isolated. Some media, such as eosin methylene blue agar (EMB) and MacConkey agar, contain lactose and a dye indicator; these are used for the growth of lactose-fermenting organisms which produce red colonies or colonies with a metallic sheen, depending on the medium. Complex selective media, such as desoxycholate citrate agar or bismuth sulfite agar, inhibit growth of most coliform bacilli and many strains of *Proteus* and permit isolation of enteric pathogens.

Bacteria may be classified as strictly aerobic or anaerobic, microaerophilic, and facultative according to their growth requirement for oxygen. The obligate aerobic organisms (for example, mycobacteria) grow well only in the presence of oxygen. Strict anaerobes grow only in the total absence of oxygen, e.g., clostridia. The lactobacilli vary from microaerophilic to anaerobic; *Actinomyces bovis, A. israelii, A. odontolyticus*, and *A. propionicus* are classed as microaerophilic or anaerobic.[11] Facultative anaerobic organisms (for example, pneumococci) grow better with about 5% carbon dioxide tension.

Based on temperature requirements for optimal growth, bacteria may be grouped into mesophiles (20 to 37°C), psychrophiles (<20°C), and thermophiles (>40°C). Most organisms are mesophilic. The mycobacteria have the spectra for temperature requirements. The saprophytic mycobacteria (e.g., *Mycobacterium phlei* and *M. smegmatis*) have a temperature range of 20 to 55°C, depending on the media and techniques employed. *M. tuberculosis* and *M. bovis* grow at 37°C and avian tubercle bacilli grow at about 45°C. Some of the atypical mycobacteria are distinguished by their ability to grow at lower temperatures (20 to 25°C).

A presumptive identification of bacteria can be made by the observation of the discrete masses of growth or colonies that can be grown on the surface of solid culture media from isolated bacteria. The colonies of certain bacterial species are characteristic with regard to size (diameter in millimeters), outline (whether they are circular and entire or indented, wavy, or rhizoid), elevation (low convex, high convex, flat, plateau-like, umbonate, or nodular), and translucency (clear and transparent, translucent, or opaque) (Figure 2.1).[12] Also, the color of the colonies (whether they are colorless, white, or otherwise pigmented) and the changes they produce in the medium (e.g., hemolysis in a blood-containing agar medium) may be useful in the tentative identification of bacteria. However, simple observation of colonial morphology may be insufficient to differentiate a species. For example, different species of *Salmonella* produce colonies that are so similar that they are not even distinguishable on ordinary media from those of other enterobacteria. On the other hand, colonies of staphylococci on solid media are opaque and confluent with unbroken edges. They may be white, yellow, or orange and may be grossly characterized by color.[13-15]

On blood agar medium, the colonies of streptococci are small, like beads, and opalescent gray with smooth or slightly rough edges. Some species are distinguished by changes produced on blood agar. Dissimilar colony forms may be seen with variants of the same strain; for example, Group A hemolytic streptococci may differentiate into matt, glossy, or mucoid colonies. The mucoid and matt forms are virulent. *Streptococcus (Diplococcus) pneumoniae* colonies may show a zone of greening. At 24-hr postincubation they resemble other streptococci (e.g., *Streptococcus viridans*); at 48-hr postincubation, pneumococcal colonies are distinctive in that they show a dense center and raised margin, giving a checkered appearance which is caused by autolysis. The growth of mycobacterial colonies requires 2 to 3 weeks incubation on Middlebrook 7H-10 agar;[10,16,17] they are first observed as dry, warty, or granular colonies. After increasing in size, the colonies have a flat irregular margin and a "cauliflower" center. The atypical mycobacteria have several distinctive colonial characteristics. Group I strains (photochromogens) (for example, *Mycobacterium kansasii*) have smooth to rough colonies. Cells arrange in cords which may be long, beaded, or banded. The most characteristic property is a bright yellow pigment which forms with young growing cultures in 6 to 24 hr after exposure to less than 1 hr of light. The same culture grown in continuous light produces a red to orange pigment. Group II strains (scotochromogens) have cells of varying size with little or no cord formation; they are usually pigmented. If grown continuously in light, colonies are red; colonies grown in absence of light are yellow to

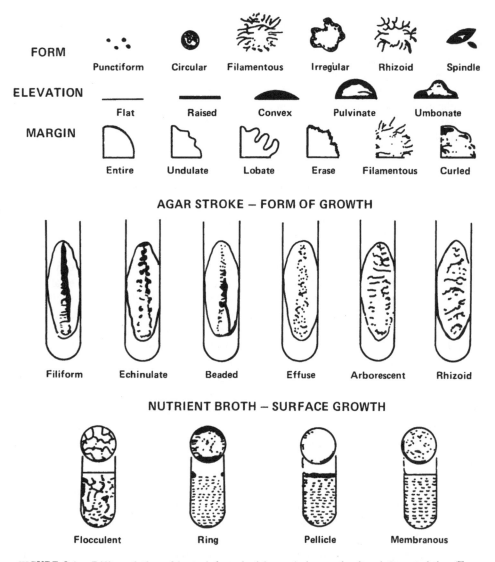

COLONIES

FORM — Punctiform · Circular · Filamentous · Irregular · Rhizoid · Spindle

ELEVATION — Flat · Raised · Convex · Pulvinate · Umbonate

MARGIN — Entire · Undulate · Lobate · Erase · Filamentous · Curled

AGAR STROKE — FORM OF GROWTH

Filiform · Echinulate · Beaded · Effuse · Arborescent · Rhizoid

NUTRIENT BROTH — SURFACE GROWTH

Flocculent · Ring · Pellicle · Membranous

FIGURE 2.1. Differentiation of bacteria by colonial morphology and cultural characteristics. (From American Society for Microbiology, *Manual of Microbiological Methods,* McGraw-Hill, New York, 1957. With permission of McGraw-Hill Book Company.)

orange. Group III strains (nonphotochromogens), the Battey-avian complex, have smooth, often gummy colonies. In liquid media there is usually no cord or pellicle formation. Cells of these strains may be coccoid and nonacid fast, but they are usually short with a bead imbedded in a weakly staining, short, tapering rod. Group IV strains of mycobacteria are distinguished by their rapid growth (3 to 5 days at 25 or 37°C). Colonies are usually smooth, with strains forming serpentine cords. The majority of these strains are stained by the Ziehl-Neelson technique[19] rather than by auramine and rhodamine.[18] For pigmentation studies, colonies are incubated in the dark. A diagnostic key to mycobacteria is given by Wayne and Doubek.[19]

However, morphological examination of bacteria for cell size, shape, and arrangement; staining properties; growth requirements; and cultural and colonial characteristics may not produce distinct differences. Therefore, further tests (biochemical, serological, or other specific tests) are often necessary to make a definitive identification of bacteria in clinical specimens.

Biochemical Tests

The most commonly used biochemical tests involve the observation of whether or not a growth of the bacterium in liquid nutrient medium will ferment particular sugars such as glucose, lactose, or mannitol. Then, acid and/or gas may be produced which may be detected by a change in color of an indicator dye present in the medium. Other tests determine whether the bacterium produces particular end products (e.g., indole, H_2S, and nitrite) when grown in suitable culture media. Many enzyme activities (such as oxidase, catalase, urease, gelatinase, collagenase, lecithinase, or lipase) are frequently measured to aid in the identification of bacteria. More elaborate procedures for the analysis of end products or cell components of bacteria may utilize special techniques such as chromatography, spectroscopy, or electrophoresis. Various biochemical tests commonly used in diagnostic bacteriology are briefly described in the following sections.

Fermentation of Sugars

The most common procedure used to determine the ability of bacteria to ferment a sugar is to add the carbohydrate or polyalcohol to be studied to a basal medium (either liquid or agar) containing an indicator capable of detecting changes in pH which develop during growth. Growth of some organisms, particularly the spore-forming anaerobes, may result in a marked reduction of the indicator, in which case the indicator must be added after (rather than before) growth; spot plate or other methods of pH determination are then used at the time of observation. Early observation of fermentation results generally eliminates any difficulty resulting from reduction of the indicator, since acidity changes usually precede reduction. Production of gas is detected in liquid media by placing Durham tubes (small inverted vials which will fill with liquid during sterilization) in the culture tubes at the time the medium is dispensed. The Durham tubes are unnecessary if a solid or semisolid medium is used. Semisolid agar is prepared by adding 0.3 to 0.5% agar to a satisfactory liquid medium and making a stab inoculation in the column of medium with a straight inoculating needle. In such a medium (or in full-strength agar), gas production is denoted by the appearance of gas bubbles and cracks; the same semisolid medium may also be used to determine motility. Full-strength (1.5%) agar should be cooled in a slanting position and inoculated on the surface of the slant. The basal medium used for fermentation tests must be free of fermentable carbohydrates and should provide the necessary nutrients for the organisms to be tested. If good growth can be obtained on 2% casein or gelatin-peptone solutions (or agar), they are preferred as media. The soluble carbohydrates or polyalcohols are added to the basal medium at a level of 0.5 to 1.0%. The indicator (commonly 1 ml of a 1.6% alcoholic solution per 1,000 ml of medium) is added before sterilization. Although litmus and Andrade's indicator (acid fuchsin decolorized with alkali) have been widely used, they do not give accurate results in terms of H-ion concentration. Thus, except for special purposes, sulfonaphthalein should be used as an indicator. Phenol red, with a pH range of 6.9 to 8.5, is useful for indication of changes on the alkaline side of neutrality and slight changes to acid; bromthymol blue has a sensitivity range extending slightly in either direction from neutrality; bromcresol purple, with a pH range of 5.4 to 7.0, is used in synthetic media and to produce pronounced pH changes in more highly buffered media.

Snell and Lapage[20] compared four methods for demonstrating the breakdown of glucose by bacteria. The ability of bacteria to metabolize carbohydrates is usually demonstrated by acid production from a carbohydrate incorporated in a nutrient medium such as a peptone-water-sugar medium (PWS), in which the source of nitrogen is provided by peptone. This is useful for detecting fermentative utilization of carbohydrate by bacteria because much acid is produced. Bacteria which metabolize glucose oxidatively (non-fermenting bacteria) may have only slight acid production, and alkali produced from peptone could neutralize the small quantities of acid. To overcome this difficulty, Hugh and Leifson[21] developed O/F medium, which utilizes a low concentration of peptone.

The Hugh-Leifson (O-F) test distinguishes the aerobic organisms rather than anaerobes. It is useful in differentiating Gram-negative oxidative rods, such as *Pseudomonas* and *Mima*, and those that are neither oxidative nor fermentative, such as *Alcaligenes* from certain facultative anaerobic Enterobacteriaceae.

Some nonfermenting bacteria remove glucose from the medium with no apparent acid production in O/F medium. Park[22] modified the

Hugh and Leifson medium (MHL) and developed a method in which removal of glucose was demonstrated enzymically using a Clinistix® reagent strip (Miles Laboratories). Clinistix reagent strip develops a blue color with glucose; absence of color development indicates removal of glucose from the medium. The ASS medium of Smith et al.,[23] with ammonium sulfate as the nitrogen source, has been used by Lapage et al.[24] to demonstrate acid production by nonfermenting bacteria.

O/F, ASS, and MHL media were more sensitive than PWS medium, but no single medium is sufficient to demonstrate all cases of nonfermentative utilization of glucose. For example, *Pseudomonas maltophilia* removed glucose from MHL medium but failed to produce acid in O/F medium. Acid production for *Acinetobacter lwoffi* was demonstrated only on ASS medium.

Lactose agar (10%) in phenol red broth base[25] is used to test for lactose fermentors. With litmus milk, formation of a pink color indicates an acid reaction caused by fermentation of lactose. Ulrich's milk[26] is the same as litmus milk except that it uses chlorphenol red as a pH indicator and methylene blue as a redox potential indicator. Both forms are available commercially from BBL, Difco, Fisher, and Oxoid.

Specific tests of fermentations of sugars by bacteria are described elsewhere in this chapter (see Specific Identification of Bacteria).

Enzyme Activities

Catalase — A catalase test may be performed by three methods.

1. Pour 1 ml of 3% hydrogen peroxide (H_2O_2) over colonies growing on agar slant.
2. Emulsify a single colony on a glass with one drop of 30% H_2O_2.
3. Add H_2O_2 directly to the liquid culture.

Appearance of gas bubbles signifies a positive test. Streptococci and pneumococci give a negative test result, whereas staphylococci and micrococci are catalase positive.

Oxidase — Kováč's method is used for the identification of *Neisseria* and *Pseudomonas* species.[27] A bacterial colony is rubbed on a filter paper strip impregnated with a fresh 1% solution of tetramethyl-*p*-phenylenediamine dihydro-

chloride. A positive reaction is indicated by purple coloring in 10 sec.[28] Oxidase test papers are available from Pathotec®.* The Ewing-Johnson modification, also known as the indophenol (cytochrome) oxidase test, may also be used.[29] In this method, 2 to 3 drops of a 1% α-naphthol solution and a 1% *p*-aminodimethylaniline hydrochloride solution are added to nutrient agar slant cultures. A positive reaction yields a blue color in 2 min. An iron-containing loop may give a false positive reaction. Reaction is delayed in an acid environment. An alternative method is to use a 1% solution of dimethyl-*p*-phenylenediamine hydrochloride dropped on a plate culture. A positive reaction is indicated by a series of colorings from pink to maroon to dark red and finally black. This reagent is less sensitive and more toxic than the tetramethyl reagent.

Coagulase — This test is used to characterize staphylococci. The tube method tests for "free" coagulase. Mix ½ ml of rabbit plasma with either a loop of bacteria from a plate or 0.1 ml of culture and check at intervals at 37°C. Any clotting indicates a positive reaction. False positive reactions may occur with mixed cultures and with some pure cultures (for example, Gram-negative rods [*Pseudomonas aeruginosa* and *Serratia marcescens*] and enterococci [*Streptococcus faecalis*]) because the citrate (anticoagulant in plasma) is used by these organisms, thereby resulting in clot formation.[30] The slide method tests for "bound" coagulase. A single colony is emulsified in 1 drop of water on a slide, and a loop of plasma is mixed with it. A positive reaction is shown by white clumping in about 5 sec. All negative slide tests should be confirmed by the tube test. The coagulase test is the single most important test in the differentiation of pathogenic and nonpathogenic staphylococci.[31] Plasma rendered incoagulable by oxalate, heparin, and ethylenediaminetetraacetic acid (EDTA) may be used to rule out false positives. Some *Staphylococcus aureus* strains may give a false negative test due to abundance of fibrinolysin production. Delayed readings of 1 to 24 hr may be required.

DNAse activity — This test correlates well with the coagulase test. An agar plate is prepared with 2 mg/ml DNA in Trypticase® soy agar containing 0.8 mg/ml anhydrous calcium chloride. After growth of the culture, the plate is flooded with 1 *N* HCl. DNAse-positive cultures have a zone of

*General Diagnostic Division, Warner Chilcott Laboratories, Morris Plains, N.J.

clearing around the streaks; the rest of the medium appears cloudy. An alternative method is to flood the plate with 0.1% toluidine blue. A bright rose pink color is produced around DNAse-producing colonies.

β-D-Galactosidase (ONPG) test — O-Nitrophenyl-β-O-galactopyranoside/PO_4 buffer pH 7.0 is added to a loop of emulsified culture from a triple sugar iron (TSI) or Kligler iron agar (KIA) slant. Toluene is added to liberate the enzyme during a 5-min incubation period at 37°C. Incubation is continued at 37°C for 18 to 24 hr. Liberation of O-nitrophenol gives a yellow color which indicates a positive reaction.[32] The original procedure utilized a lactose-glucose-SH_2 medium. Lactose fermentation requires two enzymes, β-galactosidase and a permease. Potential lactose fermenters can be detected by cultivation on a lactose-containing medium. Mutations may occur which yield mutants containing permease, whereby delayed fermentation of lactose may occur.

Other Biochemical Tests

Citrate utilization — A Simmons citrate agar slant (pH 6.9) is inoculated with a saline suspension of a young agar slant culture. The blue color from the indicator bromothymol blue means that citrate is utilized.[33] Some bacteria (for example, *Escherichia coli*) cannot utilize citrate as a sole C source. Therefore, growth in synthetic medium having nitrogen as Na-ammonium PO_4 and C as sodium citrate serves to differentiate these bacteria. Christensen citrate agar[34] is used to differentiate organisms utilizing citrate in the presence of organic nitrogen. *Shigella* cannot do this; many strains of anaerogenic, nonmotile *E. coli* can utilize Christensen citrate agar at 37°C with phenol red as the indicator. A positive reaction is indicated by a red color.

Ferric chloride test — This test is performed to detect hydrolysis of sodium hippurate[35] by mixing 8/10 ml of culture in sodium hippurate broth and 0.2 ml of 12% ferric chloride in 2% aqueous hydrochloric acid. The test is positive if after 10 to 15 min a permanent precipitate forms, indicating the presence of benzoic acid. Exact amounts of reagents must be used or false negatives may occur.

Gluconate oxidation test — This test is used in identifying nonpigmented strains of *Pseudomonas aeruginosa* which are able to oxidize dextrose or gluconate to ketogluconate.[36] The ketogluconate is detected by reduction of copper salts in Benedict's solution, with a positive reaction indicated by a color change from blue to yellow-green. A gluconate substrate tablet* in 1 ml of water is heavily inoculated with the test organism. After overnight incubation at 37°C the culture is tested for reducing ability with either Benedict's solution or a Clinitest® tablet.**

Inulin — Inulin is metabolized by the majority of pneumococcal strains. With few exceptions (e.g., *Streptococcus salivarius, S. sanguis*), α-hemolytic streptococci do not decompose inulin.

Malonate utilization — A culture inoculated in sodium malonate broth is incubated at 37°C for 24 to 48 hr. Malonate utilization is indicated by a change in color from green to blue. Some negative cultures change to yellow. Leifson's malonate broth, modified by the addition of a small amount of yeast extract and glucose, is of considerable value in the differentiation of salmonellae and species of *Arizona*. The majority of *Salmonella* cultures do not utilize malonate, whereas the majority of strains belonging to *Arizona* do. Modified malonate medium is also valuable since many strains of *Klebsiella* and *Aerobacter* utilize it, whereas *Serratia* strains do not.[37]

Hydrogen sulfide (H_2S) production[38,39] — A filter paper strip saturated with a 5% lead acetate solution is inserted between the plug and inner wall above a sulfur-containing liquid medium inoculated with the test organism. Hydrogen sulfide production causes a blackening of the lead acetate strip. Negative tests should be checked for dissolved sulfide by adding a few drops of $2 N$ HCl to the tube and closing with the plug and lead acetate strip in place as before. If sulfide is liberated, the strip will darken. *Brucella suis* Type 1 cultures produce the greatest amount of H_2S, while Biotypes 2, 3, and 4 of this species produce very little or no H_2S. *B. abortus* produces moderate amounts of H_2S.

Nitrate reduction test — Add equal drops each of a 0.8% solution of sulfanilic acid/acetic acid and a 0.5% solution of α-naphthylamine/acetic acid to a growing culture in a liquid medium. The culture is nitrite positive if a red color develops in 1 to 2 min. All negative tests should be confirmed by

*Key Scientific Products Company, Los Angeles, Cal.
**Ames Company, Elkhart, Ind.

adding zinc dust to the broth medium. If un-reduced nitrate is present, a red color will develop. This precaution is necessary since false negative results are given by those organisms that can reduce nitrate beyond the nitrite stage to nitrogen or ammonia.

Ehrlich indole test modified[40] — Gently add ½ ml of a 1% solution of paradimethylaminobenzaldehyde in acidified ethyl alcohol to a culture which was previously treated with 1 ml of ether or xylene; shake well and let stand until the solvent rises to the top of the culture surface. The reagent forms a ring between the culture medium and the solvent; this will be a brilliant red if indole is extracted into the solvent layer.

A modification of the Kovács indole test[41] — This test requires addition to the growing culture of a mixture containing five drops of a 5% solution of paradimethylaminobenzaldehyde in isoamyl alcohol to which HCl is slowly added. A deep red color indicates indole.

Voges-Proskauer test for acetyl methyl carbinol[41-44] — Add 6/10 ml of a 5% solution of α-naphthol in ethyl alcohol and 0.2 ml of a 40% KOH solution containing 0.3% creatine to 1 ml of a 48-hr culture grown in methyl red-Voges-Proskauer (MR-VP) medium; shake well and let stand for 10 to 20 min. Acetyl methyl carbinol production is indicated by formation of a bright orange-red color at the surface of the medium, gradually extending throughout the medium.

Optochin (ethylhydrocupreine hydrochloride) test — A 6-mm disk impregnated with (5 μg) Optochin is placed on a bacterial lawn on a blood agar disk in an aerobic environment. A positive reaction is indicated by an 18-mm or greater zone of growth inhibition after an overnight incubation at 37°C.[45] Pneumococci are positive; chemolytic streptococci are negative. Disks are commercially available as Taxo P Sensi-Discs®.*

Niacin test — This test differentiates *Mycobacterium tuberculosis* (human) from other mycobacteria. One method of identification is the Runyon modification of the Konns test.[46] Aniline and cyanogen bromide are added to a water or saline rinse of the colonies. A yellow color occurs almost immediately throughout the solution if niacin is present. Poor results occur if *M. tuberculosis* is grown on blood agar or Middlebrook 7H-10 agar.

Acrolein reaction — This test is used to identify clostridia. Schiff reagent (basic fuchsin and 4%

*Baltimore Biological Laboratory, Cockeysville, Md.

sodium sulfite/HCl) is added to the culture in thiogylcollate broth. A positive reaction is indicated by a magenta color developing in 15 min at room temperature.

Susceptibility to Bile and Deoxycholate

The test for susceptibility to 0.1% sodium deoxycholate may be done using either a liquid medium (e.g., thioglycollate medium in which the deoxycholate is added after autoclaving) or a solid medium. The thioglycollate broth is steamed for 15 min and cooled before inoculation. Ascitic fluid (25%) and menadione (0.5 μg/ml) are also added to the thioglycollate medium in studies with *Bacteroides melaninogenicus*. An identical tube of broth without deoxycholate is used as a control. *Fusobacterium mortiferum* and *F. varium* strains grow rapidly in broth containing 0.1% deoxycholate so that it is possible to evaluate the results of this test after overnight incubation. When deoxycholate agar is used, it is necessary to inoculate heavily using at least one loopful of colonial growth from solid media or 0.5 ml of undiluted broth culture. The tests must be incubated for 3 to 4 days in order to obtain growth of those *Fusobacterium* strains that are able to grow on this medium.

The bile test for anaerobes utilizes Difco Bactooxgall in a final concentration of 2% (equivalent to 20% bile) plus 0.1% sodium deoxycholate in either fluid thioglycollate medium or peptone-yeast extract-glucose broth in anaerobic tubes. The bile and bile acid may be added either before or after autoclaving. Control tubes without bile and bile acid are set up concurrently. All deoxycholate and bile tests (with the exception of the deoxycholate agar) are read after a period of anaerobic incubation sufficient to give adequate growth in control tubes.

Bile solubility test — A few drops of a 10% solution of sodium deoxycholate or sodium taurocholate are added to a fresh dilute culture in buffered (pH 7.4) broth. A positive reaction is seen by loss of turbidity and an increase in viscosity in 5 to 15 min. This is due to activation of autolytic enzymes causing dissolution of the cell. It is an excellent test for rapid differentiation of pneumococci (most are positive) from streptococci (most are negative).

Tween® 80 hydrolysis — Incubate ½% Tween 80 in phosphate buffer (pH 7.0) containing neutral red indicator with mycobacteria strains. Mycobacteria that degrade Tween 80 to oleic acid are detected by a change of color of the indicator.

Sodium pectinate hydrolysis — This test is used to identify *Erwinia* species.[47] A Gram-negative organism that liquefies pectinate is not considered to be a member of the Enterobacteriaceae. *Erwinia* are plant pathogens that find their way into the new lyophilized food concentrates. No other reagents are necessary to perform sodium pectinate liquefaction tests.

Hydrolysis of urea — Some bacteria can split urea into ammonia and CO_2 by means of the enzyme urease (for example, *Proteus rettgeri*). The medium becomes alkaline, and an indicator is used to see the change in color. The solid medium test utilizes a heavily inoculated slant of Christensen urea agar[48] incubated at 37°C. Negative tubes are carried for 4 days to detect delayed positive reactions. Liquid medium may also be used.[49-51] Prepared media are available from BBL, Difco, and Fisher.

Arginine hydrolysis — Organisms are incubated for 24 hr in arginine broth. Nesslers reagent (0.25 ml) is then added to the culture. A brown color develops in the culture if it is positive for arginine hydrolysis. Arginine decarboxylation as well as ornithine and lysine decarboxylation tests are used for the differentiation of species of bacteria in Enterobacteriaceae.[52]

Tellurite reduction — This test is useful in differentiating *Mycobacterium avium* and *M. intracellulare* (Battey-Avium strains) from other slow-growing mycobacteria. It is performed in mycobacterial cultures grown in Middlebrook 7H-9 liquid medium and incubated at 37°C for 7 days. After a 7-day incubation period, two drops of a sterile 0.2% solution of potassium tellurite are added to the culture. The presence of a black metallic precipitate indicates a positive result.

Ammonia test — This test is performed on a culture after incubation in nutrient or peptone broth for 5 days. Filter paper is immersed with Nesslers reagent and placed in the upper part of the culture tube. The tube is warmed in a water bath at 50 to 60°C. Ammonia is present if the filter paper turns brown or black.

Casein hydrolysis test — This test is performed by using skim milk medium and observing the clearing around colonies of casein-hydrolyzing organisms. Interference (false clearing) by acid production of lactose fermenters is prevented by pouring a 10% solution of mercuric chloride in 20% hydrochloric acid over the skim milk medium. Casein is not hydrolyzed if the cleared area disappears.

Gelatin liquefaction — This is determined using denatured gelatin plus charcoal disks (Oxoid). Add one gelatin-charcoal disk to inoculated nutrient broth and incubate at 37°C. Liquefaction is seen by release of charcoal granules which fall to the bottom of the tube. Alternatively, 1 ml of broth or 0.01 *M* calcium chloride in saline is heavily inoculated; a disk is added, placed in a 37°C water bath, and examined at 15-min intervals for 2 to 3 hr. Pour plates may be used (100 g nutrient agar and 20 g gelatin). Culture plates are flooded with saturated ammonium sulfate solution; halos form around colonies of organisms producing gelatinase.

Lecithinase activity — This may be observed by using the media of Willis and Hobbs[55] and Lowbury and Lilly.[56] Using pour plates, add 20% dextrose and 2 ml of egg yolk (per 100 ml melted nutrient agar) to the media. Lecithinase-producing colonies give a precipitate.

Levan production — This is demonstrated from sucrose by formation of large mucoid colonies on sucrose agar pour plates (5% sucrose in 100 ml melted nutrient agar is steamed for 30 min and cooled to 55°C; then, 5 ml sterile horse serum is added).

Lipolytic activity — This is monitored using tributyrin agar (Oxoid or Difco). Most of the fat-splitting organisms which cause spoilage in butter split glyceryl tributyrate. Colonies of lipolytic organisms clear the medium.

Phosphatase test — This test is positive for *Staphylococcus aureus* (negative strains have been reported). Use 1 ml of 1% phenolphthalein phosphate in 100 ml melted nutrient agar for preparation of pour plates which are kept refrigerated for at least 1 week before using. After inoculation and overnight incubation, expose the plates to ammonia vapor. Phosphatase-positive staphylococci colonies will turn pink due to free phenolphthalein.

Phenylalanine deamination — This test utilizes the deamination of phenylalanine to phenylpyruvic acid.[53,54] Add 0.2 ml of 10% ferric chloride to a growing culture in phenylalanine liquid medium or on an agar slant. If phenylpyruvic acid is formed, a green color develops. The test is used to differentiate Enterobacteriaceae.

Many other biochemical tests are occasionally used to aid in the identification of bacteria from clinical specimens. These tests, which are specific for the differentiation of certain species of bacteria, are described in a subsequent section in connection with the specific identification of bacteria.

Serological Tests

Bacterial species and types can often be identified by specific antibody reactions (to bacterial antigens) observed in serological tests. Antigens are substances that elicit production of antibodies in a foreign species and can react with those antibodies. Bacteria and some bacterial products function as antigens in a host. Specific antibodies are formed by the host to mediate the destruction and removal of the foreign substances. Infection by flagellated bacteria may give rise to host antibodies directed against the flagellar surface antigens. Serum of an infected individual mixed on a slide with a drop of culture of the particular infecting bacteria is inspected through the low-power objective; flagellar clumping is seen in a positive test. The same test performed by the tube dilution technique permits estimation of the titer of antibody. A constant amount of antigen is added to a series of tubes containing serum dilutions. After mixing, tubes are kept at 37 or 4°C, depending on the method followed. The titer is equivalent to the highest dilution of serum showing visible agglutination. A quantitative precipitin reaction yields antibody nitrogen when antigen nitrogen is subtracted from total antigen-antibody nitrogen.[57] The capillary precipitin method utilizes a column of 1 cm of serum and an equal amount of antigen containing extract in a glass capillary tube which is plugged and inserted into a clay rack. A white cloud or ring at the junction of the two liquids indicates a positive reaction and usually appears within 5 to 10 min. This is a convenient method for grouping of β-hemolytic streptococci. The slide agglutination test is used for the definitive identification of the various members of Enterobacteriaceae. Commercially available diagnostic antisera are used to identify the more common *Salmonella* and to group the *Shigella* organisms. Absorbed sera available from the National Communicable Disease Center (Atlanta) and the National *Salmonella* and *Shigella* Center (Ottawa, Canada) are required for exact antigenic analysis.

The quellung or capsular swelling reaction, also described as a capsular precipitin reaction,[3] is used for typing pneumococci. A similar procedure is used for rapid diagnosis of *Haemophilus influenzae* meningitis (using spinal fluid) and identification of *Klebsiella* types (Friedlander's bacilli). An optical phenomenon occurs when the pneumococcus is in contact with homologous anticapsular serum. The organism is seen surrounded by a halo when a loop of culture and a loop each of 1% methylene blue solution and anticapsular (typing) serum[58] are mixed on a slide and observed under the oil immersion lens. The quellung reaction is not widely used for typing pneumococci because treatment no longer depends on the identity of the specific capsular type. However, it remains a rapid diagnostic procedure utilizing direct clinical specimens. Pool and unitypic sera for typing pneumococci are available from the Center for Disease Control (Atlanta) for the first 33 capsular types. Statens Seruminstitut (Copenhagen, Denmark) produces antisera (omni-, pooled, and unitypic) that react with all 84 pneumococcal capsular serotypes. For reconciliation of American and Danish systems of nomenclature, see Lund.[59]

Bacteriophage Typing

Bacteria are susceptible to infection by bacterial viruses, bacteriophages (also known as phages) which are highly host specific. This method may be used for typing bacteria by testing the susceptibility of the culture to lysis by each of a set of type-specific lytic bacteriophages; the phage type of the culture is identified according to its pattern of susceptibility to the different phages. As a result of the specificity, phage typing is used in research studies and is widely used as an epidemiologic tool. For example, it can prove two strains to be identical or different. Among the staphylococci, only coagulase-positive organisms can be phage typed, thereby giving a most reliable means of identifying *Staphylococcus aureus* strains. Phage typing may allow subdivision of a serological entity, as in the case of serotype *Salmonella typhi* which is divisible into 80 different phage types. For further discussion of the methods of phage typing, see Blair and Williams.[60]

Animal Pathogenicity

Virulence studies in animals are sometimes used for the identification of certain bacteria. With pneumococcal infection, an animal host may be used to obtain a pure culture from a sputum specimen. The mouse usually eliminates other sputum bacteria before death occurs. One milliliter of emulsified sputum is injected into the peritoneal cavity of the white mouse. Death will occur within 96 hr if pneumococci are present; however, a few types, i.e., Type 14 pneumococcus, are not virulent in mice. Organisms are cultured from heart blood on blood agar plate and broth. Peritoneal washings may be stained and typed immediately.

Avian tubercle bacilli resemble Runyon Group III strains, but are distinguished by growth at 45°C, pathogenicity for rabbit and fowl, and allergic reaction of the infected animal to avian tuberculin.

Severity of staphylococcal infection in a patient may be gauged by giving albino mice (by intravenous route) 0.05 ml of an overnight infusion broth culture of the strain to be tested. Death of 50% or more of inoculated mice within 2 days indicates severe infection.[61] *Staphylococcus albus* is usually much less pathogenic than *S. aureus*. Strains recently isolated from saprophytic sources are nonpathogenic unless administered in large doses; death then occurs, apparently from toxemia.

Bacteria produce two classes of toxins: exotoxins which are heat-labile toxic proteins produced mainly by certain Gram-positive and a few Gram-negative bacteria, and endotoxins, which are relatively heat-stable phospholipid-polysaccharide complexes which are an integral part of the Gram-negative bacterial cell wall. They are released only if cellular integrity is altered. Although the ultimate mechanism of endotoxins in disease has yet to be entirely elucidated, they are responsible for endotoxin shock in humans. The most frequent causative bacteria are *Escherichia coli*, *Proteus*, *Pseudomonas aeruginosa*, and *Bacteroides*.

Specific exotoxins are associated with *Clostridium botulinum* (type-specific neurotoxins with botulism), *C. perfringens* (α-toxins with gas gangrene), and *C. tetani* (tetanospasmin and tetanolysin with tetanus). *Corynebacterium diphtheriae* produces diphtheria toxin which causes diphtheria. *Staphylococcus aureus* produces α-, β-, and γ-toxins and leukicidin in staphylococcal pyogenic infections; enterotoxin in staphylococcal food poisoning; and erythrogenic toxin in staphylococcal scarlet fever. *Streptococcus pyogenes* produces streptolysins O and S in streptococcal pyogenic infections and erythrogenic toxin in

streptococcal scarlet fever. The α-hemolysin test[62] utilizes the supernatant of a 48-hr culture grown in the presence of 30% CO_2. One milliliter of a 2.5% suspension of rabbits' washed red cells is added to each milliliter of a series of doubling dilutions of supernatant. A control tube contains commercial staphylococcal antitoxin added to the lowest dilution of supernate. Tubes are incubated for 1 hr at 37°C. The reciprocal of the highest dilution which produces 50% hemolysis is the activity in hemolytic units per milliliter. Leukocidin tests[63-66] measure the ability of the specimen to destroy the polymorphonuclear leukocytes.

When a specific bacterial toxin is involved in the mechanism of pathogenesis, animal pathogenicity tests are conducted by use of specific neutralizing antisera, and the pathogenic organisms are thereby identified with a high degree of specificity. For example, the final identification of a diphtheria bacillus may be made by injecting culture material intradermally into two guinea pigs; one has been protected by prior injection of diphtheria antitoxin, and the other is given partial protection with a small dose of antitoxin at the time of the test. The development of an inflammatory and necrotic skin reaction in the partially protected, but not in the fully protected, animal identifies the culture as an organism producing diphtheria toxin. The tetanus bacillus is identified similarly by a test in mice that proves its production of the specific tetanus toxin. Strains of *Clostridum welchii* may be identified by demonstration of the set of specific toxins they produce; in this case, the culture fluid of the clostridia is mixed with a specific antiserum before injection.

Antibiotic Sensitivity

The organism is tested for its ability to grow on artificial nutrient media containing different antibiotic and chemotherapeutic agents in different concentrations. The methods of antibiotic sensitivity testing include tube dilution (which is time consuming and expensive, but more accurate) and the disk method (which is rapid and easier). With the tube dilution method, the medium should not be enriched with blood or serum. Some researchers caution that Trypticase soy broth may contain substances inhibitory to cephalosporin antibiotics when testing Enterobacteriaceae.[67]

The filter paper disk method was originally described by Bondi et al.,[68] and its application has been extended by many groups.[69-71] Some pathogenic bacteria produce acid from glucose, which is the basis of one form of antibiotic disk-sensitivity test.[72,73] Disks impregnated with various antibiotics are placed in 1 ml of inoculated phenol red dextrose broth containing a 0.25% yeast extract. If effective, the antibiotic-containing tubes remain unchanged in color after incubation; otherwise, tubes are yellow if antibiotic is ineffective. If the disk method is to be accurate, care must be exercised in storage and handling of the disks. Humidity inactivates penicillin G, oxacillin, methicillin, cloxacillin, and cephalothin disks. Maximum stability is assured by storage over "Drierite®" (calcium chloride) in an evacuated desiccator at -20°C. The desiccator must be brought to room temperature before opening in order to avoid condensation on disks. Short-term storage at 4°C in an infrequently opened refrigerator is acceptable. One must take into account various generation times with mixed cultures, the synergistic action of combinations in primary cultures, and the influence of various bloods on inhibition zone size. The World Health Organization Collaborative Study[71] recommends the Mueller-Hinton medium for sensitivity testing. With SS and EMB media, metallic cations may be inhibitory. Some antibiotics (for example, streptomycin and neomycin) are inactivated by anaerobic conditions.

The methods most often followed with disks are the Bauer-Kirby-Sherris-Turck single-disk diffusion method (also referred to as the Kirby-Bauer method) and the Seattle method.[74] For Gram-positive bacteria, antibiotic disks routinely used include: penicillin, tetracycline, streptomycin, kanamycin, methicillin, erythromycin, chloramphenicol, cephalothin, lincomycin, and oleandomycin. Nitrofurantoin is used for testing bacteria from the urinary tract. For Gram-negative bacteria, antibiotic disks of choice include: chloramphenicol, tetracycline, streptomycin, kanamycin, polymyxin B, ampicillin, and cephalothin. Sulfamethizole, nitrofurantoin, and nalidixic acid are used for testing urinary tract infections.

Haltalin et al.[75] proposed the use of the agar plate dilution method, which was semiautomated by the use of an inocule-replicating apparatus.[76] With the agar dilution method of Haltalin et al., ampicillin, cephalothin, chloramphenicol, kanamycin, polymyxin B, tetracycline, and gentamicin are used to test Gram-negative organ-

isms. Gram-positive organisms are tested with penicillin G, methicillin, erythromycin, cephalothin, and gentamicin. Both sets are used to test enterococci, *Listeria* species, and unusual organisms. The replicating apparatus permits simultaneous inoculation of 32 organisms onto each plate. Final antibiotic concentrations in agar, except for gentamicin and penicillin, are 20, 10, 5, 1.25, and 0.3 μg/ml, respectively; gentamicin concentrations are 10, 5, 2.5, 1.25, and 0.3 μg/ml, respectively; and penicillin concentrations are 2.5, 1.25, 0.6, 0.2, and 0.08 units/ml, respectively. Results are reported as minimal inhibitory concentration (MIC), the lowest concentration of antibiotic that will result in complete inhibition of growth or barely visible growth haze.[77]

The Bauer disk technique is used routinely to test susceptibility of enteropathogenic *Escherichia coli* to neomycin, to test susceptibilities of *Proteus* and *Pseudomonas* species to carbenicillin, and to test susceptibilities to drugs not included in the test set. For the disk technique, the depth of agar (5 to 6 mm) is important since antibiotics diffuse downward as well as outward. This is not a factor with the agar dilution technique since the antibiotics are evenly concentrated throughout the medium. With the agar dilution, results are read either as growth or no growth for a given concentration of antibiotic. The disk technique requires precise measurements of zone diameters.

Garrod and Waterworth[78] emphasize that standard procedures must be followed in performance of antibiotic-sensitivity testing. They place emphasis on controlling the weight of inoculum and ensuring its uniform distribution. They recommend interpretation by comparison of inhibition zone sizes with those of a suitable control organism (*Staphylococcus aureus* for systemic infection and *Escherichia coli* for urinary tract infection) on the same medium.

For a further discussion, see Ericson and Sherris[79] and Gavan et al.,[80] who have evaluated the various techniques available for antimicrobial susceptibility testing.

Specific Identification of Bacteria
Aerobic and Facultatively Anaerobic Bacteria
Gram-positive Cocci
Micrococci

The family Micrococcaceae is classified into six different genera, *Micrococcus, Gaffykya, Sarcina, Methanococcus, Peptococcus,* and *Staphylococcus,* based on the arrangement of cells. Microscopically, the cells may appear singly or in pairs, clusters, tetrads, or packets of eight. Most of the species in this family are not usually parasitic to man, with the exception of *Staphylococcus aureus* and *S. epidermidis*.

Staphylococci

The genus *Staphylococcus* may be differentiated from *Micrococcus* by the ability of staphylococci and the inability of micrococci to anaerobically ferment glucose. The two genera also differ in the guanine plus cytosine (GC) content of their DNA: *Micrococcus*, 66 to 75% GC; *Staphylococcus*, 29 to 37% GC.[81-85] Differentiation in the DNA base composition is accompanied by differences in the chemical composition of the cell wall of these organisms. Schliefer and Kandler[86,87] and Schleifer[88] demonstrated that staphylococci differ from micrococci by the presence of glycine-rich peptidoglycans in their cell walls. Baddiley et al.,[89] Davison et al.,[90] and Schleifer[88] proposed the distinct types of teichoic acids in the cell walls of staphylococci. Schleifer and Kocur[91] characterized three different types, of staphylococcal teichoic acids — ribitol, glycerol, and ribitol and glycerol.

Barry et al.[92] proposed two tests that should be routinely performed for the identification of *Staphylococcus aureus*, tube coagulase and thermonuclease (heat-stable nuclease). The latter test is performed with the culture heated to 100°C for 15 min and then cooled; it uses the toluidine blue-deoxyribonucleic acid agar (TDA) technique.[92,93] The TDA plates are observed for a pink halo around each well after 1 to 4 hr at 35°C; coagulase tests are read for a definite coagulum. The inoculum for the tests is prepared by pre-incubation in a small volume of brain-heart infusion (BHI) medium.

Jay[94] proposed a plate method for the characterization of staphylococci on the basis of extracellular protein production. Staphylococci are streaked onto a 4% NZ-amine type NAK medium fortified with thiamine and niacin. The medium is solidified with 0.75% agar and incubated for 24 hr at 37°C. Growths are removed with cotton swabs, and plates are flooded with 25% trichloracetic acid (TCA). After 15 min various degrees of protein precipitate (extracellular protein, ECP) are noted under the areas of growth. Pathogenicity has been associated with the production of extracellular substances by staphylococci. While coagulase production is accepted as characteristic of pathogenic strains, not all coagulase-positive staphylococci are pathogenic. Jay[94] suggested the possibility that strains which produce maximum levels of ECP are of pathogenic significance regardless of their coagulase reaction. Although the method seems specific for staphylococci, *Salmonella typhi, S. newington*, and *S. typhimurium* also produce ECP in a similar manner.

Staphylococcus aureus can be differentiated from *S. epidermidis* by DNAse activity and the coagulase test; the latter test is simpler to perform and provides earlier results.[95] Also, *S. epidermidis* has a white colony and does not form mannitol.

Streptococci

The original classification of streptococci was based on hemolytic properties of the group observed around the colonies in poured blood agar plates.[96] The organisms were differentiated according to the type of hemolysis: α for green zone with intact cell envelope, β for clear colorless zone with cell envelope dissolved, and γ when no change in red blood cells occurred. The Lancefield system for classification of streptococci is based on the antigenic characteristics of the group-specific carbohydrate reactions.[97]

For practical purposes, streptococci may be divided into four main groups.

Pyogenic group — On blood agar, streptococci produce small grayish colonies and give characteristic reactions due to the production of hemolysins which result in hemolysis. The pyogenic group of streptococci elaborates group-specific cell wall (C) carbohydrate. In the Lancefield system of classification, the antigenic characteristics of the group-specific C substance are identified as Lancefield Groups A, B, C, D, E, F, and G. Type identification within these groups is done by the precipitin test using type-specific M substance or by the Griffith system of T-agglutination. In epidemiological investigation, typing of Group A streptococci by T-agglutination has proven to be of real value.[98,99] Group A streptococci are responsible for pyogenic infections in man, including streptococcal throat pharyngitis, pneumonias, impetigo, and boils. Rheumatic fever and glomerulonephritis are both sequelae to Group A infection. Group B streptococci cause mastitis in cattle, and certain types are responsible for puerperal fever in women and infections of the newborn. Group D streptococci and other α-hemolytic and nonhemolytic streptococci have been isolated from the blood and shown to cause septicemia and subacute endocarditis. The majority of invasive β-hemolytic streptococci pathogenic for humans fall into Group A, but strains of Groups B, C, and G are encountered in respiratory infections, bacteremia, and endocarditis.

The Center for Disease Control (CDC, Atlanta) recommends the serological grouping procedure (isolation of pure colonies, extraction of the group-specific carbohydrate, and demonstration of a serological reaction between the extracted antigen and specific grouping antiserum) as the method of choice for identifying streptococci. For laboratories unable to do this, Facklam et al.[100] recommend tests based on determination of hemolytic activity, bacitracin susceptibility for Group A streptococci, hydrolysis of sodium hippurate by Group B streptococci, hydrolysis of esculin in the presence of 40% bile for Group D streptococci, and tolerance of 6.5% NaCl broth for enterococcal Group D streptococci (Table 2.1).[100]

Viridans streptococci — These organisms do not produce soluble hemolysins or β-hemolysis on blood agar. Many species induce α-hemolysis, i. e., they turn hemoglobin green. Some species have no action on blood and are called γ-streptococci. They do not produce C carbohydrate. The colonies resemble pneumococci, but the *Streptococcus viridans* are not bile soluble. These organisms comprise the most prominent group of the normal flora of the human respiratory tract and are ordinarily associated with disease only when they settle on an abnormal heart valve (subacute bacterial endocarditis) or establish themselves in the meninges or the urinary tract of humans. They also occur widely in nature. Some synthesize large molecular weight polysaccharides such as dextran.

Enterococci (*Streptococcus faecalis*) — The enterococci produce D-specific C carbohydrate and are capable of growing at 10 and 45°C in 0.1% methylene blue milk, 40% bile agar, or 6.5% NaCl. They cause variable hemolysis. *S. faecalis* is part of the normal flora of the intestinal tract of man and animals[101] and may cause disease when introduced into tissues, bloodstream, urinary tract, or meninges. Occasionally they are associated with food poisoning. Enterococci are quite resistant to many antimicrobial agents. Penicillins often inhibit but do not kill enterococci unless an aminoglycoside is present. Fecal streptococci may be differentiated by the scheme presented in Figure 2.2.

Lactic streptococci — These organisms elaborate Group N-specific C carbohydrate and vary in hemolytic properties. The organisms grow on 40% bile agar but not at 45°C or in 6.5% NaCl, which differentiates them from enterococci. Lactostreptococci do not produce disease, but they are commonly present in milk and are often responsible for its normal coagulation.

Streptococci may be differentiated on the basis of the morphological and growth character-

TABLE 2.1

Presumptive Identification of Streptococci

Group identification

Hemolysis	Group A, β	Group B, β	Nongroup A, B, or D, β	Group D Enterococcus, β, α or none	Group D not an Enterococcus, α or none	Viridans nongroup D, α or none
Bacitracin sensitivity	+	−[a]	−[a]	−	−	V
Hippurate hydrolysis	−	+	−	V	−	−[a]
Bile-esculin hydrolysis	−	−	−	+	+	−
Tolerance to 6.5% NaCl	−	V	−[a]	+	−	−

Note: V = variable.

[a] An occasional exception occurs.

From Facklam, R. R., Padula, J. F., Thacker, L. G., Worthan, E. C., and Sconyers, B. J., *Appl. Microbiol.*, 27, 107, 1974. With permission.

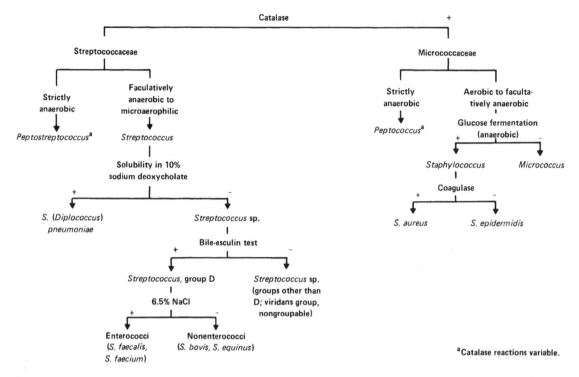

FIGURE 2.2. Scheme for identification of Gram-positive cocci. (From Washington, J.A., II, *Laboratory Procedures in Clinical Microbiology*, Little, Brown & Company, Boston, 1974. With permission.)

istics[102] summarized in Table 2.2. There are a number of physiologic growth and fermentation reactions used to differentiate certain groups and species of streptococci. Some of the tests are more reliable than others. Group B may be differentiated by its ability to hydrolyze sodium hippurate and its ability to grow on 10 and 40% bile agar. Group C is further subdivided by its fermentation reactions on trehalose and sorbitol. Those which do not ferment either of these sugars are classified as *Streptococcus equi*. Streptococci which ferment sorbitol but not trehalose are considered to be animal Group C strains, whereas those which ferment trehalose but not sorbitol are considered to be human strains. Group D streptococci present some difficulties in classification. They may exhibit α, β, or no hemolysis (γ). They grow on 10 to 40% bile agar. Further biochemical differentiation of various species of streptococci is summarized in Table 2.3.

Precipitation and agglutination tests are usually performed for classification of various groups of streptococci. Type-specific antisera are used for the precipitation reactions for specific antigenic polysaccharides of streptococci. Table 2.4 shows chemical compositions of various streptococcal polysaccharides.[103]

Antiserum is prepared in rabbits by injecting formalinized streptococcal cells in an equal volume of Freund's adjuvant. Rabbits are bled, and serum is tested against specific antigens by the precipitin test. The precipitating antigens of streptococci are prepared from the organism grown for 18 to 20 hr at 37°C in a suitable broth medium, followed by incubation in formalin for 24 hr. The cells are then harvested by centrifugation and resuspended in formalinized saline solution. Extracts are then prepared from 400 to 500 mg of cell preparation by the method described by Fuller.[104] A precipitin test is then conducted using a standard technique such as the micro-double-diffusion method.[105]

Pneumococci

The pneumococci are represented by a single species, *Streptococcus (Diplococcus) pneumoniae*, commonly known as the pneumococcus. It is classified under tribe Streptococcaceae, family Lactobacillaceae, and order Eubacteriales. There are over 80 serotypes which are differentiated on the basis of the capsular material elaborated by the organism. The organisms are normal habitants of the upper respiratory tract of humans and can cause pneumonia, sinusitis, otitis, meningitis, and

TABLE 2.2

Identification of Streptococci Based on Morphology and Growth

Species	Characteristics
All	Spherical; arrangement in pairs or chains; nonacid fast; nonmotile; nonsporing; facultative anaerobic; catalase negative; attack sugars fermentatively

Species	On blood agar		SF medium	Growth at		Bacitracin susceptibility	Sodium hippurate
	Surface colonies	Subsurface colonies		10°C	45°C		
β-Hemolytic streptococci							
Group A	White to gray opaque, hard, dry 12-mm zones of hemolysis	2- to 2.5-mm zones of hemolysis with sharply defined edges	No growth	–	–	Susceptible	Not hydrolyzed
Group B	Gray translucent narrow hemolytic zone	0.5-mm zone of hemolysis after 24 hr, 1 mm after 48 hr	No growth	–	–	Resistant	Hydrolyzed
Group C (human)	Similar to Group A	Similar to Group A	No growth	–	–	Resistant	Not hydrolyzed
Group D (enterococci)	Gray, translucent, soft zone of hemolysis wider than the colony	3- to 4-mm zones of hemolysis	Growth, acid reaction	+	+	Resistant	Not hydrolyzed
						Agar	Broth
α-Hemolytic streptococci *Streptococcus faecalis*	Relatively large, gray, shiny, translucent, no greening after 24 hr; slight greening at 48 to 72 hr	Large, nonhemolytic after 25 hr, definite greening after 48 to 72 hr; hemolysis evident after 24 hr refrigeration	Growth			Small, compact	No change

TABLE 2.2 (continued)

| Species | On blood agar | | SF medium | Colonies on 5% sucrose | |
	Surface colonies	Subsurface colonies		Agar	Broth
S. salivarius	Small, raised, convex, opaque narrow zone of hemolysis with or without greening depending on blood used	Typical α appearance, fixation of cells, greening and hemolysis after 24 to 48 hr	No growth	Large, raised mucoid	No change
S. sanguis	α-Hemolysis in 24 to 48 hr	α-Hemolysis in 24 to 48 hr	No growth	Small, compact	Gelling of medium
S. mitis group	α-Hemolysis in 24 to 48 hr	α-Hemolysis in 24 to 48 hr	No growth	Small, compact	No change
Streptococcus MG	Nonhemolytic after 24 hr, but may give α appearance after 48 hr, variations due to blood	Nonhemolytic after 24 hr, changing to α at 48 hr, refrigeration for 24 hr produces typical appearance	No growth	Small, compact, fluorescent in ultraviolet light	No change

Species	Hemolysis in sheep blood agar	6.5% NaCl broth	Gelatin liquefaction	Litmus milk
Enterococci[a]				
Streptococcus faecalis	γ to α	Growth	–	Reduced
S. faecalis var. liquifaciens	γ to α	Growth	+	Reduced
S. faecalis var. zymogenes	β	Growth	±	Reduced
S. durans	β	Growth	–	Not reduced

[a]The enterococci will grow at 10 and 45°C and will survive temperature of 56°C for 30 min. They also grow at pH 9.6 and in 0.1% methylene blue milk.

Adapted from Facklam,[102] Facklam et al.,[102] and Pavlova et al.[101]

TABLE 2.3

Summary of Biochemical Differentiation of Various Species of Streptococci

Biochemical test	S. pyogenes	S. agalactiae	S. dysgalactiae	S. equisimilis	S. equi	S. zooepidemicus	S. bovis	S. equinus	S. durans	S. faecalis	S. faecium	S. sanguis	S. hominis	S. lactis	S. cremoris	S. mitis	S. pneumoniae	S. viridans	S. uberis
Esculin hydrolysis	−	−	−	d	−	+	+	+	+	+	+	+	+	d	d	−	:	d	+
Arginine hydrolysis	+	+	+	+	+	+	−	−	+	+	+	+	−	+	−	d	:	−	+
Hippurate hydrolysis	−	+	d	−	−	−	−	−	d	d	d	−	−	d	−	−	−	+	+
Gelatin liquefaction	−	−	−	−	−	−	−	−	−	±	−	−	−	−	−	−	−	−	−
Litmus milk	A	AC	B	B	−	A	A	−	A	RAC	A	AC	RAC	RAC	RAC	RAC	AC	A	RAC
Arabinose (acid)	−	−	−	−	−	−	d	−	−	−	+	−	−	d	−	−	−	d	−
Maltose (acid)	+	+	+	+	+	d	+	+	+	+	+	+	+	+	−	+	+	+	+
Trehalose (acid)	+	+	+	+	−	−	d	−	d	+	+	+	+	d	d	−	+	+	+
Raffinose (acid)	−	−	−	−	−	+	+	−	−	−	−	d	+	d	−	+	+	d	−
Salicin (acid)	+	d	d	d	+	−	+	+	d	+	+	+	+	d	d	+	d	+	+
Glycerol (acid)	−	+	−	+	−	−	−	−	+	+	+	−	−	−	−	−	+	+	+
Mannitol (acid)	−	−	−	−	−	−	d	−	+	+	d	−	−	d	−	−	d	d	+
Sorbitol (acid)	−	−	d	−	−	+	−	−	−	+	−	−	−	−	−	−	−	d	+
Bile solubility	−	−	−	−	−	−	−	−	−	−	−	−	−	−	−	−	+	−	−

Note: A, acid; B, acid produced but clotting variable; C, clot; R, reduction of different biochemical types; ••, results not known.

Adapted from Washington,[95] Facklam et al.,[100] and Cowan and Steel.[130]

TABLE 2.4

Chemical Composition (%) of Streptococcal Polysaccharide Type Antigens

Type	Glucose	Galactose	Rhamnose	Glucosamine	Galactosamine	Fucose
I	22	46	23		13	
II	26	12	24		16	
III	33	52	15			
IV[a]	35	31	26	8		
IIIm from culture medium[a]	17	55	10			17

[a]IIIm excreted into the culture medium by strains carrying the Type III antigen in the cell wall; reacts with Antitype III serum.

From Huis In't Veld, J. H. J., Meuzelaar, H. L. C., and Tom, A., *Appl. Microbiol.*, 26, 92, 1973. With permission.

other infectious processes. They may be readily isolated from the sputum, blood, and exudates of persons with lobar pneumonia; from the cerebrospinal fluid in pneumococcal mengitis; and from patients with otitis media, peritonitis, endocarditis, and other diseases.

The pneumococci are Gram-positive diplococci, often lancet shaped or arranged in chains, and possess a capsule of polysaccharide which permits easy typing with specific antisera. They can be lysed by surface-active agents, e.g., bile salts. *Diplococcus pneumoniae* is primarily but not obligately parasitic and requires enrichment media for its cultivation. Pneumococci from blood and body fluids grow well in beef-heart infusion broth or in Trypticase soy broth. They also grow in standard thioglycollate medium. A few strains require an increased CO_2 tension on initial isolation on solid medium. Pneumococci ferment carbohydrates, with the production of acid but no gas. Most strains of pneumococci give rise to colonies with typical morphology. They are smooth, glistening, 1 to 2 mm in diameter, and α-hemolytic. As the colony ages, central autolysis gives rise to a central depression which gives the colony a checker-like appearance.[106] Colonies of Types 3 and 37 are larger and more mucoid and can be recognized by direct inspection of plates.

Pneumococcal colonies are presumptively identified by examining plates with obliquely reflected illumination under a colony or by using a dissecting microscope at a magnification of 40 to 50. The colonial and cellular morphology of α-hemolytic streptococci may resemble that of pneumococci, and several tests are sometimes performed to facilitate their differentiation (Table 2.5).[107] α-Streptococci produce small, raised,

opaque, convex colonies with or without greening and hemolysis.

Merrill et al.[108] have rediscovered the value of the capsular precipitin (quellung) reaction of the pneumococcus with homologous anticapsular serum.[109,110] Using omniserum (a pool of concentrated rabbit antiserum to 82 pneumococcal types),[58,111] 89% correlation was obtained between the quellung reaction and the results of culture. Gram stain gave a 55% correlation with culture results.

The technique used was to air-dry a sputum smear and overlay it with a loop of omniserum. A loop of methylene blue was placed on the underside of a coverslip which was then placed on the specimen. Using a light microscope, oil immersion lens, and reduced illumination, the capsular precipitin reaction could be seen immediately.

Lorian and Markovits[112] offered a new test for distinguishing pneumococci from other α-hemolytic streptococci for use in those cases where other characteristics were insufficient for an unequivocal identification. It is based on the observation[113,114] that strains of pneumococci, if grown anaerobically on horse blood agar and then exposed to air at 6°C, produce a ring of β-hemolysis surrounding the zone of inhibition formed by a methicillin disk. All 125 strains of *Diplococcus pneumoniae* and 125 strains of α-hemolytic streptococci (91 strains of *Streptococcus mitis*, 32 strains of *S. salivarius*, and 2 strains of *S. sanguis*) isolated from sputum were susceptible to methicillin and showed α-hemolysis after 24 hr of anaerobic incubation at 37°C. Following 24 hr of exposure to air at 6°C, 76% of the strains of pneumococci produced a ring of β-hemolysis 2 to 7 mm in width surrounding the

TABLE 2.5

Differentiation of *Diplococcus* and *Streptococcus* by Conventional Methods

Test	Organism										
	Diplococcus	*S. pyogenes*	*S. agalactiae*	*S. canis*	*S. dysgalactiae*	*S. equi*	*S. equisimitis*	*S. suis*	*S. uberis*	*S. zooepidemicus*	*Streptococcus group*
Hemolysis[a]	g	c	g⁻ᶜ	c	g⁻ᶜ	c	c	c	cg	c	c
Motility	−	−	−	−	−	−	−	−	−	−	−
Capsulation	+	+	+	+	+	+	+	+	+	+	+
Optochin® sensitive	+		−		−				−		
Sodium hippurate		−	+	−	−	−	−	−	+	−	−
Litmus milk reduction[b]	ac⁻	a	aᶜ	ar	acʳ		aʳ		aʳ	a	
Gelatin liquefaction	−	−	−		−	−	−		−	−	
Fermentation of[c]											
Arabinose	+(−)	+	−						−		
Dextrose	+(−)	+	+		+	+	+		+	+	
Inulin	+(−)	−	−	−	−	−	−	−	W	−	−
Lactose	+(−)	+(−)	+(−)	+	±	−	±		+	+(−)	±
Maltose		+	+		+	+	+	−	+	+	
Mannitol		−(+)	−	−	−	−	−	+	+	+(−)	−
Salicin	+	+	+(−)	+	±	+	+(−)	+	+	+	−
Sorbitol	−	−	W	−	⊄	−	+	±	+	+	−
Trehalose	+	+	W	−	+	−	+	+	+	−	−

[a] c, clear; i, incomplete; −, none; g, green.
[b] a, acid; b, alkaline (basic); r, reduction; c, coagulation; d, digestion.
[c] +, pH acid; −, pH neutral or alkaline; +(−), usually +, few −; −(+), usually −, few +; ±, positive or negative, more strains +; W, weak positive.

From Oetjen, K. A. B. and Harris, D. L., *J. Am. Vet. Med. Assoc.*, 163, 169, 1973. With permission.

zones of inhibition produced with the methicillin disk; a ring of α-hemolysis was seen in the remaining 24% of the pneumococci. After 48 hr of exposure to air at 6°C, all 125 strains of pneumococci showed a ring of β-hemolysis 3 to 8 mm in width surrounding the zone of methicillin inhibition. None of the 125 strains of α-hemolytic streptococci showed β-hemolysis after either 24 or 48 hr of exposure to air at 6°C.

As mentioned previously, the Optochin (ethylhydrocupreine hydrochloride) disk test[45] is used to distinguish pneumococci (growth inhibited) from hemolytic streptococci. Bacitracin is used as a test for Group A streptococci.[115-117] A disk saturated with a 1 μg/ml solution of bacitracin placed on a heavily inoculated blood agar plate inhibits 97% of Group A streptococci. Most other hemolytic streptococci are uninhibited. This is a presumptive test, and the Lancefield procedure of serological grouping[97] or fluorescent antibody (FA) grouping[118] is needed for final identification. Disks are commercially available as Taxo A disks (BBI) and as bacitracin differentiation disks (Difco and Case Laboratories, Chicago). Note that differential disks have 0.04 μm per disk, while sensitivity disks have 10 μm per disk. Reliability of the differential test requires pure cultures on which hemolysis was previously determined. Many α-hemolytic streptococci and pneumococci are sensitive to bacitracin. A summary of identification of Gram-positive cocci is presented in Figure 2.2.

Gram-negative Cocci
Neisseriaceae

The neisseriae are Gram-negative diplococci, many species of which normally colonize the oropharynx and the gastrointestinal and genitourinary tracts of humans. The two pathogenic species of *Neisseria*, *N. meningitidis* and *N. gonorrhoeae*, are similar in their morphological and cultural characteristics. They are coffee-bean shaped cocci that are often paired and are catalase and oxidase positive; some species of neisseria utilize carbohydrate oxidatively.

Examination with a dissecting microscope of staining *N. meningitidis*, *N. gonorrhoeae*, and *N. lactamicus* grown on translucent agar media shows the nonpathogenic *Neisseria* to be distinguishable by colonies with dentate edges and/or pigmentation.[119] Meningococcal colonies are entire and nonpigmented. Isolates of Group Y meningococcal

strains have characteristic growth in larger colony size and concentric rings. Fresh isolates of *N. meningitidis* are differentiated from laboratory stock cultures in that they are translucent and homogeneous without an "internal structure" as opposed to a frosted appearance, "internal structure," and concentric rings of growth seen in laboratory cultures. Lactose-fermenting strains show a marked increase in yellow pigment production when grown on agar media containing 0.02% cystine, whereas the meningococci do not. The difficulty of serogrouping *N. lactamicus* strains due to autoagglutination can be corrected by increasing the molarity of the saline. This will permit all lactose-fermenting *Neisseria* to be serogrouped.

Neisseria species of medical importance can be differentiated by the morphological, biochemical and serological tests summarized in Table 2.6 and Figure 2.3.

Slide and tube agglutination tests can be performed to identify *Neisseria* species. The slide method is generally used to facilitate rapid identification. The Subcommittee on Taxonomy and Nomenclature of the *Neisseriae*[120] designated four serological groups (A, B, C, and D) as the international standard of *N. meningitidis*. Group A

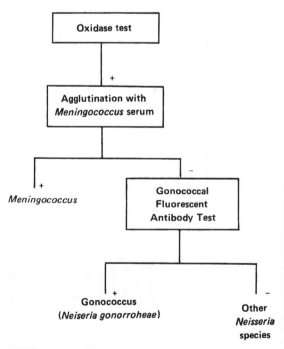

FIGURE 2.3. Identification of Gram-positive cocci. (From Wasilauskas, B. L. and Ellner, P. D., *J. Infect. Dis.*, 124, 449, 1971. With permission.)

TABLE 2.6

Differentiation of Some Species of *Neisseria*

Test	N. gonorrhoeae	N. meningitidis	N. lactamicus	N. subflava
Growth on Thayer-Martin or Martin-Lester medium	+	+	+	–
Growth on nutrient agar				
22°C	–	–	–	±
30°C	+	+	+	+
37°C	+	+	+	+
β-Galactosidase[a] (ONPG)	–	–	+	–
Carbohydrate utilization[b]				
Glucose	+	+	+	+
Maltose	–	+	+	+
Sucrose	–	–	–	–
Lactose	–	–	+	–

[a]Must be performed with colonies grown on medium containing lactose.
[b]Cystine Trypticase® (BBL) or tryptic (Difco) semisolid agar (CTA) with added 1% sugar, inoculated by stabbing three or four times with a heavy suspension of organisms in 0.5 ml of Trypticase, tryptic soy broth, or sterile distilled water; incubated at 35°C without added CO_2 and examined at 24, 48, and 72 hr.

From Washington, J. A., II, *Laboratory Procedure in Clinical Microbiology*, Little, Brown & Company, Boston, 1974, chap. 4. With permission.

is the classical epidemic strain Type I; Group B is Type II; Group C is Type II alpha, and Group D is Type IV.

Antisera are commercially available to test meningococci. Saline suspensions from overnight cultures on chocolate agar or starch agar (Difco) should be used. Antigens are made equivalent to a No. 3 MacFarland standard (900 million cells per milliliter). The antigen is added to a specific antiserum on a microscope slide and is shaken for a maximum of 3 min before being called negative. Most strains agglutinate within 1 min. Several colonies from negative or rough strains should be transferred to starch agar. Generally, they will agglutinate after one to three transfers. Many nonagglutinables from the original drug plates will agglutinate after one transfer on chocolate agar without drugs.

Serogrouping may also be performed by fluorescence microscopy. Several reports describe fluorescent antibody (FA) procedures for examination of cerebrospinal fluid (CSF) for detection and presumptive identification of *Neisseria meningitidis*.[121-123] A heat-fixed smear of CSF is overlayed with conjugates of polyvalent *N. meningitidis* (Groups A, B, and C). The cover of a large petri dish fitted with moist filter paper is placed over the slide and is allowed to stand at room temperature for 20 min. The excess reagent is removed and rinsed with PBS and then with distilled water. After mounting the slide with a coverslip (using a drop of carbonate-buffered glycerol, pH 9, as mounting medium), the slide is examined with an FA microscope and oil-immersion objective. The morphology (capsular vs. cell wall staining) and intensity of staining (± to 4+) are noted. If a specimen is positive with any one of the polyvalent reagents, additional smears and stains should be prepared with the individual components.

Neisseria gonorrhoeae species may be identified by the FA technique by a rapid method in which a small portion of the colony is emulsified in a drop of distilled water on a glass slide. The smear is air-dried for 15 to 20 min and overlayed with a working dilution of *N. gonorrhoeae* conjugate. The slide is then incubated for 5 min at room temperature in a moist atmosphere and is rinsed with distilled water. A drop of mounting fluid is placed on the smear and a coverslip is applied. The smear is then examined with a fluorescent microscope. Cells of *N. gonorrhoeae* should stain brilliantly at the working titer. However, if organisms suspected to be *N. gonorrhoeae* do not fluoresce in the above method and are Gram-negative diplococci which produce oxidase-positive colonies on Martin-Lester or Thayer-Martin medium, they may be tested by the following method (the Long method). A few colonies are suspended in PBS, and a drop of suspension is placed on a glass slide; the smear is air-dried and then heat-fixed gently. One or two drops of *N. gonorrhoeae* FA conjugate is placed over the smear. The slide is covered and incubated for 1 hr at 37°C. The slide is rinsed with PBS and then with distilled water. The smear is then examined with the fluorescent microscope as described previously. Cultures of *Staphylococcus aureus* and *Enterobacter cloacae* should be used as negative controls.

Gram-negative Bacilliae
Enterobacteriaceae

According to Ewing,[124] this family consists of 4 divisions and 11 genera, namely, *Escherichia* and *Shigella* (Division I); *Edwardsiella*, *Salmonella*, *Arizona*, and *Citrobacter* (Division II); *Klebsiella*, *Enterobacter*, and *Serratia* (Division III); and *Proteus* and *Providencia* (Division IV). Morphological differentiation of strains within the species or species within the genera is difficult because all of the organisms are aerobic, Gram-negative, asporogenous, and rod-shaped bacteria that grow well on artificial media. However, colonial characteristics of some of the members are typical, for example, pigmented colonies of *Serratia* and swarming colonies of proteus species. Most of the members of the Enterobacteriaceae family are autochthonous or transient habitants of the gastrointestinal tract of animals and humans; they sometimes produce severe gastrointestinal disease. Identification of the members of Enterobacteriaceae is made primarily by biochemical reactions and serological techniques. Nitrates are reduced to nitrite, and glucose is utilized fermentatively with the formation of acid or acid and gas by Enterobacteriaceae. Different kinds of carbohydrates may be incorporated into O-F Basal medium (Difco) for the production of acid or acid and gas. Kliger's iron agar (KIA) and triple-sugar iron agar are used for the differentiation of Gram-negative intestinal bacilli on the basis of their ability to ferment dextrose and lactose and to liberate sulfides. Fuscoe[125] described a two-tube medium based on TSI medium for the biochemical identification and differentiation of pathogenic enterobacteria. With slight modification, the medium provides for differentiation of coliforms on the basis of motility; oxidation or fermentation of glucose and mannitol; and production of indole, hydrogen sulfide β-galactosidase and phenylalanine deaminase reactions.

Elazhary et al.[126] reported a procedure for identification of selected Gram-negative bacteria based on primary isolation and differentiation of the bacteria by the use of MacConkey agar and the subsequent differentiation according to a procedure using the replica plating technique (Table 2.7). Initial differentiation into lactose fermentors (+LF, *Escherichia*, *Klebsiella*, and *Enterobacter*, the latter two able to utilize citrate as their unique source of carbon) and nonlactose fermentors (–NLF, *Pseudomonas*, *Alcaligenes*, *Aeromonas*, *Proteus*, *Providencia*, *Serratia*, *Edwardsiella*, *Arizona*, *Citrobacter*, and *Shigella*) is made on the MacConkey agar plate. The genera are further differentiated by citrate, glucose (Sellers), oxidase, and lysine decarboxylase and other tests (Figure 2.4). Brooks and Goodman[127] used a modified standard medium with phenylalanine to differentiate groups of the Enterobacteriaceae (Figure 2.5).

Iveson[128] found strontium chloride B broth incubated at 37 or 43°C and subcultured into deoxycholate citrate (DC) agar to be the best medium for isolation of *Edwardsiella* species. He noted that occasionally strains isolated by direct plating were not isolated after enrichment. Strontium selenite A broth incubated at 37 or 43°C and subcultured to a modified bismuth sulfite (BS) agar was best for the isolation of *Arizona* species. The modified BS agar was found to be superior to DC agar for isolation of *Salmonella* and *Arizona* species. *Edwardsiella* and *Shigella* species were isolated exclusively on DC agar.

Roberts[129] adapted a scheme from the tables of Cowan and Steel[130] and Wilson and Miles[131] to identify the species of most Gram-negative bacilli within 24 hr after the first isolation. It is based on the following tests: swarming colony characters, sugar fermentation, indole production, growth in potassium cyanide, urease production, oxidase, motility, hydrogen sulfide production, size and pigmentation of colonies, etc. (Figure 2.6).

Enterobacteriaceae strains do not produce cytochrome oxidase. Many filter strips or disks are available for quick assay of oxidase; if blackening

TABLE 2.7

Different Tests Used in the Identification of Selected Gram-negative Bacteria

Test or substrate	Medium
Lactose	MacConkey agar
Rhmnose	MacConkey agar base + 0.5% rhamnose
Glucose[a]	Seller's differential agar
Glucose	MacConkey agar base + 0.5% glucose
Citrate	Simmons' citrate
Malonate	Malonate base + 0.03% bactopeptone
Urease[a,b]	Urea agar base
PPA [a,c]	Phenylalanine agar
Lysine decarboxylase[a]	Lysine decarboxylase broth + 0.1% lysine + 2.0% Ionager® (Oxoid®)
Cytochrome oxidase[c]	MacConkey agar

[a]Incubation under anaerobic conditions.

[b]The phenylpyruvic acid production is shown by flooding the plates with the following reagent: ammonium sulfate, 2.0 g; sulfuric acid (10%), 1.0 ml; half-saturated ferric alum, 5.0 ml. After 1 or 2 min an intensive blue color developing in the colonies indicates a positive reaction.

[c]Cytochrome oxidase activity is shown by flooding the plates with a freshly prepared 1.0% solution of tetramethy-p-phenylenediamine. Oxidase-positive colonies develop a pink color which successively becomes maroon, dark red, and black in 10 to 30 min.[27]

From Elazhary, M. A. S. Y., Saheb, S. A., Roy, R. S., and Lagace, A., *Can. J. Comp. Med.*, 37, 43, 1973. With permission.

of these materials occurs, the Gram-negative organisms do not belong in Enterobacteriaceae. Instead, *Aeromonas, Pseudomonas, Alcaligenes,* and *Vibrio* should be considered. Based on the oxidase test, and followed by phenylalanine deaminase, urease, indole, and other tests, Wasilauskas and Ellner[132] presented a scheme for presumptive identification of Gram-negative bacilli within 4 hr of primary isolation of the bacteria (Figure 2.7).

Serological Tests

An agglutination test using commercially available antisera against enteropathogenic *Escherichia coli* (EEC), *Shigella,* and *Salmonella* is frequently performed for the identification of these species.[133,134] The EEC responsible for neonatal diarrhea has been shown to be biochemically identical to commensal *E. coli*; for example, the IMViC test results are similar for both EEC and *E. coli.* The characteristics are as follows: a Gram-negative rod that ferments lactose in 48 hr and is indole positive, methyl red positive, Voges-Proskauer negative, and citrate negative. Because of the high mortality associated with EEC out-

breaks in maternity wards, rapid diagnosis is important. Therefore, the serological test is used to differentiate EEC from *E. coli*; the EEC possess heat-labile B antigens that are invariably associated with the O lipopolysaccharide complex. They cause inability to agglutinate the O antigen with homologous antiserum. In most hospital laboratories, the serologic analysis of EEC consists of performing slide agglutination tests with typical colonies of *E. coli.*

At present, there are nine serogroups of *E. coli* which have been implicated as etiologic agents of infant diarrhea in the United States (026:B6, 055:B5, 0111:B4, 0127:B8, 086:B7, 0119:B14, 0125:B15, 0126:B16, and 0128:B12). To test EEC, a dense suspension of bacteria is prepared in isotonic solution. The test organism is obtained with a wire loop from either TSIA or a nutrient agar slant. However, portions of ten colonies resembling *E. coli* on BAP and EMB may be tested directly in poly A and poly B antiserum. Poly A antiserum contains agglutinins for 026:B6, 055:B5, 011:B4, and 0127:B8 serogroups. Poly B antiserum contains agglutinins for 086:B7,

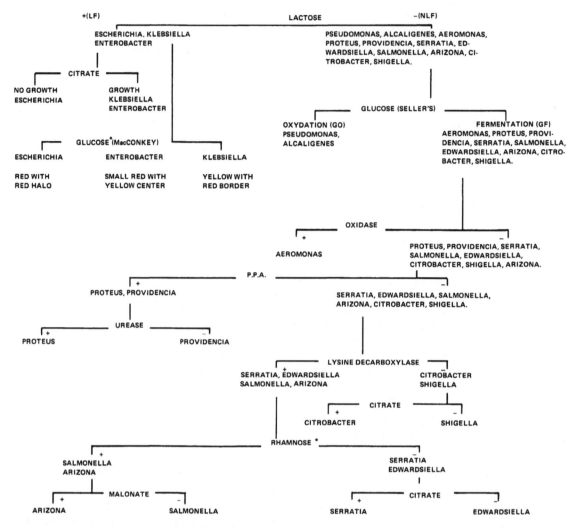

FIGURE 2.4. Differentiation of Gram-negative bacilli based on lactose fermentation test. (From Elazhary, M. A. S. Y., Saheb, S. A., Roy, R. S., and Lagacé, A., *Can. J. Comp. Med.*, 37, 43, 1973. With permission.)

019:B14, 0124:B17, 0125:B15, 0126:B16, and 0128:B12 serogroups. A drop of bacterial suspension is placed at one end of the rectangle of a 100-mm petri dish, and a drop of antiserum (poly A or poly B) is placed at the other end. The bacterial suspension and the antiserum are mixed with a wire loop along an oval tract. The plate is tilted back and forth to mix the antigen and antiserum. The antigen is tested by placing a drop of the saline suspension in a rectangle without antiserum. If no agglutination appears after 1 min of mixing with antiserum, the test is considered to be negative. Absence of agglutination should exclude EEC. If agglutination occurs in one of the antiserum pools, the bacterial suspension is heated

in a boiling water bath for 15 to 30 min and the agglutination test is repeated with the same antiserum in which agglutination previously occurred.

Similarly, agglutination tests may be performed for *Shigella* with *Shigella* Groups A, B, C, or D antiserum and *Salmonella* Groups A, B, C_1, C_2, D, and E antiserum. *Salmonella typhi* does not usually agglutinate in polyvalent O or Group D antiserum. In the unheated or live state, *S. typhi* will usually agglutinate with Vi antiserum. After heating in a boiling water bath, *S. typhi* will no longer agglutinate with Vi antiserum but will agglutinate with polyvalent O and Group D antiserum.[95]

Fluorescent antibody (FA) testing is also used

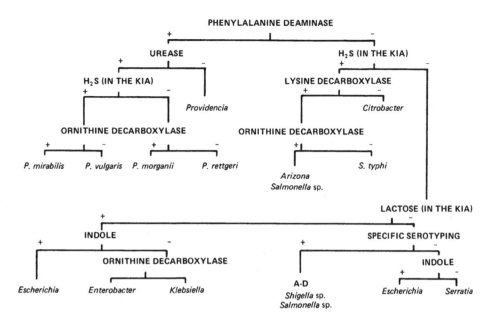

FIGURE 2.5. Differentiation of Gram-negative bacilli based on phenylalanine test. (From Brooks, W. F., Jr. and Goodman, N. L., *Am. J. Med. Technol.*, 38, 429, 1972. With permission of American Society for Medical Technology.)

routinely in many diagnostic bacteriology laboratories to identify EEC.[18,135] FA staining with OB serum (heat-labile somatic antigen B) provides results which are at least as reliable as those obtained with slide agglutination tests. FA techniques are nearly 100% reliable in excluding the presence of these bacteria (EEC) in fecal material.

Pseudomonadaceae

The two genera, *Aeromonas* and *Pseudomonas,* in the family Pseudomonadaceae are Gram-negative bacilli that morphologically resemble organisms of other genera, such as *Vibrio, Achromobacter,* and the enteric bacilli. Among the tests commonly used for differentiation of pseudomonads and aeromonads are the presence and type of flagella (polar or peritrichous), fermentation or oxidation of sugars, oxidase, demonstration of pigments (pyocynine, fluorescine, or both), ability to grow at a low temperature (e.g., 4°C) or high temperature (42°C), and serological methods. Morphological and biochemical characteristics of *Aeromonas* species are summarized in Table 2.8.

Pseudomonads are Gram-negative, asporogenous, motile, straight or slightly curved rods which produce catalase; nonmotile species are also included in this genus. When flagellated, they possess one polar flagellum or several polar flagella. Molecular oxygen is used by these organisms to degrade glucose, arabinose, ribose, and xylose. *Pseudomonas aeruginosa* does not utilize dulcitol, inulin, maltose, milezitose, raffinose, salicin, or sorbitol. Gas is not produced from carbohydrates. Some species respire in the presence of nitrate and grow under anaerobic conditions in the presence of nitrate or arginine. Indole, methyl red, and acetyl methyl carbinol tests are negative. Pseudomonads produce dense turbidity in neutral peptone broth in 18 to 24 hr at 30°C and fail to grow in brain-heart infusion broth at pH 4.5 Differential characteristics of *Pseudomonas* are summarized in Table 2.9.

Verder and Evans[143] proposed a serological method using somatic antigen of pseudomonads in the agglutination test. Since there are many serotypes within a species of pseudomonad, the agglutination test is usually useful only for the determination of serotypes within a species and not for identification at the species level. Extracellular soluble antigens may be used for the differentiation of *Pseudomonas* species by the method described by Elek.[144] A filter paper soaked in antiserum is placed in a tryptone glucose extract agar (Difco) that has been poured into a petri dish and has not yet solidified. After the agar solidifies, the organism is streaked perpendicularly to the paper strip. After a suitable incubation period (1 to 2 days at 37°C for *P. aeruginosa*), a

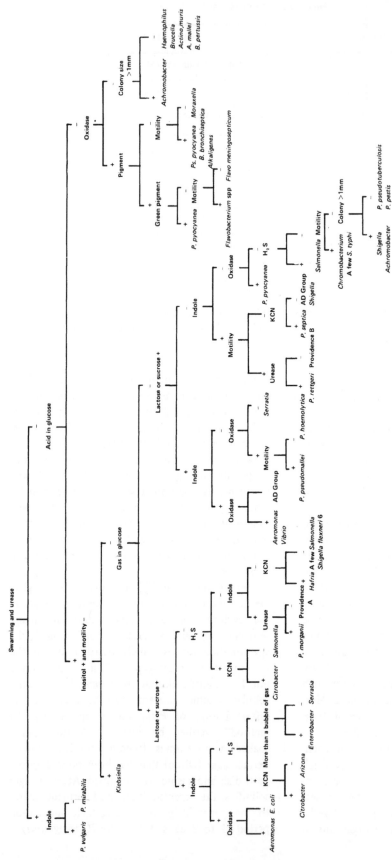

FIGURE 2.6. Scheme for identification of Gram-negative bacilli. (Adapted from Roberts, G., *Med. Lab. Technol.*, 28, 382, 1971. With permission of the Institute of Medical Laboratory Sciences.)

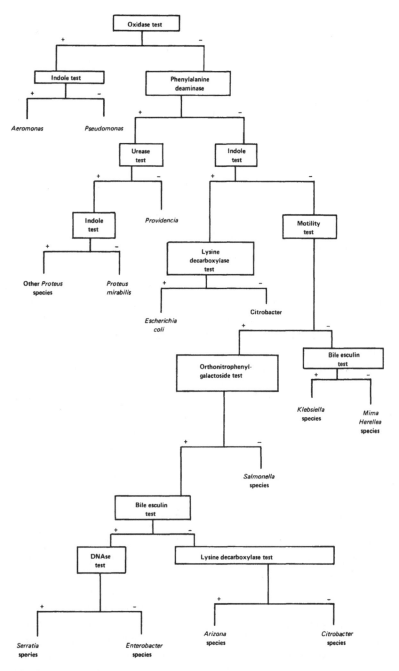

FIGURE 2.7. Scheme for identification of Gram-negative bacilli based on oxidase test. (From Wasilauskas, B. L. and Ellner, P. D., *J. Infec. Dis.*, 124, 499, 1971. With permission.)

positive reaction is seen as an arrow-shaped line of precipitation on both sides of the bacterial colony. This test can be modified and performed on a microscope slide.[145]

Brucellaceae

Haemophilus — The genus *Haemophilus* con-sists of minute, rod-shaped organisms which are sometimes thread-forming and pleomorphic cells. They are Gram-negative, nonmotile, strict para-sites, growing only in the presence of certain growth accessory factors. Most members of this genus are normal inhabitants of mucous mem-branes of the upper respiratory tract of man.

TABLE 2.8

Some Characteristics for the Differentiation of *Aeromonas* Species

Characters	*A. liquefaciens*	*A. shigelloides*	*A. salmonicida*
Number of flagella	1	>2	0
Growth at 37°C	+	+	–
Brown pigment	–	–	+
Indol	+	+	+
Methyl red	+	+	+
Voges-Proskauer	+	–	–
Simmons citrate	+	–	–
Inositol	–	+	–
Gelatin	+	–	+

From Liu, P. V., *Rapid Diagnostic Methods in Medical Microbiology,* Graber, C. D., Ed., Williams & Wilkins, Baltimore, 1970, 116. With permission.

Primary and secondary invaders cause sinusitis, conjunctivitis, acute epiglottis (croup), obstructive laryngitis, pneumonia, meningitis, and pyarthrosis in children. Typical biochemical characteristics of *Haemophilus* include variable fermentation actions, reduction of nitrates, variable production of indole, and the hemolyzing of blood in some species (Table 2.10). Encapsulated strains of *H. influenzae* may be identified by capsular swelling or precipitin tests.

Bordetella — The genus *Bordetella* contains *B. pertussis, B. parapertussis,* and *B. bronchiseptica* species. Most important for humans is *B. pertussis,* a small Gram-negative organism responsible for whooping cough in children. *B. pertussis* will not grow initially on blood agar, whereas *B. parapertussis* and *B. bronchiseptica* will. *B. pertussis* does not ferment carbohydrates and produces no significant chemical reactions. *B. parapertussis* splits urea and utilizes citrate, while *B. pertussis* does not. *B. bronchiseptica* is an animal pathogen occasionally causing respiratory infection resembling pertussis in animal handlers. It requires neither the X (hemin) nor the V (nicotinamide adenine dinucleotide, NAI) factors for growth, whereas *B. pertussis* requires X and V factors for primary isolation. *Bordetella* may be differentiated by fluorescent antibody testing (immunofluorescent procedures with known fluorescein-conjugated *B. pertussis* antiserum) or by bacterial serology.

Brucella — The species of *Brucella* are small, nonmotile, nonsporing, Gram-negative rods, usually coccobacillary, 0.4 to 1.5 μm in length; they occur singly or in pairs and rarely in short chains. Capsules, if present, are small. Growth is enhanced in an atmosphere of 5 to 10% CO_2 and will not occur in an anaerobic medium. There are three classical species of *Brucella,* each of which has several recognized biotypes. These organisms are quite host specific, but man is susceptible to all three, and under the appropriate conditions any of these species may induce brucellosis in a wide range of other hosts as well. Differentiation of *Brucella* species is based on the pattern of inhibition by dyes, production of H_2S, hydrolysis of urea, metabolic studies, serological tests, and animal inoculation studies (Table 2.11).

Pasteurella, Yersinia, and *Francisella* — Members of these genera are Gram-negative coccobacilli, occurring singly or in pairs; they vary in size from 0.2 to 0.5 × 0.25 to 1.25 μm. Members of the genus *Pasteurella* are oxidase positive and include *P. multocida, P. pneumotropica, P. haemolytica,* and *P. ureae. Yersinia* species are Gram negative; clinically important organisms include, *Y. pestis, Y. enterocolitica,* and *Y. pseudotuberculosis. Francisella (Pasteurella) tularensis* is a small, nonmotile, Gram negative coccobacillus which grows best at 35°C on cystine-glucose blood agar. Colonies may require as long as 10 to 14 days to develop. Biochemical characteristics of *F. tularensis* include fermentation of glucose, maltose, and mannose with production of acid; no gas; variable fermentation of glycerol, levulose, and dextrin; and production of H_2S in cystine medium. Slide agglutination, tube agglutination, or fluorescent antibody staining techniques may be used to identify this organism. However, rapid slide agglutination with specific antiserum is sufficient as a routine identification test in clinical bacteriology. Differentiation based on morphologi-

TABLE 2.9

Differential Characteristics of *Pseudomonas*

Species	O-F Glucose	O-F Maltose	Nitrite	Motility	Oxidase	Growth at 4°C	Growth at 41°C	Pyocanine	Fluorescein	L-Lysine decarboxylation	L-Arginine dehydrolysis	Hydrolysis Gelatin	Hydrolysis Starch	Comment
P. aeruginosa	+	-	+ (gas)	+	+	-	+	+	+	-	+	+	-	May be apyocyanogenic
P. fluorescens	+	-	d	+	+	+	-	-	+	-	+	+	-	Polar or bipolar tuft of flagella
P. putida	+	-	-	+	+	d	-	-	+	-	+	-	-	
Comamonas terrigena	-	-	+	+	+	-	-	-	-	-	-	-	-	Polar monotrichous flagella
P. alcaligenes	-	-	-	+	+	-	+	-	-	-	(+)	-	-	Polar monotrichous flagella
P. stutzeri	+	+	+ (gas)	+	+	-	d	-	-	-	-	-	+	Rough, wrinkled, dry, coherent, yellow colony
P. maltophilia	+	+	-	+	-	-	-	-	-	+	-	+	-	
P. pseudomallei	+	+	+ (gas)	+	+	-	+	-	-	-	+	+	-	
P. mallei	+	d	d	-	+	-	+	-	-	+	+	-	-	
P. cepacia, formerly P. multivorans or P. kingii (EO-1)	+	+	-	+	+	-	d	-	-	d	-	d	-	Yellow pigment on iron-containing media (TSIA); polar tuft of flagella

Note: d = different biochemical types.

From Washington, J. A., II, *Laboratory Procedure in Clinical Microbiology*, Little, Brown & Company, Boston, 1974, chap. 4. Boston, 1974, ch 1 p. 4.

TABLE 2.10

Differentiation of *Haemophilus* Species

Organism	Growth factor		Reaction		
	X	V	Indole	Nitrate	Hemolysis
H. influenzae	+	+	±	+	–
H. parainfluenzae	–	+	±	+	–
H. aegyptius[a]	+	+	–	+	–
H. hemolyticus	+	+	±	+	+
H. parahemolyticus	–	+	±	+	+
H. aphrophilus	+[b]	–	–	+	–
H. ducreyi	+	–	–	–	Slight

[a]Hemagglutinates red blood cells strongly; *H. influenzae* gives weaker hemagglutination.
[b]Requires X factor only when grown under CO_2.

Adapted from Young, V. M., in *Manual of Clinical Microbiology,* 2nd ed., Lennette, E. H., Spaulding, E. H., and Truant, J. P., Eds., American Society for Microbiology, Washington, D.C., 1974, 304. With permission.

cal and biochemical characteristics of *Pasteurella* and *Yersinia* species is summarized in Table 2.12.

Vibrio

The genus *Vibrio* of the family Spirillaceae has more than 30 species, but only a few are pathogenic, including *V. comma* and *V. El Tor* (the causative agents of cholera), *V. parahaemolyticus* (encountered in gastroenteritis in Japan), and the microaerophilic *V. fetus* (primarily an animal pathogen but occasionally isolated from a few clinical conditions in humans). The organisms are Gram-negative, slightly curved rods, 0.3 to 0.6 × 1 to 5 μm in size, occur singly or joined end to end, are actively motile, possess a single polar flagellum, and are noncapsulated; they do not form spores. Vibrios grow best aerobically on alkaline medium at 37°C. The main biochemical characteristics of vibrios are indole positive, reduction of nitrate, cholera red positive, Voges-Proskauer negative, fermentation of sucrose and mannose with production of acid, but no gas, and no fermentation of arabinose. They are also oxidase positive and have lysine and ornithine decarboxylase activities, but no arginine decarboxylase activity. Vibrios may be presumptively identified by slide agglutination testing with Inaba and Ogawa Group O antiserum. All pathogenic species are O-Group I. The differential reactions of various species of cholera and noncholera vibrios have been de-scribed by Ewing,[148] Balows et al.,[149] and Bokkenheuser.[150]

Miscellaneous Gram-negative Bacilli

There are numerous Gram-negative bacilli that are poorly categorized, sometimes unnamed or unclassified; thus, they pose problems of identification. King[151] characterized and placed into groups many such organisms according to their similarities. Hugh[152] and Tatum[153] discussed various Gram-negative bacteria including *Mimeae, Moraxella, Achromobacter, Flavobacterium,* and related species. Differential characteristics of some of these bacterial species are summarized in Table 2.13.

Gram-positive Bacilliae
Corynebacterium (family Corynebacteriaceae)

Corynebacteria are pleomorphic, Gram-positive, noncapsulated, nonmotile bacilli that do not form spores. Members of this genus are widely distributed on the mucous membranes and skin of humans and animals, in soil, and on plants. Most of the species are nonpathogenic except *C. diphtheriae, C. ulcerans,* and *C. haemolyticum.* Corynebacteria grow aerobically in ordinary medium, but potassium tellurite medium is used for their isolation. The organisms are slender, often clubbed at one or both ends, and may contain metachromatic granules. They often occur in palisades,

TABLE 2.11
Differentiation of *Brucella* Species and Their Biotypes[a]

Species	Bio-type	CO₂ require-ment	H₂S produc-tion	Growth on dyes[a] Thionin® a	b	c	Basic fuch-sin a	b	c	Agglutination in monospecific sera B. abortus	B. melitensis	Metabolic tests Gluta-mic acid	Ornithine	Ribose	Lysine
B. melitensis	1	–	–	–	+	+	+	+	+	–	+	+	–	–	–
	2	–	–	–	+	+	+	+	+	+	+	+	–	–	–
	3	–	–	–	+	+	+	+	+	+	+	+	–	–	–
B. abortus	1	+(–)	+	–	–	–	+	+	+	+	–	+	–	+	–
	2	+	+	–	–	–	–	–	–	+	–	+	–	+	–
	3	–(+)	+	–	–	+	+	+	+	+	–	+	–	+	–
	4	+(–)	+	–	–	–	+	+	+	–	+	+	–	+	–
	5	–	–	–	+	+	+	+	+	–	+	+	–	+	–
	6	–	–	–	+	+	+	+	+	+	–	+	–	+	–
	7	–	±	–	+	+	+	+	+	+	–	+	–	+	–
	8	+	–	–	+	+	+	+	+	–	+	+	–	+	–
	9	±	+	–	+	+	+	+	+	–	+	+	–	+	–
B. suis	1	–	++	+	+	+	–	–	–	+	–	–(+)	+	+	+
	2	–	–	–	+	+	–	–	–	+	–	+	+	+	–
	3	–	–	+	+	+	+	+	+	+	–	+	+	+	+
	4	–	–	+	+	+	+	+	+	+	+	+	+	+	+

Note: –(+), usually negative but positive varieties can occur; +(–) usually positive but negative varieties can occur.

[a]Species differentiation is obtained on tryptose agar with the following graded concentrations of dyes. Thionin concentration: a, 1:25000; b, 1:50000; c, 1:100000. Basic fuchsin concentration: b, 1:50000; c, 1:00000.

From *WHO Tech. Rep. Ser.*, 289, 50, 1964. With permission.

TABLE 2.12
Differentiation of *Pasteurella* and *Yersinia* Species

Test	P. multocida	P. pneumotropica	Y. pestis	Y. enterocolitica	Y. pseudo-tuberculosis
Indole	+	+			
Motility at					
25°C	–	–	–	+	+
37°C	–	–	–	–	–
Urea, Christensen's	–	+	–,d	+	+
Growth on					
McConkey agar	–	–	+	+	+
β-D-Galactosidase	–	–	+	+	+
Oxidase	+	+	–	–	–
Acid produced from					
Sucrose	+	+	–	+	–
Lactose	–	d	–	–	–
Rhamnose	0	0	–,d	–	+
Melibiose	0	0	–	–	+
Trehalose	0	0	–	±	+
Cellobiose	0	0	–	+	–
Maltose	–	+	0	0	0

Note: +, growth or positive; –, no growth or negative; 0, not definite; d, some strains vary within species.

Adapted from Washington, J. A., II, *Laboratory Procedure in Clinical Microbiology,* Little, Brown & Company, Boston, 1974, chap. 4; Wetzler, T. F., Rosen, M. N., and Marshall, J. D., Jr., in *Rapid Diagnostic Methods in Medical Microbiology,* Graber, C. D., Ed., Williams & Wilkins, Baltimore, 1970. With permission.

"chinese letters," V and L forms, or as short rods.

The main biochemical characteristics of corynebacteria are that they are catalase positive and gelatin or urea negative, nitrate is usually reduced to nitrite, glucose and maltose are fermented in 48 hr with production of acid but no gas, sucrose is very rarely fermented, and starch and glycogen are fermented only by gravis strains. Differential characteristics within the genus *Corynebacterium* are summarized in Table 2.14.

Corynebacterium diphtheriae may be identified with antisera to OK and O antigens using the fluorescent antibody test. However, the FA test does not differentiate between toxigenic and nontoxigenic strains, and cross reactions occur with streptococci and staphylococci. Toxin production may be determined either by an in vivo guinea pig virulence test or by an in vitro gel diffusion test. The test may be performed with a thick suspension of *C. diphtheriae* (obtained by washing from a Loeffler's slant) by injecting 0.2 ml of suspension intradermally into the shaved side of a guinea pig. After 4 hr the animal is injected intraperitoneally with 500 units of diphtheria antitoxin. Thirty minutes later, 0.2 ml of the cell suspension is injected into the opposite side of the guinea pig. If the suspect strain of *C. diphtheriae* is toxigenic, the site which received the injection before the antitoxin will show a characteristic necrotic lesion after 48 to 72 hr. There may be some nonspecific inflammatory reaction by 24 to 48 hr, but this can be easily distinguished from the characteristic lesion. Chicks and rabbits may be used instead of guinea pigs for this testing.

An in vitro gel diffusion test[144] may be performed by pouring peptone maltose lactose agar containing 20% horse serum into a petri dish. Before the agar hardens, a sterile strip of the filter paper impregnated with diphtheria antitoxin is placed in the petri dish. After the surface is dried, a culture of *C. diphtheriae* is streaked across the plate perpendicular to the paper strip and incubated at 37°C for 24 to 48 hr. A toxigenic organism can be detected by the formation of a visible line of antigen-antibody precipitate radiating from the intersection of the growth in front of antitoxin diffusing from the paper.

TABLE 2.13

Differential Characteristics of Some Commonly Isolated Nonfermenting Gram-negative Bacilli

Organism	O-F Glucose	Nitrite	Motility	Oxidase	10% lactose	Growth on nutrient agar	Gelatin	Comment
Alcaligenes faecalis	−	d	+	+	−	+	d	Arginine dihydrolase negative; peritrichous flagella
Mima polymorpha[a]	−	−	−	−	−	+	∓	Gram negative; coccoid, diplococcoid, bacillary
M. polymorpha var. *oxidans*[b]	−	−	−	+	−	+	−	Gram negative; coccoid, diplococcoid, bacillary
Acinetobacter calcoaceticus[c]	+	−	−	−	+	+	d	Gram negative; coccoid, diplococcoid, bacillary
Moraxella spp.	−	±	−	+	−	∓	∓	Gram negative; large diplobacillary; grows on 0.005% toluidine blue (DNase)
Flavobacterium	∓	−	−	+	−	+	+	Gram negative; long, thin bacilli with bulbous ends and thin centers; colonies may be buff to deep yellow; indole produced after 48-hr incubation

Note: d, different biochemical types.

[a] Also known as *Achromobacter lwoffi* or *Acinetobacter lwoffi.*

[b] Regarded by some as nitrate-negative *Moraxella osloensis.*

[c] Also known as *Achromobacter anitratum, Acinetobacter anitratum, Bacterium anitratum,* or *Herellea vaginicola.*[151]

Reproduced from Washington, J. A., II, *Laboratory Procedure in Clinical Microbiology,* Little, Brown & Company, Boston, 1974, chap. 4.

TABLE 2.14

Differential Characteristics of Corynebacteria

Test	C. diphtheriae			C. ulcerans	C. equi[a]	C. haemolyticum	C. aquaticum	C. hofmannii	C. xerosis	C. avis	C. pyogenes
	gravis	mitis	inter-medius								
Catalase	+	+	+	+	+	−	+	+	+	+	−
Motility	−	−	−	−	−	−	+	−	−	−	−
Hemolysis	−	(+)[d]	−	(+)	−	+		−	−	d	+
NO₃ reduction	+	+[e]	+	−	d	−	(−)[d]	(+)[d]	+	+	−
Gelatin	−	−	−	+[f]	−	−	d	−	−	−	+
Urease	−	−	−	+	d	−	(−)[d]	+	−	+	−
Acid produced from											
Glucose	+	+	+	+	−	+	+[b]	−	+	+	+
Maltose	+	+	+	+	−	+[c]	+	−	+	+[d]	+
Sucrose	−	−	−	−	−	(+)[d]	+	−	+		+

aColonies characteristically pink and moist.
bAerobically but not anaerobically.
cSlowly.
dUsually positive (+) or usually negative (−); d = variable.
eExcept belfanti strain.
fMore rapid at 25°C.

Modified from Hermann, G. J. and Weaver, R. E., in *Manual of Clinical Microbiology*, Blair, J. E., Lennette, E. H., and Truant, J. P., Eds., American Society for Microbiology, Washington, D.C., 1970, 93. With permission.

Listeria

The genus *Listeria* is a member of the family Corynebacteriaceae. *Listeria monocytogenes* is responsible for causing meningitis, meningoencephalitis (in neonates and persons above 40 years), flu-like low-grade septicemia in gravida, and perinatal septicemia. The organisms are diphtheroid-like bacilli, usually occurring in pairs, and may resemble diplococci. *Listeria* are Gram positive, motile, noncapsulated, asporogenic, and 1.0 to 2.0 × 0.5 μm in size. Blood agar and tryptone agar may be used for primary isolation. *Listeria monocytogenes* produces narrow zones of β-hemolysis on blood agar, and the blue-green colonies may be observed with oblique light transmission. *Listeria* is catalase positive and motile. The organism may be confused with *Corynebacterium aquaticum* and *Erysipelothrix rhasopathiae* (Table 2.15). Further characteristics of *Listeria* and different tests have been reported.[157-160]

Erysipelothrix

Erysipelothrix rhusiopathiae (causative agent of erysipilas of swine and turkeys) is a member of Corynebacteriaceae. The organisms are straight or curved slender rods, 0.2 to 0.4 × 0.5 to 2.5 μm in size; they also occur in filamentous forms 4 to 15 μm in length. They are Gram positive but easily decolorized and are nonmotile, noncapsulated, without spores, and are not acid fast. *Erysipelothrix rhusiopathiae* may be differentiated from *Corynebacterium aquaticum* and *Listeria* by the tests listed in Table 2.15.

Bacillus

This genus comprises a large group of Gram-positive, aerobic, spore-forming rods that are abundant in soil and are commonly found as laboratory contaminants. Members of the genus sporulate only aerobically. *Bacillus* species are catalase positive, most are motile, some produce capsules, and some are thermophilic. Many species of *Bacillus* have been recognized; all but one (*B. anthracis*) are saprophytic. Differentiation of *B. anthracis* from *B. cereus* and *B. megaterium* may cause confusion. However, the three species can be characterized by the tests summarized in Table 2.16.

Lactobacillus

The genus *Lactobacillus* includes a number of nonpathogenic species encountered in the oral cavity, intestine, and vagina. The bacteria are Gram positive, occur singly or in chains (with a tendency to form palisades), nonacid fast, nonspore forming, nonmotile, and often long and slender but sometimes pleomorphic; most of the species are microaerophilic, but a few are strictly anaerobic and grow best on glucose-containing medium or other acidic media such as Rogasa's selective agar.

Anaerobic Bacteria
Gram-positive Sporulating Bacilli – Clostridia

A number of known species of the genus *Clostridium* are widely distributed in soil, plants, animals, and humans. Some of the *Clostridium*

TABLE 2.15

Differential Characteristics of *Corynebacterium aquaticum*, *Listeria monocytogenes*, and *Erysipelothrix rhusiopathiae*

Test	C. aquaticum	L. monocytogenes	E. rhusiopathiae
Catalase	+	+	−
Motility	+	+	−
Methyl red	−	+	−
Voges-Proskauer	−	+	−
Oxid-ferm (O-F)	O	F	F
Acid produced from			
Glucose	+	+	+
Maltose	+	(+)	+
Sucrose	+	d	d
Lactose	−	(+)	+

Note: (+), positive after 48 hr; d, variable.

Based on data of Leifson,[157] Killinger,[158] and Weaver.[159]

TABLE 2.16

Differentiation of *Bacillus anthracis* from Other Bacilli

Test	B. anthracis	B. cereus	B. megaterium
Motility in agar	−	+	+
Hemolysis on sheep's blood agar (24 hr)	−	+	−
Peptonization of litmus milk (72 hr)	−	+	±
Fermentation of salicin	−	+	−
Growth on penicillin agar (10 μm/ml)	−	+	−
Growth on PEA medium[a]	−	+	+
Growth on CH medium[b]	−	+	+
Capsulation	+	−	−
Gelatin	−	+	+
Immunofluorescence	+	−	−
Animal pathogenicity	+	−	−

Note: All cultures incubated at 35 to 37°C.

[a] 0.3% phenylethyl alcohol in heart infusion agar.
[b] 0.25% chloral hydrate in heart infusion agar.

Modified from Williams, R. P. and Wende, R. D., in *Rapid Diagnostic Methods in Medical Microbiology,* Graber, C. D., Ed., Williams & Wilkins, Baltimore, 1970. With permission.

species are pathogenic to humans and animals, causing gas gangrene, wound infections, tetanus, and food poisoning. Clostridia are usually Gram positive, but some are rapidly decolorized and appear Gram negative. Many are fat with uniform diameter, some are thin, and some are quite pleomorphic with widely variable length in a single strain. They may be motile or nonmotile. All but three species (*C. carnis, C. histolyticum,* and *C. tertium*) are obligate anaerobes. Clostridia may be identified by morphological, biochemical, and toxicological studies (Table 2.17). Members of the genus *Clostridium* are generally considered easy to speciate. All cultures of clostridia should be tested for production of toxin by animal inoculation tests. Most of these toxins can be detected in chopped meat dextrose subcultures incubated for 18 to 24 hr at 37°C. However, the conditions for the elaboration of toxin may vary depending on the bacterial species. For further details on the biochemical characteristics of clostridia and their toxins, see Dowell and Hawkins,[162] Holdeman and Moore,[163] and Finegold et al.[164]

Gram-positive Nonsporulating Bacilli — Bifidobacterium and Propionibacterium

Bifidobacterium species (including organisms formerly called *Lactobacillus bifidus, Actinomyces eriksonii,* and *A. parabifidus*) are a predominant part of the human fecal flora. In clinical material they usually occur in association with other organisms. Production of gas in reduced medium differentiates these bacterial species from species of anaerobic lactobacilli, but not from species of *Eubacterium.* Some strains show multiple short branching, often with spherical bulbous ends. Cells of bifidobacteria are generally thicker, and those of actinomyces grow more rapidly.

Propionibacteria are "diphtheroids," delicate, club-shaped rods that often do not stain uniformly. One end is often rounded; the other end is tapered or pointed and is stained less intensely. Some cells may be coccoid, bifid (a split club shape), or even branched (one or more side arms which themselves may have side arms). Cells usually are arranged singly, in pairs of "V" and "Y" configurations, and in clumps often described

TABLE 2.17

Differential Characteristics of Some Pathogenic (or Toxigenic) and Nonpathogenic Species of the Genus *Clostridium*

Species	Hemolysis	Lecithinase	Lipase	Urease	Aerobic growth	Gelatin	Milk	Meat	Glucose	Mannitol	Lactose	Sucrose	Maltose	Glycerol	NO$_3$ reduction	Indole	Motility	Spores	Toxicity for Mice	Volatile end products[a] (glucose culture)
Pathogenic species																				
C. botulinum A, B, F	+	−	+	−	−	D	D(−)	D	+	−	−	−	+	V	−	−	+	OST	+	Ac, P, Ib, B, Iv, (Ic)
C. botulinum B, C, D, E	+	−	+	−	−	V	−	−	+	−	−	V	V	−	−(+)	−	+(−)	OST	+	Ac, P, B
C. botulinum C, D	+	−	+	−	−	D	A	−	+	−	−	−	V	−	−	−(+)	+	OST	+	Ac, P, B
C. tetani	+	−	−	−	−	D(−)	−	V	−	−	−	−	−	−	−	V	+	RT	+	Ac, P, B, Bu
C. chauvoei	+	−	−	−	−	V	C	−	+	−	+	V	+	−	V	−	+	OST	V	Ac, (P), (Ib), (B), (Iv), (Ic)
C. haemolyticum	+	+	−	−	−	D	−	−	+	−	−	−	−	−	−	+	+	OST	+	Ac, P, B
C. histolyticum	+	−	−	−	+	D	D	D	−	−	−	−	−	−	−	−	+	OST	+	Ac
C. novyi A	+	+(−)	+	−	−	D	C(D)	G	+	−	−	−	V	+	−(+)	−	+	OST	V	Ac, P, B
C. novyi B	+	+	−	−	−	D	V	G	+	−	−	−	V	+	+	−	+	OST	+	Ac, P, B
C. perfringens	+[b]	+	−	−	−	+	C	−	+	−	+	+	+	+(−)	+	−	−	OST	V	Ac, (P), B
C. septicum	+	−	−	−	−	D	C	G	+	−	+	−	+	−(+)	V	−	+	OST	+	Ac, (P), B
C. sordellii	+	V	−	+	−	D	D	D	+	−	−	−	+	+	−	+	+	OST	−[c]	Ac, P, Ib, (B), Iv, (Ic), Et, Pr
Nonpathogenic species																				
C. bifermentans	+	+	−	−	−	D	D	D(−)	+	−	−	−	+	+(−)	−	+(−)	+	OST	−	Ac, P, Ib, B, Iv, (Ic), Et, Pr
C. butyricum	−	−	−	−	−	−	C	−	+	+	+	+	+	−	V	−	+	OST	−	Ac, (P), B
C. sporogenes	+	−	+	−	−	D	D	D	+	−	−	+	+	V	−	−	+	OST	−	Ac, P, Ib, B, Iv, (Va), (Ic)
C. tertium	−	−	−	−	+	−	C	G	+	+	+	+	+	−	V	−	+	OT	−	Ac, (P), B

Note: +, positive; −, negative; A, acid; C, coagulated; D, digested; G, gas; V, variable; OST, oval subterminal; OT, oval terminal; RT, round terminal; ( ) occasional reactions.

[a]Measured by gas-liquid chromatography (GLC); Ac, acetic; P, propionic; Ib, isobutyric; B, butyric; Iv, isovaleric; Va, valeric; Ic, isocaproic; Bu, butanol; Et, ethanol; Pr, propanol; ( ), variable end products.
[b]Pathogenic for guinea pigs.
[c]Characteristic double zone of hemolysis (as early as 24 hr).

Data from Dowell and Hawkins,[162] Holdeman and Moore,[163] and Finegold et al.[169]

as "Chinese alphabets." *Propionibacterium* species may be differentiated from the species of *Coryne-bacterium, Ramibacterium,* or *Actinomyces* by tests for catalase production, esculin hydrolysis, and acids produced from fermentation of glucose.

Biochemical characteristics which may be helpful in identification of the nonsporulating, anaerobic Gram-positive bacilli and related species are listed in Table 2.18.

Gram-negative Bacilli – Bacteroides and Fusobacterium

The bacilli of *Bacteroides* and *Fusobacterium* are motile, nonspore-forming, anaerobic organisms having peritrichous flagella. They are somewhat difficult to differentiate from other Gram-negative bacilli such as anaerobic vibrios, spirillae, and selenomonads. The anaerobic vibrios, seleno-monads, and spirillae usually are curved or S-shaped. Spirillae have tufts of short polar flagella whereas the flagellar tufts of *Selenomonas* species are located laterally, usually on the concave side of the cell. The main characteristics by which these organisms can be differentiated are listed in Table 2.19.

Anaerobic Cocci – Peptococcus, Peptostreptococcus, Sarcina, and Veillonella

In general, these bacteria are most difficult to speciate, primarily because of the present state of confusion as to their classification. Morphological methods and end-product analysis of GLC are helpful in the differentiation of the species of anaerobic cocci. Biochemical characteristics of some of these bacterial species are summarized in Table 2.20.

Actinomycetaceae

The family Actinomycetaceae consists of heterogenous groups of filamentous organisms which resemble fungi. However, they are closely related to true bacteria in terms of size, absence of nuclear membrane and mitochondria, resistance to antifungal agents but susceptibility to antibacterial agents, and cell wall components. Three genera, one anaerobic (*Actinomyces*) and two aerobic (*Nocardia* and *Streptomyces*) which are medically important are briefly described in this section. For further details on the Actinomycetaceae, see Ajello et al.,[167,168] Beneke and Rogers,[169] Holdeman and Moore,[163] Moore and Holdeman,[170] Pine and Georg,[171] and Gordon.[172]

Actinomyces

Among many species of *Actinomyces*, three species (*A. israelli, A. naeslundii,* and *A. bovis*) are medically important. *Actinomyces* in humans is primarily the agent of a disease of the skin and subcutaneous tissue of the head and neck (cervico-facial) with possible spread to the abdomen, thorax, and other areas of the body. *Actinomyces bovis,* the etiologic agent of bovine actinomycosis (lumpy jaw), has not been reported to be transmitted to humans. *Actinomyces* are Gram positive, nonacid fast, nonmotile, anaerobic, filamentous organisms which display characteristic branching. The filaments break easily into short bacillary fragments in the form of "V" or "Y" when observed in a Gram-stained smear. "Sulfur granules" forming in tissues consist of a central mass of Gram-positive filamentous mycelia. In addition, a peripheral "halo" consisting of swollen Gram-negative "clubs" may be present. Typical morphology (microscopic and colonial characteristics) and biochemical reactions may be used to identify the members of *Actinomyces*. Microscopically, the organisms appear as long branched filaments, and colonies on brain-heart infusion plates appear as spider form after 24 to 48 hr growth or molar tooth shape after 7 days growth. Biochemical reactions of actinomyces include fermentation of glucose, lactose, maltose, sucrose, and mannitol with acid production; no fermentation of glycerol; reduction of nitrate (usually); no change (or acid change) in milk, no liquefaction of gelatin; and catalase negative. By gas chromatography, variable end products are found in glucose broth cultures of *Actinomyces* species; however, succinic, acetic, formic, and lactic acids are commonly present.

Nocardia and Streptomyces

Members of *Nocardia* and *Streptomyces* are filamentous, branching, less than 1 μm in diameter (some species tend to fragment irregularly into bacillary and coccoid forms), and Gram-positive, but stain irregularly; some species are weakly and partially acid fast, nonmotile, and noncapsulated. They grow aerobically on ordinary bacterial and mycological media, e.g., beef infusion blood agar and Sabourad dextrose agar (SDA), and particularly well on Lowenstein-Jensen medium, but are inhibited on media that contain antibacterial antibiotics. Growth is slow (colonies of *N. asteroides* are usually apparent within 3 days, but

TABLE 2.18

Differential Characteristics of Nonsporulating Anaerobic Gram-positive Bacilli

Species	H_2S	Aerobic growth	Catalase	Gas	Milk	Gelatin	NO_3	Indole	Glucose	Mannitol	Lactose	Sucrose	Maltose	Glycerol	Esculin	GLC of end products Volatile	GLC of end products Methylated
Bifidobacterium bifidum	–	–	–	+(–)	C	–	–	–	A	–	A	–	–	–	–(+)	Ac	La, (S)
B. eriksonii	–	–	–	V	C	–	–	–	A	V	A	A	A	–	+(–)	Ac	(Py), La, (S)
B. longum	–	–	–	V	C	–	–	–	A	–	A	A	A	–	–(+)	Ac	
Catenabacterium filamentosum (Eubacterium filamentosum)	–(+)	–	–	+	C	–	–	–	A	V	A	A	A	–	+	Ac	(Py), La, (S)
Eubacterium lentum	–(+)	–	–	V	–	–	–(+)	–	–	–	–	–	–	–	–(+)	Ac	(Py), (La), (S)
E. limosum	+(–)	–	+	V	C	D	+	+(–)	A	A(–)	–	–	–	–	V	Ac, (P), (Ib), (B), (Iv)	
Propionibacterium acnes	–	+	+	V	–	–	+	+(–)	A	V	–	–	–	A	–	Ac, P, (Iv)	
P. freudenreichii (ss freudenreichii)	–	+	+(–)	+	–	–	–	–	A	A	–	–	–	A	V		
P. granulosum	–	–	+	V	C(–)	V	–	–	A	–	–	–	–(+)	A	–		
P. jensenii	–	+	–(+)	+(–)	C(–)	–	–(+)	–(+)	A	–(A)	–	A	A	V	V	Ac, P	
P. pentosaceum	–	+	+	+	C	–	V	–	A	A	V	A	V	A(–)	+(–)		
P. shermanii	–	+	+	V	C	–	–	–	A	V	A	A	A(–)	V	–(+)		
(P. freudenreichii ss shermanii)									A	–	A	–(A)	–(A)	A	–		
Ramibacterium alactolyticum (Eubacterium alactolyticum)	–	–	–	V	–	–	–	–	A	V	–	–	–	–	–	Ac, (P), (Ib), (B), Ca	

Note: La, lactic acid; S, succinic acid; Ca, caproic acid; Py, pyruvic acid; for other symbols, see Table 2.17.

From Washington, J. A., II, *Laboratory Procedure in Clinical Microbiology*, Little, Brown & Company, Boston, 1974, chap. 4. With permission.

TABLE 2.19

Definite Characteristics of Commonly Encountered *Bacteroides* and *Fusobacterium* Species[a]

Species	Amygdalin	Arabinose	Cellobiose	Glucose	Lactose	Levulose	Maltose	Mannitol	Mannose	Melezitose	Raffinose	Rhamnose	Sucrose	Trehalose	Xylose	Esculin	Gelatin	Starch	20% bile + 0.1% deoxycholate	0.1% deoxycholate	Reduction of nitrate	Production of indole	Conversion of lactate to propionate	Conversion of threonine to propionate	Fatty acids produced[b]
Bacteroides																									
corrodens	–	–	–	–	–	–	–	–	–	–	–	–	–	–	–	–	–	–	N	N/S	+	–	–	–	A, L, S
fragilis																									
ss[c] distasonis	w+	+–	+/–	+w	+w	+w	+w	–	+w	+/–	+w	+/–	+w	w+	+w	+	–+	+/–	S	I	–	–	–	–	A, P, S (IB, IV, L)
ss fragilis	w+	+–	–w	+w	w+	+w	+w	–	+w	+/–	+w	+/–	+w	–	+w	+	–w	+	S	I	–	–	+–	–	A, P, S (IB, IV, L)
ss ovatus	w+	w+	w+	+w	+w	+w	+w	w–	+w	–w	+–	+/–	+w	+/–	+w	+	–w	+	S	I	–	+	+–	–	A, S (P, IB, IV, L)
ss thetaiotaomicron	+w	+w	+–	+	+w	+w	+w	–	+w	+/–	+w	+/–	+w	+–	+w	+	+–	+–	S	I	–	+	+–	–	A, S (P, IB, IV, L)
ss vulgatus	+/–	+–	–w	+w	+w	+w	+w	–	+w	–	+w	+w	+w	+–	+w	+–	+–	+–	S	I	–	–	–	–	A, S (P, IB, IV, L)
melaninogenicus																									
ss asaccharolyticus	–	–	–	–	–	–	–	–	–	–	–	–	–	–	–	–	+	–	I	I	+–	+–	–	+–	A, IB, B, IV (P, L, S)
ss intermedius	w–	–w	–	+w	–+	+/–	+/–	–	+/–	–	–+	–	+/–	–w	–w	–	+	+–	S	I	–	+–	–	+–	A, IB, IV, S
ss melaninogenicus	+–	w–	+–	+	+	+	+	–	+	–	+	–w	+	–w	–w	+–	+	+	S	I	–	–	–	–	L, S (A, B, IV)
oralis	+w	–	+–	+w	+w	+w	+–	–	+w	–w	+w	–+	+w	–w	–w	+–	+–	+–	S	I	–	–	–	–	A, S (IB, IV, L)
Fusobacterium																									
mortiferum	–w	–	w–	+w	w+	w+	w–	–	w+	–	+/–	–	+/–	–w	–+	+	+/–	–w	N	N	–	–	–	–	A, P, B (IV, L, S)
necrophorum	–	–	–	–w	–	–w	–	–	–	–	–	–	–	–	–	–	–	–	I[n]	I	–	+	+	+	A, P, B (L, S)
nucleatum	–	–	–	–w	–	w–	–	–	–	–	–	–	–	–w	–+	–	–w	–	N	N	–	+	–	+	A, P, B, S (L)
varium	–	–	–	w+	–	w	+–	–	w	–	+w	+	+w	–w	–w	–	–w	+	N	N	–	+–	+	+	A, B, L (P, S)

Note: –, negative reaction of pH above 6.0 in carbohydrate fermentations; w, weak reaction or pH 5.6 to 6.0 in carbohydrate fermentations; +, positive reaction or pH 5.5 or less in carbohydrate fermentations; S, stimulated growth; I, inhibited growth (not necessarily complete inhibition); N, no effect on growth (see Anaerobe Laboratory).

[a]Material for this table is taken from Anaerobe Laboratory, Outline of Clinical Methods in Anaerobic Bacteriology, Virginia Polytechnic Institute, Blacksburg, 1972; Sutter, V. L., Attebery, H. R., Rosenblatt, J. E., Bricknell, K., and Finegold, S. M., *Anaerobic Bacteriology Manual*, Extension Division, University of California, Los Angeles, 1972.

[b]Fatty acids: A, acetic; P, propionic; IB, isobutyric; B, butyric; IV, isovaleric; L, lactic; S, succinic; ( ), variable. No quantitation of fatty acids is implied.

[c]Subspecies.

From Sutter, V. L., Attebery, H. R., and Finegold, S. M., in *Manual of Clinical Microbiology*, 2nd ed., Lennette, E. H., Spaulding, E. H., and Truant, J. P., Eds., American Society for Microbiology, Washington, D.C., 1974, 393. With permission.

TABLE 2.20

Differential Characteristics of Anaerobic Cocci[a]

Group	Gram reaction	Gas	Catalase	NO₃	Indole	Gelatin	Glucose	Lactose	Maltose	GLC of end products Volatile	GLC of end products Methylated
Peptococcus, CDC Group 2	+	–	+	+(–)	–	–	A	–	–	Ac, (P), B, (Ib), (Iv), (Ic)	
Peptostreptococcus, CDC Group 1 (*Peptococcus asaccharolyticus*)	+	+	–	–	+	–	–	–	–		
Peptostreptococcus, CDC Group 2 (*Peptococcus prevotii*)	+	V	–	–(+)	–	–(D)	–(A)	–	–	Ac, (P), (B), (Iv)	(La)
Peptostreptococcus, CDC Group 3 (*Peptococcus anaerobius* and others)	+	V	–	–(+)	–	–	A	–(A)	A(–)	Ac, (P), Ib, Iv, (Ic)	
Sarcina species	+	V	–	–	–(+)	–(D)	–	–	–	Ac, (B)	
Veillonella alcalescens	–	V	+	+	–	–	–	–	–	Ac, P	
Veillonella parvula	–	+(–)	–	+	–	–	–	–	–	Ac, P	

Note: For symbols, see Table 2.18.

[a]Based on data of and adapted from Dowell et aL[166] and Holdeman and Moore.[163]

From Washington, J. A., II, *Laboratory Procedure in Clinical Microbiology*, Little, Brown & Company, Boston, 1974, chap. 4.

55

others may take a week or more), and colonies of *N. asteroides* and *N. brasiliensis* vary in appearance from chalky white to orange to pink. Differential characteristics that are considered to be helpful in the differentiation of *Nocardia* and *Streptomyces* are listed in Table 2.21.

Mycobacteria

The mycobacteria comprise a large group of acid-fast bacilli ranging in size from 0.3 to 0.6 X 0.5 to 4.0 μm. They are nonmotile, nonsporogenous, and do not produce a capsule in susceptible animals or on culture. They vary in morphology from coccoid to long filamentous forms depending on the strain, culture medium, and environment. The most rapidly growing species require 2 to 3 days on a simple media at a temperature of 20 to 40°C, and most pathogens require 2 to 6 weeks on complex media. About a dozen species are pathogens, producing slowly developing, destructive granulomas that may necrose with ulceration or cavitation or heal with possible disfiguration. Tuberculosis (caused by *Mycobacterium tuberculosis*) of the lungs may disseminate to other parts of the body via the bloodstream. Pathogenicity for experimental animals varies with the mycobacterial species. Animals may be useful for both primary isolation (e.g., *M. leprae*) and identification (e.g., *M. tuberculosis* and *M. bovis*). Acid-fast bacteria commonly occur in soil, water, and milk and in the alimentary tract of healthy animals and man. All are important to the medical microbiologist because most of the pathogenic and nonpathogenic mycobacteria stain similarly. The mycobacteria other than leprosy bacilli and tubercle bacilli are called atypical because they are unlike tubercle bacilli. Runyon[46,178] grouped the acid-fast bacilli on the basis of growth rate and pigmentation of grossly visible colonies as follows:

Group I — Photochromogens; grow slowly (2 to 3 weeks); colonies nonpigmented in the dark but turning yellow after exposure to light.

Group II — Scotochromogens; grow slowly (2 to 3 weeks); colonies yellow to orange when grown in light or dark.

Group III — Nonphotochromogens; grow slowly (2 to 3 weeks); colonies have no marked color but are usually slightly buff colored.

Group IV — Rapid growers (2 to 7 days); pigmentation varies.

Mycobacteria may be identified by a number of tests summarized in Table 2.22. Main species of *Mycobacterium* which are of medical importance are briefly described in Table 2.23.

Treponemes

The treponemes are members of the family Spirochaetaceae. They are fine, filamentous, helical cells, usually occurring singly and exhibiting three types of vigorous motility [rotation about the long axis, nonpolar (to and fro) locomotion, and flexion]. The cell body of the treponemes has a diameter of 0.1 to 0.25 μm with a coiled length of 5 to 15 μm; in a liquid environment, the spirals usually have uniform depth and amplitude (up to 0.5 μm), but may taper toward each end. These organisms can be examined microscopically with usual bacterial or tissue staining methods because the cell diameter of treponemes is below the resolution of most light microscopes. However, most treponemes are readily observed in wet preparation under dark-field illumination. Genus *Treponema* contains the organisms that cause venereal syphilis (*T. pallidum*), nonvenereal syphilis (bejel), the widespread tropical scourge yaws (*T. pertenue*), pinta (*T. carateum*), and some other diseases of humans and animals.

Treponema pallidum — This is a relatively fragile, highly parasitic organism. It has never been cultivated in a virulent form in artificial media, chick embryos, or tissue cultures, although it may be maintained alive and virulent without multiplication for several days in certain complex artificial media. Syphilis, caused by *T. pallidum,* is ordinarily transmitted by sexual contact. The organism may be present in either external lesions or in discharge from deeper in the genitourinary tract. Organisms most commonly penetrate the mucous membrane, but may be introduced through breaks in the skin. Syphilis is characterized by the formation of a primary inflammatory lesion known as a chancre, followed by a generalized systemic skin rash (secondary lesion). Untreated patients may recover from secondary lesions, but often enter a tertiary stage in which symptoms of the disease are difficult to recognize. Identification of *T. pallidum* is made by dark-field or fluorescent microscopy of the specimen obtained from primary or secondary lesions. Flocculation or complement-fixation tests for Wasserman antibody, as well as a screening test for syphilis, are used for diagnosis. The Venereal Disease Reasearch

TABLE 2.21

Physiological Characteristics of Pathogenic *Nocardia*, *Actinomadura*, and *Streptomyces* Species

Species	Cell wall type[a]	Decomposition[b] of							Acid from		Growth at 10°C
		Casein	Tryosine	Xanthine	Starch	Gelatin	Bromcresol purple milk	Urea (12, 24)	Lactose	Xylose	
N. asteroides	IV	−	−	−	−[c]	−[a]	−[e]	+	−	−	
N. brasiliensis	IV	+	+	−	−[c]	+	+	+	−	−	
N. caviae	IV	−	−	+	−[c]	−	−[e]	+	−	−	
A. dassonvillei	III	+	+	+	−			± 35% +	−	+	−
A. madurae	III	+	+ 14% neg	−	+	+	+	−	± 55% +	+	−
A. pelletieri	III	+	+	−	13% +	+	+	−	−	−	
S. somaliensis	I	+	+	−	±	+	+	−	−	−	
S. paraguayensis	I	+	+	+		+	+	+	−	−	
Streptomyces spp. (nonpathogenic)	I	−	+	+	(+)[f]	(+)[f]	(+)[f]	± 50% +	+	+	+

Note: See Gordon,[173] Mackinnon and Artagaveytia-Allende,[173a] and Mariat.[174]

[a] Cell walls of all actinomycetes contain glucosamine, muramic acid, glutamic acid, and alanine. In addition, major amounts of the following components are found in the respective groups: (I) LL-diaminopimelic acid (DAP) and glycine; (III) *meso*-DAP; (V) *meso*-DAP, arabinose, and galactose (see Becker et al.[175]).

[b] Within 2 weeks at 27°C.

[c] About 50% of strains positive by different method (see Gordon and Mihm[176]).

[d] Some strains reportedly liquefy certain gelatin media (see Gonzalez-Mendoza and Mariat[177]).

[e] Usually turns alkaline.

[f] Most species give positive tests.

From Gordon, M. A., in *Manual of Clinical Microbiology*, Lennette, E. H., Spaulding, E. H., and Truant, J. P., Eds, American Society for Microbiology, Washington, D.C., 1974, 176. With permission.

TABLE 2.22

Tests for Identifying Clinically Significant Mycobacteria[a]

Species or subgroup	Niacin	Nitrate reduction		Catalase activity			Tween® hydrolysis (days)		Tellurite reduction in 3 days	Pigment formation	
		>1+	>3+	<40 mm	>50 mm	68°+	<5	<10		Dark	Light
M. tuberculosis	++	++	++	++	-	-	NA[e]	∓	-	-	-
M. bovis	-	-	-	++	-	-	NA	-	-	-	-
Group I											
M. kansasii	-[f]	++	++	++	++	++	++	++	-	-	++
M. marinum	-	-	-	±	∓	-	++	++	-	-	++
Group II											
M. sp. scrofulaceum	-	∓	-	++	++	++	-	-	-	++	++
M. sp. aquae	-	-	-	++	++	++	+	++	-	++	++
M. flavescens	-	++	+	++	++	++	++	++	-	++	++
M. xenopi	-	-	-	++	-	++	-	-	-	+g	+g
Group III											
M. avium[i]	-	-	-	++	-	++	-	-	±	-	∓
M. intracellulare[i]	-	-	-	++	-	+	-	-	+	-	-
M. gastri	-	-	-	++	-	-	++	++	-	-	-
M. terrae complex	-	++	+	++	++	++	++	++	-	-	-
"v" (*M. triviale*)	-	++	+	++	++	++	+	-	-	-	-
Group IV											
M. fortuitum	-	±	±	++	++	++	-	∓	++	-	-
M. smegmatis	-	++	∓	++	++	++	++	++	++	-	-
M. phlei	-	++	∓	++	++	++	++	++	++	++	++
M. vaccae	-	++	+	++	++	++	±	+	+	++	++
M. borstelense	V	-	-	++	++	++	-	-	-	-	-
M. rhodochrous	-	++	±	++	++	++	-	++	±	++	++

Note: d= different biochemical types.

[a]Key to percentage of strains reacting as indicated: (++) 85% or more; (+) 75–84%; (±) 50–74%; (∓) 15–49%; (−) 15%; (V) variable.

[b](R) Rough, wrinkled surface; (S) smooth, glistening surface; (V) variable (colonies may be smooth when first seen but become wrinkled and matted).

[c]Granular to flocculent growth in clear medium; (T) turbid growth, usually with sediment that suspends on shaking.

[d](INH) Isoniazid; (FAS) p-aminosalicylic acid; (TCH) thiophene-2-carboxylic acid hydrazide.

[e]Not applicable; test is not necessary.

[f]Positive in rare instances.

[g]Pigment increases with age.

[h]May require 7 to 10 days for a positive test. g, in 3 days or less.

[i]With tests listed, the pairs of organisms so indicated can not be separated; colonial morphology on 7H10 may be helpful in the case of *M. phlei-M. vaccae.*

Table 2.22 (continued)
Tests for Identifying Significant Mycobacteria

Species or subgroup	Growth on 5% NaCl	Growth in less than 7 days	Arylsulfatase in 3 days	Growth on MacConkey agar	Colonial morphology L-J[b]	Growth in PB-S[c]	Cord formation	Growth inhibition[d] INH 1.0 μg/ml	PAS 1.0 μg/ml	TCH 10.0 μg/ml
M. tuberculosis	−	−	NA	−	R	G	+	+	+	−
M. bovis	−	−	NA	−	R	G	+	+	+	+
Group I										
M. kansasii	−	−	NA	−	V	T	±	V	−	NA
M. marinum	−	±	NA	−	V	T	±	V	−	NA
Group II										
M. sp. *scrofulaceum*	−	−	NA	−	S	T	−	−	−	NA
M. sp. *aquae*	−	−	NA	−	S	T	−	−	−	NA
M. flavescens	±	−	NA	−	S	T	−	+	−	NA
M. xenopi	−	−	+[h]	−	S	T	−	+	−	−
Group III										
M. avium[i]	−	−	V	−	S	T	−	−	−	−
M. intracellulare[i]	−	−	−	±	S	T	−	−	−	−
M. gastri	−	−	−	−	S	T	−	−	−	−
M. terrae complex	−	−	V	−	S	T	−	−	−	−
"V" (*M. triviale*)	++	−	V	−	R	T	−	−	−	−
Group IV										
M. fortuitum	++	++	++[j]	++	R	G	−	−	−	NA
M. smegmatis	++	++	−	−	R	G	−	−	−	NA
M. phlei	++	++	−	−	R	G	−	−	−	NA
M. vaccae	++	++	−[j]	−	V	G	−	−	−	NA
M. borstelense	−	++	++[j]	±	R	G	−	−	−	NA
M. rhodochrous	++	++	−	−	R	G	−	−	−	NA

[j] 3 days or less for a positive test.

From Washington, J. A., II, *Laboratory Procedure in Clinical Microbiology*, Little, Brown & Company, Boston, 1974, ch. 1 p. 4. With permission.

TABLE 2.23

Differential Characteristics of Certain Mycobacterial Species

Species and references	Growth characteristics	Unique features
Mycobacterium tuberculosis (tubercle bacillus)[179]	Slow growers; nonpigmented; colony appears in 2 weeks or longer on L-J medium – rough or matte colonies – and becomes wrinkled and white; on 7H10, colonies are flat, wrinkled, or white; in PB-S, growth is granular in 1 week and flocculent with a pellicle and sediment in 2 to 3 weeks, medium not turbid	Niacin positive; cord formation; catalase negative at 68°C; nitrate reduction; growth inhibited by INH, PAS, and rifampin at 5 μg/ml; causes fatal disease in guinea pigs but not in rabbits
M. bovis (bovine tubercle bacillus)[180,181]	Slow growers; nonpigmented; colonies usually smaller than those of *M. tuberculosis*, appear in 3 weeks or longer on L-J medium, scanty white or buff-colored growth; no growth on 7H10 medium; in PB-S, granular sediment in 2 weeks or more, seldom with a pellicle	Niacin negative; inhibited by TCH at 0.2 μg/ml; cord formation; catalase negative at 68°C; negative nitrate reduction test; growth inhibited by INH and rifampin at 0.5 μg/ml; causes fatal disease in guinea pigs and rabbits
M. kansasii (Group I, photochromogen or yellow bacillus)[182]	Slow growers, colonies appear in 2 weeks or longer on L-J and on 7H10; growth raised, matte, or wrinkled and buff colored if inoculated in darkness, yellow pigmented on exposure to light; in PB-S growth is granular and turbid in 1 week with sediment and pellicle in 2 weeks; these suspend readily	Growth becomes yellow when exposed to light (photochromogenic); cells elongated, broad, and beaded in sputum, tissue, and young cultures; loose cord formation occasionally; catalase positive at 25 and 68°C; nitrate reduction; niacin formed rarely; Tween® 80 hydrolysis in <5 days; resistant to PAS
M. marinum (*M. balnei*, Group I, photochromogen or yellow bacillus)[183]	Slow growers, colonies appear in 2 weeks or longer at 24 to 30°C on L-J or 7H10; colonies smooth or matte but may become wrinkled; growth buff colored in darkness, yellow in light; in PB-S, turbid growth and granular in 1 week with viscous sediment which suspends on shaking	Photochromogenic; optimal growth at 24 to 30°C and poor or no growth at 37°C when originally isolated; loose cords may be formed; negative nitrate reduction test; Tween® 80 hydrolyzed in less than 5 days; niacin rarely formed; not susceptible to IMH or PAS
M. scrofulaceum (*M. marianum, M. scrofula*, Group II, scotochromogen)[184,185]	Slow growers, growth appears in 2 weeks or longer on L-J and 7H10; rounded smooth, glistening, yellow colonies, pigmented in darkness and in light; subcultures grow on L-J and 7H10 in 1 week or longer to form smooth, moist, yellow growth; in PB-S, growth in 1 week with turbidity and yellow sediment which suspends on shaking	Smooth yellow colonies; no cord formation; niacin negative; no hydrolysis of Tween® 80 in 14 days or less; negative nitrate reduction test; negative arylsulfatase in 3 days; resistant to INH and PAS
M. gordonae (*M. aquae*, Group II, tap water scotochromogen)[184,185]	Slow growers, colonies appear in 3 weeks or longer on L-J and 7H10, usually as single or few isolated colonies; subcultures develop in 1 to 2 weeks as smooth yellow growth in darkness and in light; in PB-S turbid growth in 1 week with a yellow sediment which suspends easily	Raised, rounded, smooth yellow growth; hydrolyzes Tween® 80 in 5 days or less; negative nitrate reduction test; niacin negative; negative arylsulfatase in 3 days; resistant to INH and PAS

TABLE 2.23 (continued)

Differential Characteristics of Certain Mycobacterial Species

Species and references	Growth characteristics	Unique features
M. xenopi (Group II, slow growers, scoto-chromogens; yellow glistening colonies)[186]	Growth appears in 4 or more weeks on L-J and 7H10 medium to form small yellow colonies in darkness and in light; in PB-S, granular to turbid growth in 3 weeks	Slow growers at 45 and 37°C; small yellow colonies with radiating filaments; niacin negative; nitrate reduction test negative; catalase positive at 68°C; Tween® 80 not hydrolyzed; arylsulfatase positive in 4 to 7 days; growth inhibited by 1 µg/ml INH
M. avium complex (complex includes at least 21 different serotypes, e.g., avian tubercle bacilli of Serotypes 1, 2, and 3; *M. intracellulare*, *M. battey*, Battery bacilli of Serotypes 4 through 21; Avian Battey-swine complex, avian-like mycobacteria, nonphotochromogens Group III)[186,187]	Slow growers, colonies appear in 2 or more weeks when originally isolated and 1 or more weeks on subculture; nonpigmented or buff-colored colonies; growth may be matte or wrinkled on L-J or 7H10 but the surface is usually moist, early growth colorless on 7H10; good growth at 40°C; in PB-S forms a uniformly turbid suspension in 1 week, a viscous sediment formed in 1 to 2 weeks which suspends readily	Smooth, glistening, nonphoto-chromogenic colonies; buff to yellow colonies; does not hydrolyze Tween® 80; no cord formation; niacin negative; catalase positive at 68°C; negative nitrate reduction test; negative arylsulfatase in 3 days
M. terrae (Group III, slow growers, nonpigmented)[188]	Growth appears in 2 or more weeks on L-J and 7H10 as rounded, smooth, glistening, nonphotochromogenic, colorless to buff colonies; subcultures appear in 1 week or longer as smooth growth; in PB-S, growth in uniform turbidity with sediment but without a pellicle	Superficially resembles *M. avium* complex; nonphotochromogenic; nitrate reduction; Tween® 80 hydrolysis; catalase positive at 68°C; niacin negative; does not form cords
M. triviale (Group III, nonpigmented)[189,190]	Growth appears in 2 or more weeks on L-J and 7H10 as single, rounded, matte, nonphotochromogenic colonies; in PB-S, growth turbid with granular sediment but no pellicle	Superficially resembles *M. tuberculosis* and *M. avium* complex; does not form cords; niacin negative; nitrate reduction; catalase positive at 68°C; Tween® 80 hydrolysis in 5 days or less; growth in medium containing 5% NaCl
M. gastri (Group III, nonpigmented)[181,191]	Growth appears in 2 or more weeks on L-J and 7H10 medium as smooth to rough nonphotochromogenic colonies; subcultures on solid medium form colorless to buff colonies in 1 week or longer; in PB-S, turbid growth with sediment but no pellicle in 1 week	Superficially resembles *M. avium* complex; nonphotochromo-genic; Tween® 80 hydrolysis in 5 days or less; catalase negative at 68°C; arylsulfatase negative in 3 days; niacin negative; no cord formation; negative nitrate reduction test

TABLE 2.23 (continued)

Differential Characteristics of Certain Mycobacterial Species

Species and references	Growth characteristics	Unique features
M. fortuitum (*M. ranae*, Group IV, rapid grower)[192,193]	Rapid grower, wrinkled or matte buff colonies in 7 days or less; on subculture luxuriant growth in 3 days; in PB-S, growth granular to flocculent in 3 days with a pellicle and heavy sediment in 7 days; grows well between 24 and 37°C but not at 45°C; growth on MacConkey's agar in 5 days, radiating filaments in colonies on cornmeal or 7H10 agar examined at 100°C	Rapid growth on L-J and 7H10 medium; arylsulfatase formation; niacin negative; nitrate reduction in 5 days; catalase positive at 68°C; resistant to INH and PAS at 10 μg/ml; focal necrosis of renal cortex in mice (i.v. injection of 0.1 mg)
M. abscessus, *M. chelonei*, *M. fortuitum* subsp. *abscessus*, *M. runyonii* (Group IV, rapid growers)[193,194]	Rapid growers, form visible colonies in 3 days or less; rounded smooth or matte buff colonies within 7 days on L-J and 7H10 medium; on subculture, luxuriant growth in 3 days; in PB-S, growth granular to flocculent in 3 days with sediment; grows well between 24 and 37°C but not at 45°C; growth on MacConkey's agar in 5 days; colonies on cornmeal agar or 7H10, entire or scalloped edges	Resembles *M. fortuitum;* does not reduce nitrate; niacin negative; no cord formation; catalase positive at 68°C; arylsulfatase positive; negative nitrate reduction; not susceptible to INH or PAS at 10 μg/ml
M. leprae (leprosy bacillus, Hansen's bacillus)[195]	No growth in vitro; growth on mouse foot pads; abundant leprosy bacilli in lipromatous leprosy but scant in tuberculoid lesions	Cannot be cultured on artificial media; acid-fast bacilli in tissue sections or smears in intracellular bundles (globi); bacilli range in size from 0.2 to 0.5 × 1.5 to 8 μm with parallel sides and rounded ends, often in clumps or rounded masses or in groups, side by side, rarely bent or curved
M. ulcerans[196]	Slow growers, growth in 3 to 4 weeks (at 30°C) on L-J and egg yolk agar in the form of small rounded, matte colonies; in PB-S, growth slightly turbid with sediment in 2 weeks; colonies colorless initially, buff to slightly yellow in 5 to 6 weeks (nonphotochromogenic)	Growth at 30°C but not at 37°C; no cord formation; niacin production variable; negative nitrate reduction test; no hydrolysis of Tween® 80; catalase positive at 68°C; resistant to INH, PAS, and TCH at 10 μg/ml; resembles *M. marinum*

Laboratory (VDRL) test was developed by Harris et al.[197] The test has a high degree of sensitivity and specificity. Patients with negative serology and lesions that are suspect should be retested 1 to 2 weeks following a negative test. The best serologic confirmatory test for the diagnosis of syphilis is the fluorescent treponemal antibody absorbed (FTA-ABS) test.[198] The FTA-ABS is an indirect fluorescent antibody test which combines the rabbit-virulent Nichols strain of *T. pallidum* with the patient's serum (the unknown) and allows them to react. This is followed by washing and the application of fluorescein-labeled antihuman globulin. If the patient has antibodies that react with the virulent organism, the conjugated anti-human globulin reacts with this bound serum. When reviewed under ultraviolet light, the organisms fluoresce. If the serological test for syphilis (STS) is negative and the *Treponema* immobilization test (TPI) is positive, FTA-ABS test should be conducted.

Treponema pertenue — This organism is morphologically and serologically indistinguishable from *T. pallidum*. It causes yaws (frambesia), a nonvenereal tropical syphilis resembling *T. pallidum* syphilis except for the character of the lesion produced. A primary lesion (mother yaws) appears in 3 to 4 weeks as a painless red papule surrounded by a zone of erythema. Secondary lesions are similar to primary lesions, but appear several weeks later. The lesions on the soles of the feet may become "crab yaws," hyperkeratotic lesions. Disfigurement of the nose and face is common. Since *T. pertenue* shows a positive Wassermann reaction and other serological reactions, diagnosis to differentiate this disease from venereal syphilis is based on character of the chancre, mode of transmission (common mode of transmission is contact between individuals or by flies), and frequency of visceral and tertiary lesions.

Treponema carateum — This organism causes pinta (mal de pinto in Mexico or carate in Colombia), a nonvenereal type and clinically a skin disease. Skin may undergo dyschromic changes (gray, bluish gray, or pinkish, but eventually becoming white). Spirochetes may be demonstrated in serous fluids from lesions or from fissures in plantar keratosis. It is negative for TPI test, but the serological reaction of Kahn is identical to that of syphilis. The Wassermann reaction is negative for the primary stage, but positive for the secondary stage.

Leptospires

Members of the genus *Leptospira* (family Spirochaetaceae) occur in fresh natural water, animals, and humans. One single species, *Leptospira interrogans,* comprises two major groups, the so-called saprophytic and pathogenic leptospires (*L. icterohemorrhagiae* causes acute, febrile, systemic disease of man and other mammals). The organisms are helicoidal, usually 6 to 20 μm in length, and approximately 0.1 μm in diameter with semicircular hooked ends (occasionally one or both ends straight). They are motile; by electron microscopy, the cytoplasm is seen as helicoidally wound about an axial filament. The quick flexion, the active motility in the line of the long axis, and alteration of this direction clearly distinguish (for the experienced observer) leptospires from floating filaments of fibrin or cellular material. Leptospires may be differentiated from other spirochetes, the treponemes, and the borreliae largely by the very thin body, the tightness of their spiral, and the turned ends. The treponemes are about the same length as the leptospires, but the body is thicker. Both the excussion of the spiral and the wavelength are two to three times that of the leptospires. The borreliae are the largest of the pathogenic spirochetes. Of these three genera, only the borreliae can be seen microscopically when fixed and stained with dyes. The treponemes and leptospires require silver impregnation to be outlined in a fixed section of tissue. The leptospires grow aerobically in medium containing 10% serum (preferably rabbit) or serum components at pH 6.8 to 7.8 and at 30°C. Incubation time for optimal growth ranges from a few days to 4 weeks or longer (usually 6 to 14 days).

Identification of the leptospires is based on serological tests and animal pathogenicity; they are not distinguishable on the basis of biochemical characteristics. Serotypes are identified by microscopic-agglutination and agglutinin-adsorption tests with serotype-specific rabbit antiserum. Macroscopic plate agglutination test may be performed on a piece of glass with one drop of serum and one drop of antigen. A clouding of the mixture indicates a positive reaction.[199,200]

The leptospires may be identified by animal pathogenicity test since they produce lethal to subclinical infections in hamsters, guinea pigs, gerbils and weanling rabbits on intraperitoneal inoculation. Further informatin on the leptospires may be obtained from publications by Alexander,[201] Baker and Cox,[202] WHO,[203] and Galton et al.[204]

Mycoplasma

Mycoplasma are members of the family Mycoplasmataceae. They are highly pleomorphic organisms without a rigid cell wall; thus, they are fragile and plastic. They are the smallest organisms (125 to 250 nm in size) capable of growing in a cell-free medium. Also called "pleuropneumonia-like organisms" (PPLO), there are pathogenic as well as nonpathogenic species. *Mycoplasma pneumoniae* is pathogenic to humans, associated with primary atypical pneumonia and bronchitis. *M. hominis* is suspected in nongonococcal urethritis and salpingitis, occasionally invading the bloodstream. There are also atypical forms of *Mycoplasma* species (T-strains) isolated from genital regions, aborted fetal membranes, and the pharynx. *Mycoplasma* may be seen as filamentous and coccoid forms, Gram negative (stain poorly), nonmotile, and without capsules or spires. They grow aerobically or facultative anaerobically on several media enriched with peptone and serum, ascitic fluid, whole blood, or egg yolk. They grow selectively on media containing bacterial and fungal inhibitors (e.g., penicillin, thallium acetate, amphotercin B) and multiply in chick embryo and yeast cell components. Pathogenic species (*M. pneumoniae*) require a 37°C growth temperature, giving distinctive "fried egg" colonies on solid medium with dark, circular centers that penetrate agar. *Mycoplasma* species may be differentiated by the physical, biochemical, and serological characteristics summarized in Table 2.24. Further details may be found in the many publications on this subject.[205-211]

TABLE 2.24

Differential Characteristics of Mycoplasmas of Human Origin

Reaction or characteristic	*M. pneumoniae*	*M. hominis* I	*M. salivarium*	*M. pharyngis* (*orale* I)	*M. orale* II	*M. fermentans*	T strains
Tetrazolium reduced aerobically	+	−	−	−	−	−	?
Growth on 0.002% methylene blue agar	+	−	−	−	−	−	?
Dextrose fermented without gas	+	−	−	−	−	+	−
Arginine hydrolyzed with production of NH₃	−	+	+	+	+	+	−
Methylene blue reduced	+	±	±	?	?	+	?
Urea hydrolyzed	−	−	−	−	−	−	+
Resistant to 1:2,000 thallium acetate	+	+	+	+	+	+	−
Turbidity in broth (G = granular: D = diffuse; F = faint)	G	F	D	D	D	D	D
"Fried egg" colonies typically produced	−	+	+	±	+	+	
Spots or film on normal horse serum agar	−	−	+	−	−	+	
Hemolysis of erythrocytes							
Sheep	β	γ	α	γ	?	α	α′
Guinea pig	β	α′γ	α	γ	α	α	β
Horse	αβ	α′γ	α′	γ	?	α	γ
Human O	α	γ	γ	γ	?	α	α′
Hemagglutination of guinea pig erythrocytes	+	−	−	−	−	−	−

From Crawford, Y. E., in *Manual of Clinical Microbiology,* Blair, J. E., Lennette, E. H., and Truant, J. P., Eds., American Society for Microbiology, Washington, D.C., 1970, 252. With permission.

Rickettsiae

The rickettsiae are members of the family Ricketsiaceae characterized by coccobacillary or bacillary microorganisms measuring 250 to 500 nm in size. They are obligate intracellular parasites and are sensitive to the tetracyclines and chloramphenicol. They have rigid cell walls, contain both RNA and DNA, and reproduce by binary fission. Their natural reservoirs are certain insects (lice, fleas, and ticks) and, with the exception of Q fever, they are all transmitted to many by these insects. They appear pleomorphic in shape and are found singly, in pairs, in short chains, or in filamentous forms. They stain well with the Giemsa, Castaneda, and Macchiavello methods. Rickettsial diseases of man can be classified into four major groups, namely, typhus fever (*Rickettsia prowazeki, R. mooseri,* and *R. tsutsugamushi*); spotted fevers (*R. rickettsii, R. akari,* and *R. conorii*); Q fever (*Coxiella burnetii*); and trench fever (*R. quintana*).

Identification of rickettsial agents may be achieved either by isolating and identifying the organisms or by serologic procedures. Rickettsia grow well in guinea pigs and mice, causing fever, scrotal swelling, and sickness. They also grow in the yolk sacs of 5- to 7-day-old chick embryos and in certain cell cultures. Serologically, the Wiel-Felix (not specific for each disease), CF (specific), and agglutination (specific) tests are used. Table 2.25 summarizes serological test reactions in the rickettsial diseases.

METHODS IN THE PUBLIC HEALTH LABORATORY

Public health laboratories are concerned with the identification of bacteria causing endemic or epidemic diseases. In many instances the organisms are transmitted through water supplies, food, air, or individual contacts. From the public health point of view, it is important to screen the population for signs of disease, to identify and classify the etiologic agent, to determine the source of the disease-causing organism, and to apply preventive measures in order to control spread of the disease among the masses. Generally, bacteriological contamination control in the public health laboratory seeks the lowest possible level of bacterial load in the water treatment system, producing potable water and seeking elimination of pathogens in the environment, food, and water. To accomplish these ends, microbiological contamination control tests are performed, which include the following:

1. Microbial air sampling with devices such as air impaction samplers, liquid impingers, and settling plates.
2. Particle size sampling devices such as membrane filters.
3. Performance of surface sampling with cotton swabs or Roda® plates.
4. Collection of accumulated surface contamination on small strips and assay of same for viable organisms.

TABLE 2.25

Time of Appearance of Antibodies in Serum from Humans

Disease	Complement fixation (days)	Weil-Felix (days)	Rickettsial agglutination (days)
Spotted fever group[a]	8–10	5–12[a]	Unknown
Q fever	8–14	None	5–8
Epidemic and murine typhus	7–9	7–14	5–7
Scrub typhus	Unknown	10–14	Unknown

[a]Except for rickettsialpox in which Weil-Felix antibodies are not found.

From Ormsbee, R. A., in *Manual of Clinical Microbiology,* 2nd ed., Lennette, E. H., Spaulding, E. H., and Truant, J. P., Eds., American Society for Microbiology, Washington, D.C., 1974, 812. With permission.

5. Component system testing.

6. Food (milk, eggs, meat, vegetables, etc.) sampling and testing for bacterial counts.

7. Screening tests for the detection and identification of bacteria on urine, feces, sera, or other clinical specimens from a group of people in a suspected population.

Morphological, biochemical, serological, etc. tests used in the diagnostic bacteriology laboratory may also be employed in the public health laboratory for the identification of the etiologic agents. However, certain tests are more frequently used in public health laboratories, e.g., the coliform test in water supplies, serological tests for venereal diseases, bacteriophage typing for staphylococci or *Salmonella* food poisoning, etc. A brief review of the tests commonly used in the public laboratory is presented in the following section.

The American Public Health Association and affiliated societies have recommended methods for routine detection and enumeration of coliforms and other indicator organisms; these methods are used daily in every health department laboratory.

Standard plate count test — This test is designed to enumerate the total viable bacterial population as a guide to determining the efficiency of water treatment methods such a sedimentation, flocculation, filtration, and disinfection. Properly treated water should yield fewer than 100 colonies per milliliter. Plate counts are not usually done in the routine examination of drinking water at the present time. However, the test is useful in the examination of new supplies where daily sampling is necessary.

Coliform test — A presumptive test, confirmed test, and completed test are carried out in systematic order according to the results of each step. Fermentation tubes containing 0.5% lactose, or tryptone broth containing 0.5% lactose and sodium lauryl sulfate (which is inhibitory to noncoliforms), are inoculated with measured samples and incubated for 24 (±2) hr at 35 (±0.5)°C. If gas appears, the presumptive test is positive and further tests are made to confirm the coliform organisms. Coliform organisms may be enumerated by two methods. The presumptive test procedure may be used in estimating the number of coliform organisms present in a given sample. In the most probable number (MPN) method, multiple tubes of lauryl tryptone broth are inoculated with different volumes of samples, such as 10 ml

(5 tubes), 1 ml (5 tubes), and 0.1 ml (5 tubes). If the number of positive and negative tubes in each dilution is known, it is possible to calculate the probable number of coliforms present in a given volume of water. A combination of the numbers of positive and negative tubes provides an index of pollution which is usually expressed at the most probably number (MPN) of coliform bacteria per 100 ml of sample. The number of organisms present is not absolute but a statistical estimate, i.e., the index represents the number of coliforms which, more frequently than any other number, gives the observed result.

Determination of Bacteria in Water Samples — Membrane Filter Method

One hundred volumes of water samples (two separate samples) are passed through 47-mm membrane filters. If the supply is known or is expected to contain more than 100 coliform bacilli per 100 ml, 10 ml of water diluted with 90 ml of ¼-strength Ringer solution should be used. Sterile Whatman No. 17 absorbent pads are placed in sterile petri dishes, and 2.5 to 3 ml of enriched Teepol® broth[213] is placed over the surface. Chlorinated samples may be tested using one membrane for 6 hr at 25°C followed by 18 hr at 35°C for the presumptive count and one membrane for 6 hr at 25°C followed by 18 hr at 44 ± 0.2°C for the *Escherichia coli* count. Unchlorinated samples may be tested using one membrane for 4 hr at 30°C followed by 14 hr at 35°C for the presumptive count and one membrane for 4 hr at 30°C followed by 14 hr at 44 ± 0.2°C for the *E. coli* count. Yellow colonies are counted and reported as presumptive coliform and *E. coli* counts per 100 ml of water. Membrane counts may be higher than most probably counts (MPN) because they include all organisms producing acid, as well as those producing acid and gas. Plate counts are not usually done in the routine examination of drinking water at the present time. However, plate count information is useful in the examination of new supplies where daily sampling is necessary.

Clostridium welchii (*Perfringens*) in water may be tested by the litmus milk method or black tube method using differential reinforced clostridial medium containing 70 μg/ml polymyxin to inhibit facultative anaerobes. Fecal streptococci in water may be tested using azide dextrose broth.

It is not possible to assess the potability of any

water supply by a single examination. Presumptive coliforms and *Escherichia coli* should be absent from 100 ml. Enterococci and *Clostridium perfringens* in the absence of coliforms suggest contamination at a remote time. These organisms persist longer in water than coliforms.

Determination of Bacteria in the Atmosphere

Bacteria of the atmosphere may be collected, enumerated, and identified by the standard bacteriological techniques including the use of selective media. Microorganisms in the air at low levels may be collected by the simple method of allowing dust to settle on an open petri dish containing nutrient agar. This is useful in enclosed spaces. Dust and microorganisms may also be collected by drawing air through a tube containing a filter of wet sand or cotton. The cotton or sand is then shaken in broth. There are many other types of devices for collecting microorganisms in the air. One, an impingement device, consists of an agar-covered cylinder rotating slowly around its axis vertically. An air stream carrying dust and microorganisms impinges on the sticky agar surface as the drum rotates. There are also bubbling, atomizing, and electrostatic devices. Also, the membrane filter is adaptable to direct collection by filtration of air. Among the airborne pathogenic bacteria that could transmit diseases are *Streptococcus pyogenes* (β-hemolytic type), *Corynebacterium Diphtheriae, Mycobacterium tuberculosis, Streptococcus (Pneumococcus) pneumoniae* (Gram positive), *Klebsiella pneumoniae* (Gram positive), *Neisseria meningitidis, Yersinia (Pasteurella) pestis, Bordetella pertussis,* and *Haemophilus influenzae.*

A number of investigators found no relationship between the concentration of viable bacteria in air and such parameters as temperature, humidity, and wind speed. Skali[214] reported no correlation between two air pollutants, SO_2 and total suspended particulate matter, on the viability of airborne bacteria.

Using newer, sophisticated techniques, Lee et al.[215] reassessed the relationship of air pollutants and meteorological parameters with airborne bacteria. In Cincinnati during the summer of 1969, samples were collected for 1-hr periods with Andersen cascade impactors.[216] Six different size fractions were collected which permitted the determination of concentration as well as size of airborne bacteria. The particles were implanted on

Trypticase soy agar with added fungicide and incubated for 24 hr. Estimation of aerodynamic size of viable bacteria in terms of equivalent spheres of unit density was accomplished using the method of Flesch et al.[217] The analytical procedures followed were those described by Jutze and Tabor[218] and used in continuous air monitoring projects (CAMP). Sulfur dioxide was measured colorimetrically using pararosaniline following collection in an aqueous solution of tetrachloromercurate.[219] Iodine released from a neutral buffered KI solution and determined colorimetrically gave the total oxidant concentration.[220] The flame ionization technique was used to determine total hydrocarbon concentration.[221] Carbon monoxide was determined by nondispersive infrared absorption. The colorimetric method of Saltzman[222] was used to quantitate nitrogen dioxide. The coefficient of haze (Coh) as determined with an AISI tape sample was used to yield a soiling index (ASTM standards);[223] particulate concentrations were estimated from the Coh index values according to the formula of Lee et al.[224]

Results indicated $\mu g/m^3$ = 35.3 + 89.1 Coh, a significant correlation of concentration of viable airborne bacteria with CO and total hydrocarbons. There was a negative correlation of size of bacteria with relative humidity and CO concentration.

Several sampling methods are used for determining bacterial contamination of the air. With impactors, air jets strike a solid obstacle on which the particles settle. With impingers, the bacteria settle directly on a liquid medium. With the Andersen sampler, settling is carried out on nutrient medium. Settling methods of sampling combined with the use of dyes would permit determination of the total number of microorganisms and the proportion of living and dead constituents.[225]

Enrichment in selenite F broth and plating on deoxycholate citrate agar are suitable for isolation of Salmonellae from feces, but are inefficient when sewage is examined. The sewage may contain large numbers of *Proteus* and *Pseudomonas* that are not inhibited by the technique used with feces.

Ryan[226] readily isolated Salmonellae by plating selenite F-enrichment cultures on deoxycholate citrate agar with incubation anaerobically in the presence of 10% (v/v) carbon dioxide. Salmonellae produced jet-black colonies; pseudomonads were suppressed, and the growth of *Proteus* was markedly inhibited. To satisfactorily recover *Salmonella typhi,* double-strength selenite medium and deoxycholate citrate agar without lactose were used.

Collins and Willoughby[227] found casein-peptone-starch (CPS) medium to be the best for counting aquatic bacteria in unpolluted lake water. Staples and Fry[228] compared CPS with other media for counting heterotrophs in polluted and unpolluted surface waters. Media studied were yeast extract medium,[229] iron-peptone medium,[230] sodium caseinate,[231] CPS,[232] nutrient agar, and plate count agar (PCA) as described by the American Public Health Association.[233] Results showed that for clean and polluted rivers, CPS is the most suitable medium for counting fresh-water bacteria.

Most methods for studying soil microbes are indirect in nature. Ideally, studies should be performed on an undisturbed environment. Direct microscopic observations of microbes in soil are difficult, and most techniques require removal or disturbance of the soil in some way. Fluorescent antibody techniques have met with some success.[234-239] Infrared photography has been used to visualize bacteria in the presence of soil particles.[240] Electron microscopy has been used to elucidate the morphology of soil microbes.[241-243] Soil sectioning techniques have been used to study spatial distribution of microbes in natural environments.[244-248] Selective substrates of immersion tubes may be rapidly overgrown by species having shorter generation time or a constitutive enzyme complement capable of utilizing these media. This would exclude colonization by slower growing species.

A modification of the immersion method is favored by many groups. In the pedoscope method, pedoscope of Perfil'ev and Gabe,[249,250] microcapillaries are inserted for various times in soil, mud, and water and then examined microscopically. Replica soil plate technique can be adapted to a study of density and changes in distribution of microbes in natural soils[251] and provide information on the generation time of microbes in soil.[252] Respiratory measurements are widely used for studying microbes in soil and correlate well with other indices of microbial activity such as transformations of carbon, nitrogen, phosphorus and sulfur; accumulation of metabolic intermediates; and pH changes.[251] Methods of measuring soil respiration range from manometry and titrimentry to infrared spectro-

scopy, gas chromatography, and polarography.[251,253] For further discussion of soil microbiology, see Stotzky.[254]

Davis[255] discussed some of the microbial aspects of pollution. When the natural decomposition of food or material takes place, a particular type of flora is established. Microbial activities may be favorable or unfavorable in relation to pollution. Sulfate-reducing bacteria produce H_2S, blacken heavy metals and their solutions, and reduce the Eh value of water to kill fish and poison the respiratory systems of higher forms of life. Some microorganisms intensify pollution by converting mercury compounds to the toxic dimethyl derivative or oxidizing nonacidic substances to acids.

Methods for Testing Milk and Other Dairy Products

The Phosphatase Test

This test is used to provide information concerning the adequacy of pasteurization of milk. It is based on the destruction by pasteurization of the heat-sensitive enzyme phosphatase, normally present in fresh milk. The test for presence or absence of phosphatase is based on the ability of phosphatase to liberate phenol from phosphoric-phenyl ester added to a sample of milk and held at 40°C for 15 min. If the milk is not pasteurized (or is insufficiently pasteurized), the phosphatase enzyme normally present in the milk will be active and will decompose the added phenyl phosphate, liberating the phenol. Phenol turns blue if 2,6-dichloroquinone chloroimide with $CuSO_4$ as catalyst is added to the mixture. The appearance of a blue color (indophenol blue) indicates the presence of free phenol liberated by undestroyed phosphatase in the milk. The indophenol blue is extracted by shaking with neutral *n*-butyl alcohol. The color content of the tested sample of milk is then compared with the color of standards containing known amounts of phenol and treated with the same reagents. When milk is pasteurized at 143°F for 30 min, 96% of the enzyme is destroyed. Only a trace of blue color should appear. Heating at 145°F or above for 30 min insures complete inactivation of the phosphatase.

The Oxidation-reduction Test

Most actively growing bacteria cause a lowered oxidation-reduction (O-R) potential in their medium. The presence in milk of large enough numbers of growing bacteria to produce a significantly lowered O-R potential can be detected by the use of various dyes which undergo color change when reduced or oxidized. For example, methylene blue is colored when oxidized, but the dye becomes colorless when the O-R potential of the medium is lowered to about 0.01 V. The reductase test is used principally with raw milk to obtain useful approximation of the number and kinds of living bacteria present. The test may be performed with a 10-ml milk sample in a sterile tube added to a 1-ml standard methylene blue solution (final concentration about 1:250,000). The tube is closed with a rubber stopper and slowly inverted three times to mix. It is immediately placed in a water bath at 36°C. At the end of each hour during the test, the tube is inverted once. Observations should be made after 30 min, 1 hr, and later. The results are expressed in MBRT (methylene blue reduction time); the shorter the MBRT, the greater the number of active bacteria in the milk and the lower its bacteriological quality.

The degrees of reduction can be estimated by using resazurin dye, which changes color from slate blue to pink to colorless (with a range of colors in between). Therefore, resazurin permits reading of methylene blue.

The number of bacteria in a milk sample may be estimated by standard plate count methods. Sharpe and Jackson[256] proposed the use of poured plates over surface-inoculated plates of Baird-Parker's medium[257] for enumeration of *Staphylococcus aureus*. They suggested the use of their automatic diluting, inoculating, and pour plate preparing machine[258] for use in routine testing for *S. aureus* and other bacteria in milk.

Keller et al.[259] reported that when testing for bacteria in milk or acidified milk (pH 4.0), *Streptococcus lactis*, but not coliform bacteria, were affected by the diluent medium when counted in acidified milk. Standard methods for examination of dairy products recommend (APHA) phosphate-buffered diluent water or sodium citrate for total bacterial count determinations of milk or fermented products.[260] Using milk or two fermented products, Keller et al.[259] found a diluent of 0.1% tryptone to give the highest total counts of the three diluents.

Bhadsavle et al.[261] isolated from raw milk one species of *Clostridium* capable of growing at 0°C which was anaerobic, aerotolerant, and had characteristics similar to *C. carnis*, and one species whose minimal growth temperature was 4°C, was anaerobic, and had characteristics similar to *C. hastiforme*. Psychrophilic species of *Bacillus* were present in up to 35% of the samples of milk examined by Shehata and Collins.[262] Numbers of *Pseudomonas* sp. and other groups identified among bacteria isolated from milk have been reported by Juffs[263] (Table 2.26). The ability of such bacteria to grow at low temperatures increases in importance as the dairy industry tries to extend the shelf life of its products.

To isolate *Brucella* organisms by direct culture from milk, soil and feces, it is necessary to use a selective medium. Jones and Morgan[264] and Morgan[265] found that media containing bacteriostatic dyes were inhibitory to strains of *Brucella abortus* Biotype 2 and other fastidious strains. They also found serum-dextrose-antibiotic (SDA) medium to be the only selective medium capable of supporting the growth of all biotypes of *Brucella.* Their findings were confirmed by Painter et al.[266]

Other selective media for isolation of *Brucella* organisms have been described by Ryan[267] and Farrell.[268] Farrell and Robertson[269] reported on a study comparing Farrell's medium, Ryan's medium, Mair's medium,[270] and SDA medium. The SDA medium is recommended by the World Health Organization.[271] High isolation rates were obtained on Farrell's medium from herd samples of milk, which compared favorably with the lengthy guinea pig inoculation (6 weeks). Strains of *Brucella abortus* Biotype 2 were inhibited with Ryan's medium and with Mair's medium. Both Mair's medium and SDA medium had low isolation because of inability to suppress growth of contaminant organisms. Farrell and Robertson suggested that a combination of Farrell's and Ryan's media be used for routine isolation and that two plates of each medium should be inoculated.

METHODS IN FOOD BACTERIOLOGY

The International Commission for Microbiological Specification of Foods (ICMSF)[272] describes analytical methods for enumeration of bacteria in foods. Results of ICMSF studies reported by Erdman[273] on isolation of *Salmonella* from artificially infected raw meats showed that more positive samples were identified when enrichment broth was incubated at 43°C than when

TABLE 2.26

Numbers of the *Pseudomonas* spp. and Other Groups Identified Among the Bacteria Isolated from Milk

	No. of *Pseudomonas* spp. and other groups[a]				
	From raw milk		From pasteurized milk		
Species or group	On penicillin agar (30°)	On BCP agar (7°)	On penicillin agar (30°)	On BCP agar (7°)	Total
P. fluorescens	60	54	4	3	121
P. aeruginosa	12	4	0	0	16
P. putida	12	0	0	0	12
P. maltophilia	5	2	2	0	9
P. cepacia	2	0	0	1	3
P. pseudoalcaligenes	5	0	0	0	5
P. alcaligenes	0	1	0	0	1
All *Pseudomonas* spp.	96	61	6	4	167
Flavobacteria	22	3	5	0	30
Achromobacter, Alcaligenes, Acinetobacter	10	3	2	1	16
Fermentative Gram-negative rods	51	9	5	2	67
Gram-positive organisms	20	3	24	3	50
All isolates	199	79	42	10	330

[a]Subtotals are provided to indicate the number of isolations made from each type of milk and by each selective procedure.

From Juffs, H. S., *J. Appl. Bacteriol.*, 36, 585, 1973. With permission.

incubated at 35°C. It was found that selenite brilliant green sulfa broth was the best enrichment medium for *S. schwarzengrund, S. typhimurium,* and *S. dublin. S. seftenberg* was detected best on selenite cystine broth. Bismuth sulfite agar gave a large number of false-positive isolates when used with *Salmonella*-negative samples. The studies pointed out the need for each laboratory to use its own methods, as results were best when laboratories chose their own selective agars, indicating that familiarity with a medium plays a significant role in its suitability for the isolation of *Salmonella.*

Mossel[274-276] stresses that current methods of examination of foods for coliforms (bacteria of the coli-aerogenes group) may lead to serious underestimation of prevailing health risks; in the tests for coli-aerogenes bacteria, lactose-negative types (i.e., *Salmonella* and slow lactose fermenters equally often enteropathogenic) are missed. Based on fermentation of glucose, Mossel[277] and Henriksen[278] recommend replacement of such tests by enumeration of the entire Enterobacteriaceae group.

The proportion of salmonellae among the total Enterobacteriaceae microflora is almost always small. On this basis, instead of testing 60 × 25 g samples for salmonellae (as is standard in many places), a test for Enterobacteriaceae in a smaller sample size is likely to confer greater protection. Employing this rationale, Drion and Mossel[279] examined dried foods in 1-g aliquots for Enterobacteriaceae and found that they gave at least the same level of consumer protection as the salmonella examination scheme of 60 samples of 25 g, suggested by the National Academy of Sciences.

For control systems for foods, the testing of 60 individual 25-g subsamples with negative results was recommended by the National Research Council (NRC) Committee on *Salmonella.*[280] Silliker and Gabis[281] studied 26 low-moisture food samples naturally contaminated with salmonellae to compare the efficiency of 60 25-g, 15 100-g, and 3 500-g subsamples in recovery of salmonellae. Wet compositing was studied by pooling in groups of five preenrichment broths from the 25-g and 100-g subsamples. The wet-compositing results indicated that the method was as sensitive as the analysis of individual subsamples in detecting salmonellae. The overall results of Silliker and Gabis indicate that salmonellae may be detected with equal facility by any of the sampling and analytical methods tested.

Detection of salmonellae in high-moisture foods (meats, poultry, and eggs) is difficult because of large numbers of competing microorganisms. Many procedures specify direct selective enrichment in contrast to preenrichment which is used in analysis of dried foods.[282-285] In contrast, some reports indicate that preenrichment of meats in a non-selective medium results in greater sensitivity of *Salmonella* detection than does direct selective enrichment.[286] Gabis and Silliker[287] reported the efficacy of lumping subsamples in conjunction with the influence of preenrichment as compared to direct selective enrichment. Results paralleled those of the previous study on dried foods,[281] that is, lumping of 25-g subsamples to produce larger samples for analysis is as sensitive for detection of positive samples as is the analysis of individual 25-g subsamples. It was suggested that high-moisture foods should be preenriched (with lactose) in a nonselective medium rather than directly enriched in tetrathionate broth for the detection of salmonellae. Idziak et al.[288] studied 16 contemporary methods used for detection of *Salmonella* in meringue powder; they pointed out the variation in ability to isolate *Salmonella* from the specimens and the need for collaborative assays to define such methodology.

Palumbo et al.[289] studied the microbiology of the frankfurter process. *Salmonella seftenberg* 776 W, which is very heat resistant,[290,291] was chosen for this study. They investigated the total (nonsalmonellae) bacterial flora and salmonellae during the heating step of frankfurter processing. Most of the killing occurred after the product temperature reached 43.3°C (110°F). Gram-negative rods with some micrococci predominated at the beginning of the heating, and most were destroyed by the time the product temperature reached 60°C. Above 60°C, only micrococci were found. For salmonellae, little thermal destruction occurred before 57.2°C; no viable salmonellae were found in products heated to 68.3°C and above. Since the normal heating of frankfurters in processing both destroyed the salmonellae and reduced the total flora, the product was made safe from a public health view, thus extending the shelf life.

Using a method modified from Casman,[292]

Gilbert et al.[293] used gel diffusion to examine extracts of food samples for presence of entero-toxin. To extract endotoxin, 50 g of food was homogenized, extracted, centrifuged, and reextracted in buffered saline. This was followed by chloroform extractions concentrated to dryness in polyethylene glycol (PEG). After a second chloroform extraction, the volume of extract was adjusted with 0.005 M phosphate buffer, pH 5.6, and was passed on a column of carboxymethyl cellulose. Any toxin adsorbed was eluted with 100 ml of 0.2 M phosphate buffer (pH 7.4), concentrated to dryness in PEG, resuspended in 0.2 ml of 0.02 M phosphate-buffered saline (pH 7.4), containing 10% BHI broth and 0.02% thiomersal, and examined by gel diffusion for the presence of enterotoxin.

The slide agar double-diffusion test for entero-toxin is a modification of the method described by Šimkovičová and Gilbert.[294] Outer wells contain extracts of food or culture filtrate; one well is reserved for reference enterotoxin. The center well contains antiserum of the corresponding type. After several days incubation, slides are stained by immersion in 0.1% thiazine red in 1% glacial acetic acid to enhance precipitin lines. The presence of one or more of the enterotoxins A, B, C, and D in a sample is verified by coalescence of its precipitation line with the reference line of the corresponding type. A faint precipitin line forms with as little as 0.25 μg/ml of toxin.

Clostridium perfringens food poisoning refers to a generally mild, self-limited gastroenteritis that is quite common.[295-301] Strong et al.[302] found that 16% of samples of American meat, poultry, and fish were contaminated with *C. perfringens*. Smith and Gardner[303] recovered the organism from eight types of soil thought not to have had fecal contamination for several years. It has also been isolated from an assay of foodstuffs including milk, fruit and vegetables, spices, dehydrated soups, and spaghetti. After initial heating, the contaminated food has usually had a period of inadequate cooling and was served cold or warmed but not sufficiently reheated. The vegetative form is killed in the initial heating process, but not all of the heat-resistant spores are destroyed; these are left in a relatively anaerobic environment and germinate during the cooling or warming process.

Hauschild and Hilsheimer[304] tested four solid media for suitability of enumeration of food-borne

Clostridium perfringens. The media tested were egg yolk (EY)-free agar containing 0.04% D-cycloserine (D-CS) tentatively designated EY-free TSC (tryptone-sulfite-cycloserine) agar,[305] SFP (Shahidi-Ferguson perfringens) agar,[306] TSC agar,[307] and OPSP (oleandomycin-polymyxin-sulfadiazine perfringens) agar.[308] Only TSC and EY-free TSC agars were sufficiently selective to ensure confirmatory tests without interference from facultative anaerobes (Table 2.27). Because of the disadvantages associated with the use of egg yolk, Hauschild and Hilsheimer[305] recommend the use of EY-free TSC agar for enumeration of *C. perfringens* in foods.

Many strains of *Clostridium perfringens* biosynthesize in the enteric lumen, a low molecular weight proteinaceous toxic factor that leads to diarrhea. The standard method for selective enumeration of this anaerobic spore-bearing organism from foods is that of Marshall et al.,[310] using sulfite iron polymyxin neomycin agar at 46°C. Harmon et al.[307] suggested replacing polymyxin and neomycin by cycloserine. Mossel and Pouw[309] incorporated this suggestion into a study of pure cultures of *C. perfringens* and food, feed, and sewage samples. The advantage of the new medium is that colonies are smaller in size and much more discreet than in the Marshall medium.

TABLE 2.27

Enumeration of *Clostridium perfringens* from Various Foods in EY-free TSC and OPSP Agars[a]

		Count (cells/g)	
Food	Organism	Egg yolk-free tryptose-sulfite-cycloserine agar	Oleandomycin-polymyxin-sulfadiazine perfringens agar
Barbecued chicken	*C. perfringens*	5.6×10^6	$5.7 \times 10^6 \cdot$
	Nonspecific	$10^7 - 10^8$	$>10^9$
Barbecued pork	*C. perfringens*	2.6×10^2	2.8×10^2
	Nonspecific	$10^6 - 10^7$	$>10^7$
Turkey in gravy	*C. perfringens*	3.8×10^7	3.4×10^7
	Nonspecific	$<10^6$	$>10^9$
Beef roast	*C. perfringens*	9.0×10^7	8.2×10^7
	Nonspecific	$<10^5$	$>10^9$
Beef in gravy	*C. perfringens*	6.4×10^6	6.1×10^6
	Nonspecific	$<10^5$	$>10^9$
Pork in gravy	*C. perfringens*	2.8×10^4	1.5×10^4
	Nonspecific	$10^4 - 10^5$	$>10^9$

[a]Foods listed enumerated after 4 weeks of storage in 10% glycerol at 4°C.

From Hauschild, A. H. W. and Hilsheimer, R., *Appl. Microbiol.,* 27, 521, 1974. With permission.

REFERENCES

1. Lillie, R. D., *Conn's Biological Stains*, 8th ed., Williams & Wilkins, Baltimore, 1969.
2. Conn, H. J., Darrow, M. A., and Emmel, V. M., *Staining Procedures*, Williams & Wilkins, Baltimore, 1960.
3. Austrian, R., *Streptococcus pneumoniae (Pneumococcus)*, in *Manual of Clinical Microbiology*, 2nd ed., Lennette, E. H., Spaulding, E. H., and Truant, J. P., Eds., American Society for Microbiology, Washington, D.C., 1974, 113.
4. Chapman, G. H., The significance of sodium chloride in studies of Staphylococci, *J. Bacteriol.*, 50, 201, 1945.
5. Zebovitz, E., Evans, J. B., and Niven, C. F., Jr., Tellurite-glycine agar: a selective plating medium for the quantitative detection of coagulase-positive staphylococci, *J. Bacteriol.*, 70, 686, 1955.
6. Greer, J. E. and Menard, R. R., A new selective and differential medium for *Staphylococcus aureus*, *Am. J. Public Health*, 49, 685, 1959.
7. Blair, J. E., Borman, E. K., Bynoe, E. T., Updyke, E. L., and Williams, R. E. O., Hospital Acquired Staphylococcal Disease, Recommended Procedures for Laboratory Investigation, U.S. Department of Health, Education and Welfare, Public Health Service, Atlanta, 1958.
8. Reed, R. W. and Morgante, O., Recent advances in the laboratory diagnosis of tuberculosis, *Am. J. Med. Sci.*, 231, 320, 1956.
9. *Tuberculosis Laboratory Methods*, Armed Forces Cooperative Study on the Chemotherapy of Tuberculosis, Veterans Administration Department of Medicine and Surgery, Central Office, Washington, D.C., 1958.
10. Middlebrook, G. and Cohn, M. L., Bacteriology of tuberculosis: laboratory methods, *Am. J. Public Health*, 48, 844, 1958.
11. Dowell, V. R. and Hawkins, T. M., Laboratory methods in anaerobic bacteriology, *U.S. Public Health Serv. Publ.*, No. 1803, 1968.
12. American Society for Microbiology, *Manual of Microbiological Methods*, McGraw-Hill, New York, 1957.
13. World Health Organization Expert Committee, Streptococcal and staphylococcal infections, *WHO Tech. Rep. Ser.*, 394, 1, 1968.
14. Recommendations of the National Conference on Hospital Acquired Staphylococcal Disease, U.S. Department of Health, Education and Welfare, Public Health Service, Atlanta, 1958.
15. Elek, S. D., *Staphylococcus pyogenes and its Relation to Disease*, E. and S. Livingstone, Edinburgh, 1959.
16. Vestal, A. L. and Kubica, G. P., Differential colonial characteristics of mycobacteria on Middlebrook and Cohn 7H10 agar-base medium, *Am. Rev. Respir. Dis.*, 94, 247, 1966.
17. Kubica, G. P. and Dye, W. E., Laboratory methods for clinical and public health mycobacteriology, *U.S. Public Health Serv. Publ.*, No. 1547, 1967.
18. Wolinsky, E., Identification and classification of mycobacteria other than *Mycobacterium tuberculosis*, in *Handbook of Tuberculosis Laboratory Methods*, Hobby, G. L., Ed., Veterans Administration, U.S. Government Printing Office, Washington, D.C., 1962.
19. Wayne, L. G. and Doubek, J. R., Diagnostic key to mycobacteria encountered in clinical laboratories, *Appl. Microbiol.*, 16, 925, 1968.
20. Snell, J. J. S. and Lapage, S. P., Comparison of four methods for demonstrating glucose breakdown by bacteria, *J. Gen. Microbiol.*, 68, 221, 1971.
21. Hugh, R. and Leifson, E., The taxonomic significance of fermentative versus oxidative metabolism of carbohydrates by various gram negative bacteria, *J. Bacteriol.*, 66, 24, 1953.
22. Park, R. W. A., A comparison of two methods for detecting attack on glucose by pseudomonads and achromobacters, *J. Gen. Microbiol.*, 46, 355, 1967.
23. Smith, N. R., Gordon, R. E., and Clark, F. E., Aerobic spore-forming bacteria, *U.S. Dept. Agric. Monogr.*, 18, 35, 1952.
24. Lapage, S. P., Hill, L. R., and Reeve, J. D., *Pseudomonas stutzeri* in pathological material, *J. Med. Microbiol.*, 1, 195, 1968.
25. Chilton, M. L. and Fulton, M., A presumptive medium for differentiating paracolon from *Salmonella* cultures, *J. Lab. Clin. Med.*, 31, 824, 1946.
26. Ulrich, J. A., New indicators to replace litmus in milk, *Science*, 99, 352, 1944.
27. Kovács, N., Identification of *Pseudomonas pyocyanea* by the oxidase reaction, *Nature*, 178, 703, 1956.
28. Steel, K. J., The oxidase reaction as a taxonomic tool, *J. Gen. Microbiol.*, 25, 297, 1961.
29. Ewing, W. H. and Johnson, J. G., The differentiation of Aeromonas and C27 cultures from Enterobacteriaceae, *Int. Bull. Bacteriol. Nomencl. Taxon.*, 10, 223, 1960.
30. Bayliss, B. G. and Hall, E. R., Plasma coagulation by organisms other than *Staphylococcus aureus*, *J. Bacteriol.*, 89, 101, 1965.
31. Quinn, E. L., Cox, F., and Fisher, M., The problem of associating coagulase negative staphylococci with disease, *Ann. N.Y. Acad. Sci.*, 128, 428, 1965.
32. Lubin, A. H. and Ewing, W. H., Studies on beta-D-galactosidase activities of *Enterobacteriaceae*, *Public Health Lab.*, 22, 84, 1964.
33. Simmons, J. S., A culture medium for differentiating organisms of typhoid-colon aerogenes groups and for isolation of certain fungi, *J. Infect. Dis.*, 39, 209, 1926.

34. Christensen, W. B., Hydrogen Sulfide Production and Citrate Utilization in the Differentiation of Enteric Pathogens and Coliform Bacteria, Res. Bull. No. 1, Weld County Health Department, Greeley, Colo., 1949, 3.
35. Ayers, S. H. and Rupp, P., Differentiation of hemolytic streptococci from human and bovine sources by the hydrolysis of sodium hippurate, *J. Infect. Dis.*, 30, 388, 1922.
36. Haynes, W. C., *Pseudomonas aeruginosa* — its characterization and identification, *J. Gen. Microbiol.*, 5, 939, 1951.
37. Ewing, W. H. and Davis, B. R., Media and Tests for Differentiation of *Enterobacteriaceae*, Center for Disease Control, Atlanta, 1970.
38. Morse, M. L. and Weaver, R. H., Rapid microtechnics for identification of cultures. III. Hydrogen sulfide production, *Am. J. Clin. Pathol.*, 20, 481, 1950.
39. Clarke, P. H., Hydrogen sulphide production by bacteria, *J. Gen. Microbiol.*, 8, 397, 1953.
40. Ewing, W. H., *Enterobacteriaceae* — biochemical methods for group differentiation, *U.S. Public Health Serv. Publ.*, No. 734, 1960.
41. Arnold, W. M., Jr. and Weaver, R. H., Quick microtechniques for the identification of cultures. I. Indole production, *J. Lab. Clin. Med.*, 33, 1334, 1948.
42. Coblentz, L. M., Rapid detection of the production of acetyl-methyl-carbinol, *Am. J. Public Health*, 33, 815, 1943.
43. Barritt, M. M., The intensification of the Voges-Proskauer reaction by the addition of 4-naphthol, *J. Pathol. Bacteriol.*, 42, 441, 1936.
44. O'Meara, R. A. Q., A simple delicate and rapid method of detecting the formation of acetylmethylcarbinol by bacteria fermenting carbohydrate, *J. Pathol. Bacteriol.*, 34, 401, 1931.
45. Bowen, M. K., Theile, L. C., Stearman, B. D., and Schaub, I. G., The optochin sensitivity test: a reliable method for identification of pneumococci, *J. Lab. Clin. Med.*, 49, 641, 1957.
46. Runyon, E. H., Anonymous mycobacteria in pulmonary disease, *Med. Clin. North Am.*, 43, 273, 1959.
47. Starr, M. P., The classical agent of bacterial root and stem disease of guayule, *Phytopathology*, 37, 291, 1947.
48. Christensen, W. B., Urea decomposition as a means of differentiating *Proteus* and paracolon cultures from each other and from *Salmonella* and *Shigella* types, *J. Bacteriol.*, 52, 461, 1946.
49. Clarke, P. H. and Cowan, S. T., Biochemical methods for bacteriology, *J. Gen. Microbiol.*, 6, 187, 1952.
50. Hormaeche, E. and Munilla, M., Biochemical tests for the differentiation of *Klebsiella* and *Cloacae*, *Int. Bull. Bacteriol. Nomencl. Taxon.*, 7, 1, 1957.
51. Stuart, C. A., van Stratum, E., and Rustigian, R., Further studies on urease production by *Proteus* and related organisms, *J. Bacteriol.*, 49, 437, 1945.
52. Ewing, W. H. and Edwards, F. P., The Principal Divisions and Groups of Enterobacteriaceae, revised, National Communicable Disease Center, Atlanta, 1962.
53. Shaw, C. and Clarke, P. H., Biochemical classification of *Proteus* and *Providence* cultures, *J. Gen. Microbiol.*, 13, 155, 1955.
54. Report of the Enterobacteriaceae Subcommittee of the Nomenclature Committee of the International Association of Microbiological Societies, *Int. Bull. Bacteriol. Nomencl. Taxon.*, 8, 25, 1958.
55. Willis, A. T. and Hobbs, G., Some new media for the isolation and identification of Clostridia, *J. Pathol. Bacteriol.*, 77, 511, 1959.
56. Lowbury, E. J. L. and Lilly, H. A., A selective plate medium for *C. welchii*, *J. Pathol. Bacteriol.*, 70, 105, 1955.
57. Kabat, E. A. and Mayer, M. M., *Experimental Immunochemistry*, Charles C Thomas, Springfield, Ill., 1961.
58. Lund, E. and Rasmussen, P., Omni-serum: a diagnostic serum, reacting with the 82 known types of *Pneumococcus*, *Acta Pathol. Microbiol. Scand.*, 68, 458, 1966.
59. Lund, E., Laboratory diagnosis of *Pneumococcus* infections, *Bull. WHO*, 23, 5, 1960.
60. Blair, J. E. and Williams, R. E. O., Phage typing of staphylococci, *Bull. WHO*, 24, 771, 1961.
61. Cohen, J. O., *The Staphylococci*, John Wiley & Sons, New York, 1970.
62. Anderson, K., A survey of toxicity in staphylococci, *J. Clin. Pathol.*, 9, 257, 1956.
63. Panton, P. N. and Valentine, F. C. O., Staphylococcal toxin, *Lancet*, 1, 506, 1932.
64. Jackson, A. W. and Little, R. M., Leucocidal effect of staphylococcal S-lysin, *Can. J. Microbiol.*, 3, 101, 1957.
65. Marks, J. and Vaughan, A. C. T., Staphylococcal S-haemolysin, *J. Pathol. Bacteriol.*, 62, 597, 1950.
66. Laskowski, L. and Pexa, M., Newer Concepts and Techniques in Clinical Diagnostic Microbiology, The Catholic Hospital Association of the U.S.A., St. Louis, 1959.
67. Ronald, A. R. and Turck, M., Factors influencing in vitro susceptibility to the cephalosporins and clinical trial of an oral cephalosoprin, cephaloglycin, *Antimicrob. Agents Chemother.*, p. 82, 1966.
68. Bondi, A., Spaudling, E. H., Smith, E. D., and Dietz, C. C., A routine method for the rapid determination of susceptibility to penicillin and other antibiotics, *Am. J. Med. Sci.*, 214, 221, 1947.
69. Ericsson, H., Rational use of antibiotics in hospitals, *Scand. J. Clin. Lab. Invest.*, 12, 1, 1960.
70. Ericsson, H., The Paper Disc Method in Quantitative Determination of Bacterial Sensitivity to Antibiotics, Karolinska Sjukhuset, Stockholm, 1961.
71. Standardization of methods for conducting microbic sensitivity tests; preliminary report of a working group of the International Collaboration Study by the World Health Organization, *WHO Tech. Rep. Ser.*, 289, 50, 1964.
72. Lorian, V., *Antibiotics and Chemotherapeutic Agents in Clinical and Laboratory Practice*, Charles C Thomas, Springfield, Ill., 1966.

73. Rogers, M. A., Ryan, W. L., and Severens, J. M., A new method for the rapid determination of bacterial sensitivity, *Antibiot. Chemother.*, 5, 382, 1955.

74. Bauer, A. W., Kirby, W. M. M., Sherris, J. C., and Turck, M., Antibiotic susceptibility testing by a standardized single disk method, *Am. J. Clin. Pathol.*, 45, 493, 1966.

75. Haltalin, K. C., Markley, A. H., and Woodman, E., Agar plate dilution method for routine antibiotic susceptibility testing in a hospital laboratory, *Am. J. Clin. Pathol.*, 60, 384, 1973.

76. Steers, E., Foltz, E. L., and Graves, B. S., An inoculareplicating apparatus for routine testing of bacterial susceptibility to antibiotics, *Antibiot. Chemother.*, 9, 307, 1959.

77. Anderson, T. G., Testing of susceptibility to antimicrobial agents and assay of antimicrobial agents in body fluids, in *Manual of Clinical Microbiology*, 1st ed., Blair, J. E., Lennette, E. H., and Truant, J. P., Eds., American Society for Microbiology, Washington, D.C., 1970, 299.

78. Garrod, L. P. and Waterworth, P. M., A study of antibiotic sensitivity testing with proposals for simple uniform methods, *J. Clin. Pathol.*, 24, 779, 1971.

79. Ericsson, H. M. and Sherris, J. C., Antibiotic sensitivity testing: Report of an international collaborative study, *Acta Pathol. Microbiol.*, Sect. B Suppl., 217, 1, 1971.

80. Gavan, T. L., Cheatle, E. L., and McFadden, H. W., Jr., Antimicrobial Susceptibility Testing, Commission on Continuing Education, American Society of Clinical Pathologists, Chicago, 1971.

81. Silvestri, L. G. and Hill, L. R., Agreement between deoxyribonucleic acid base composition and taxometric classification of gram-positive cocci, *J. Bacteriol.*, 90, 136, 1965.

82. Auletta, A. E. and Kennedy, E. R., Deoxyribonucleic acid base composition of some members of the *Micrococcaceae*, *J. Bacteriol.*, 92, 28, 1966.

83. Rosypal, S., Rosypalová, A., and Hořejš, J., The classification of micrococci and staphylococci based on their DNA base composition and Adansonian analysis, *J. Gen. Microbiol.*, 44, 281, 1966.

84. Garrity, F. L., Detrick, B., and Kennedy, E. R., Deoxyribonucleic acid base composition in the taxonomy of *Staphylococcus*, *J. Bacteriol.*, 97, 557, 1969.

85. Kocur, M., Bergan, T., and Mortensen, N., DNA base composition of gram-positive cocci, *J. Gen. Microbiol.*, 69, 167, 1971.

86. Schleifer, K. H. and Kandler, O., Amino acid sequence of the murein of *Planococcus* and other *Micrococcaceae*, *J. Bacteriol.*, 103, 387, 1970.

87. Schleifer, K. H. and Kandler, O., Peptidoglycan types of bacterial cell walls and their taxonomic implications, *Bacteriol. Rev.*, 36, 407, 1972.

88. Schleifer, K. H., Chemical composition of staphylococcal cell walls, in *2nd Int. Symp. Streptococcus and Staphylococcus Infections*, Jeljaszewicz, J., Ed., Karger, Basel, 1973.

89. Baddiley, J., Brock, J. H., Davison, A. L., and Partridge, M. D., The wall composition of micrococci, *J. Gen. Microbiol.*, 54, 393, 1968.

90. Davison, A. L., Baddiley, T., Hofstad, T., Losnegard, N., and Oeding, P., Serological investigations on teichoic acids from the walls of *Staphylococci, Nature*, 202, 872, 1964.

91. Schleifer, K. H. and Kocur, M., Classification of Staphylococci based on chemical and biochemical properties, *Arch. Mikrobiol.*, 93, 65, 1973.

92. Barry, A. L., Lachica, R. V. F., and Atchison, F. W., Identification of *Staphylococcus aureus* by simultaneous use of tube coagulase and thermonuclease tests, *Appl. Microbiol.*, 25, 496, 1973.

93. Lachica, R. V. F., Genigeorgis, C., and Heoprich, P. D., Metachromatic agar-diffusion methods for detecting staphylococcal nuclease activity, *Appl. Microbiol.*, 21, 585, 1971.

94. Jay, J. M., Use of a plating method to estimate extracellular protein production by staphylococci, *Infect. Immun.*, 3, 544, 1971.

95. Washington, J. A., II, Identification of bacteria, in *Laboratory Procedure in Clinical Microbiology*, Little, Brown & Company, Boston, 1974, chap. 4.

96. Brown, J. H., The Use of Blood Agar for Study of Streptococci, Monogr. No. 9, Rockefeller Institute for Medical Research, New York, 1919.

97. Lancefield, R. C., A serological differentiation of human and other groups of hemolytic streptococci, *J. Exp. Med.*, 57, 571, 1933.

98. Moody, M. D., Padula, J., Lizana, D., and Hall, C. T., Epidemiological characterization of group A streptococci by T-agglutination and M-precipitation tests in the public health laboratory, *Health Lab. Sci.*, 2, 149, 1965.

99. Wilson, E., Zimmerman, R. A., and Moody, M. D., Value of T-agglutination typing of group A streptococci in epidemiological investigations, *Health Lab. Sci.*, 5, 199, 1968.

100. Facklam, R. R., Padula, J. F., Thacker, L. G., Worthan, E. C., and Sconyers, B. J., Presumptive identification of Group A, B, and D Streptococci, *Appl. Microbiol.*, 27, 107, 1974.

101. Pavlova, M. T., Brezenski, F. T., and Litsky, W., Evaluation of various media for isolation, enumeration and identification of fecal streptococci from natural sources, *Health Lab. Sci.*, 9, 289, 1972.

102. Facklam, R. R., Recognition of group D streptococcal species of human origin by biochemical and physiological tests, *Appl. Microbiol.*, 23, 1131, 1972.

103. Huis In't Veld, J. H. J., Meuzelaar, H. L. C., and Tom, A., Analysis of Streptococcal cell wall fractions by Curie-Point gas-liquid chromatography, *Appl. Microbiol.*, 26, 92, 1973.

104. **Fuller, A. T.,** The formamide method for the extraction of polysaccharides from haemolytic streptococci, *Br. J. Exp. Pathol.,* 19, 130, 1938.

105. **Shuman, R. D. and Wood, R. L.,** Swine abscesses caused by Lancefield group E streptococci. III. Application of precipitin test in selected groups of market pigs, *Cornell Vet.,* 57, 356, 1967.

106. **Austrian, R.,** The pneumococci, in *Rapid Diagnostic Methods in Medical Microbiology,* Graber, C. D., Ed., Williams & Wilkins, Baltimore, 1970.

107. **Oetjen, K. A. B. and Harris, D. L.,** Scheme for systematic identification of aerobic pathogenic bacteria, *J. Am. Vet. Med. Assoc.,* 163, 169, 1973.

108. **Merrill, C. W., Gwaltney, J. M., Jr., Hendley, J. O., and Sande, M. A.,** Rapid identification of pneumococci, Gram stains vs. the quellung reaction, *N. Engl. J. Med.,* 288, 510, 1923.

109. **Neufeld, F.,** Ueber die Agglutination der Pneumokokken und über die Theorieen der Agglutination, *Z. Hyg. Infektionskr.,* 40, 54, 1902.

110. **Neufeld, F.,** Etinger-tulczynska: Nasale Pneumokokkeninfectionen und Pneumokokkenkeimträger im Teirversuch, *Z. Hyg. Infektionskr.,* 112, 492, 1931.

111. **Lund, E.,** Polyvalent, diagnostic pneumococcus sera, *Acta Pathol. Microbiol. Scand.,* 59, 533, 1963.

112. **Lorian, V. and Markovits, G.,** Disk test for the differentiation of Pneumococci from other alpha-hemolytic Streptococci, *Appl. Microbiol.,* 26, 116, 1973.

113. **Lorian, V. and Popoola, B.,** Pneumococci producing beta hemolysis on agar, *Appl. Microbiol.,* 24, 44, 1972.

114. **Lorian, V., Waluschka, A., and Popoola, B.,** A pneumococcal beta hemolysis produced under the effect of antibiotics, *Appl. Microbiol.,* 25, 290, 1973.

115. **Maxted, W. R.,** The use of bacitracin for identifying group A hemolytic streptococci, *J. Clin. Pathol.,* 6, 224, 1953.

116. **Levinson, M. L. and Frank, P. F.,** Differentiation of group A from other beta hemolytic streptococci with bacitracin, *J. Bacteriol.,* 69, 284, 1955.

117. **Jelinková, J. and Rotta, J.,** The bacitracin test for recognition of group A streptococci, *Int. J. Syst. Bacteriol.,* 17, 297, 1967.

118. **Cherry, W. B. and Moody, M. D.,** Fluorescent-antibody techniques in diagnostic bacteriology, *Bacteriol. Rev.,* 29, 222, 1965.

119. **Ivler, D.,** Aids in classification and characterization of *Neisseria* species, *J. Gen. Microbiol.,* 69, 10, 1971.

120. **Branhan, E. E.,** Serological relationships among meningococci, *Bacteriol. Rev.,* 17, 175, 1953.

121. **Biegeleisen, J. Z., Jr., Mitchell, M. S., Marcus, B. B., Rhoden, D. L., and Blumberg, R. W.,** Immunofluorescence techniques for demonstrating bacterial pathogens associated with cerebrospinal meningitis. I. Clinical evaluation of conjugates on smears prepared directly from cerebrospinal fluid sediments, *J. Lab. Clin. Med.,* 65, 976, 1965.

122. **Cherry, W. B.,** Fluorescent-antibody techniques, in *Manual of Clinical Microbiology,* Blair, J. E., Lennette, E. H., and Truant, J. P., Eds., American Society for Microbiology, Washington, D.C., 1970, 693.

123. **Fox, H. A., Hagen, P. A., Turner, D. J., Glasgow, L. A., and Connor, J. D.,** Immunofluorescence in the diagnosis of acute bacterial meningitis: A cooperative evaluation of the technique in a clinical laboratory setting, *Pediatrics,* 43, 44, 1969.

124. **Ewing, W. H.,** Enterobacteriaceae: Taxonomy and Nomenclature, National Communicable Disease Center, Atlanta, 1966.

125. **Fuscoe, F. J.,** A dual composite medium for the differentiation of the pathogenic enterobacteria and coliform organisms, *Med. Lab. Technol.,* 29, 261, 1972.

126. **Elazhary, M. A. S. Y., Saheb, S. A., Roy, R. S., and Lagacé, A.,** A simple procedure for the preliminary identification of aerobic gram negative intestinal bacteria with special reference to the enterobacteriaceae, *Can. J. Comp. Med.,* 37, 43, 1973.

127. **Brooks, W. F., Jr. and Goodman, N. L.,** Rapid differential testing of the Enterobacteriaceae using modified standard media, *Am. J. Med. Technol.,* 38, 429, 1972.

128. **Iveson, J. B.,** Enrichment procedures for the isolation of *Salmonella, Arizona, Edwardsiella* and *Shigella* from faeces, *J. Hyg.,* 71, 349, 1973.

129. **Roberts, G.,** The rapid identification of Gram-negative bacilli, *Med. Lab. Technol.,* 28, 382, 1971.

130. **Cowan, S. T. and Steel, K. J.,** *Manual for the Identification of Medical Bacteria,* Cambridge University Press, Cambridge, 1965.

131. **Wilson, G. S. and Miles, A. A.,** Topley and Wilson's *Principles of Bacteriology and Immunity,* Edward Arnold, London, 1964.

132. **Wasilauskas, B. L. and Ellner, P. D.,** Presumptive identification of bacteria from blood cultures in four hours, *J. Infect. Dis.,* 124, 499, 1971.

133. **Edwards, P. R. and Ewing, W. H.,** *Identification of Enterobacteriaceae,* 3rd ed., Burgess, Minneapolis, 1922, 1.

134. **Martin, W. J.,** *Enterobacteriaceae,* in *Manual of Clinical Microbiology,* Blair, J. E., Lennette, E. H., and Truant, J. P., Eds., American Society for Microbiology, Washington, D.C., 1970, 151.

135. **Cherry, W. B., Thomason, B. M., Pomales-Lebrón, A., and Ewing, W. H.,** Rapid presumptive identification of enteropathogenic *Escherichia coli* in faecal smears by means of fluorescent antibody. III. Field evaluation, *Bull. WHO,* 25, 159, 1961.

136. **Liu, P. V.,** The Pseudomonads and the Aeromonads, in *Rapid Diagnostic Methods in Medical Microbiology,* Graber, C. D., Ed., Williams & Wilkins, Baltimore, 1970, 116.

137. Stainer, R. Y., Palleroni, N. J., and Doudoroff, M., The aerobic pseudomonads: A taxonomic study, *J. Gen. Microbiol.*, 43, 159, 1966.

138. Gilardi, G. L., Diagnostic criteria for differentiation of pseudomonads pathogenic for man, *Appl. Microbiol.*, 16, 1497, 1968.

139. Gilardi, G. H., Characterization of EO-1 strains (*Pseudomonas kingii*) isolated from clinical specimens and the hospital environment, *Appl. Microbiol.*, 20, 521, 1970.

140. Sutter, V. L., Identification of *Pseudomonas* species isolated from hospital environment and human sources, *Appl. Microbiol.*, 16, 1532, 1968.

141. Hugh, R. and Leifson, E., The taxonomic significance of fermentative versus oxidative metabolism of carbohydrates by various gram-negative bacteria, *J. Bacteriol.*, 66, 24, 1953.

142. Ballard, R. W., Palleroni, N. J., Doudoroff, M., Stainier, R. Y., and Mandel, M., Taxonomy of the aerobic pseudomonads: *Pseudomonas cepacia, P. marginata, P. alliicola* and *P. caryophylli, J. Gen. Microbiol.*, 60, 199, 1970.

143. Verder, E. and Evans, J., A proposed antigenic schema for the identification of strains of *Pseudomonas aeruginosa, J. Infect. Dis.*, 109, 183, 1961.

144. Elek, S. D., The recognition of toxigenic bacterial strains *in vitro, Br. Med. J.*, 1, 493, 1948.

145. Liu, P., A modification of Elek's technique to be performed on microscope slides, *Am. J. Clin. Pathol.*, 36, 471, 1961.

146. Young, V. M., Haemophilus, in *Manual of Clinical Microbiology*, 2nd ed., Lennette, E. H., Spaulding, E. H., and Truant, J. P., Eds., American Society for Microbiology, Washington, D.C., 1974, 304.

147. Wetzler, T. F., Rosen, M. N., and Marshall, J. D., Jr., The Pasteurella, the Yersinia and the Francisella, in *Rapid Diagnostic Methods in Medical Microbiology*, Graber, C. D., Ed., Williams & Wilkins, Baltimore, 1970.

148. Ewing, W. H., *Vibrio cholerae*, in *Diagnostic Procedures for Bacterial, Mycotic and Parasitic Infections*, 5th ed., Bodily, H. L., Updyke, E. L., and Mason, J. O., Eds., American Public Health Association, New York, 1970, 565.

149. Balows, A., Herman, G. J., and DeWitt W. E., The isolation and identification of *Vibrio cholera* – a review, *Health Lab. Sci.*, 8, 167, 1971.

150. Bokkenheuser, V., *Vibrio fetus* infection in man. I. Ten new cases and some epidemiologic observations, *Am. J. Epidemiol.*, 91, 400, 1970.

151. King, E. O., *The Identification of Unusual Pathogenic Gram-negative Bacteria*, preliminary revision of Weaver, R. E., Tatum, H. W., and Hollis, D. G., National Communicable Disease Center, Department of Health, Education, and Welfare, Atlanta, 1972.

152. Hugh, R., A practical approach to the identification of certain nonfermentative gram-negative rods encountered in clinical specimens, *J. Conf. Public Health Lab. Dir.*, 28, 168, 1970.

153. Tatum, H. W., Miscellaneous Gram-negative bacteria, in *Manual of Clinical Microbiology*, Blair, J. E., Lennette, E. H., and Truant, J. P., Eds., American Society for Microbiology, Washington, D.C., 1970, 191.

154. Samuels, S. B., Pittman, B., and Cherry, W. B., Practical physiological schema for the identification of *Herellea vaginicola* and its differentiation from similar organisms, *Appl. Microbiol.*, 18, 1015, 1969.

155. Sonnenwirth, A. C., Gram-negative bacilli, vibrios, and spirilla, in *Gradwohl's Clinical Laboratory Methods and Diagnosis: A Textbook on Laboratory Procedures and Their Interpretation*, Vol. 2, 7th ed., Frankel, S., Reitman, S., and Sonnenwirth, A. C., Eds., C. V. Mosby, St. Louis, 1970, 1269.

156. Hermann, G. J. and Weaver, R. E., Corynebacterium, in *Manual of Clinical Microbiology*, Blair, J. E., Lennette, E. H., and Truant, J. P., Eds., American Society for Microbiology, Washington, D.C., 1970, 93.

157. Leifson, E., The bacterial flora of distilled and stored water. III. New species of the genera *Corynebacterium, Flavobacterium, Spirillum*, and *Pseudomonas, Int. Bull. Bacteriol. Nomencl. Taxon.*, 12, 161, 1962.

158. Killinger, A. H., *Listeria monocytogenes*, in *Manual of Clinical Microbiology*, 2nd ed., Lennette, E. H., Spaulding, E. H., and Truant, J. P., Eds., American Society for Microbiology, Washington, D.C., 1974, 135.

159. Weaver, R. E., *Erysipelothrix*, in *Manual of Clinical Microbiology*, 2nd ed., Lennette, E. H., Spaulding, E. H., and Truant, J. P., Eds., American Society for Microbiology, Washington, D.C., 1974, 140.

160. Wetzler, T. F., Freeman, N. R., French, M. L. V., Renkowski, L. A., Eveland, W. C., and Carver, O. J., Biological characterization of *Listeria monocytogenes, Health Lab. Sci.*, 5, 46, 1968.

161. Williams, R. P. and Wende, R. D., *Bacillus anthracis* and its differentiation from other species of *Bacillus*, in *Rapid Diagnostic Methods in Medical Microbiology*, Graber, C. D., Ed., Williams & Wilkins, Baltimore, 1970.

162. Dowell, V. R., Jr. and Hawkins, T. M., Laboratory Methods in Anaerobic Bacteriology: CDC Laboratory Manual, Publ. (HSM) 73-8222, Center for Disease Control, Department of Health, Education, and Welfare, Atlanta, April 1973.

163. Holdeman, L. V. and Moore, W. E. C., *Anaerobe Laboratory Manual*, Virginia Polytechnic Institute and State University, Blacksburg, Va., 1972.

164. Finegold, S. M., Sutter, V. L., Cato, E. P., and Holdeman, L. V., Anaerobic bacteria, in *Rapid Diagnostic Methods in Medical Microbiology*, Graber, C. D., Ed., Williams & Wilkins, Baltimore, 1970.

165. Sutter, V. L., Attebery, H. R., and Finegold, S. M., Gram-negative nonspore forming anaerobic bacilli, in *Manual of Clinical Microbiology*, 2nd ed., Lennette, E. H., Spaulding, E. H., and Truant, J. P., Eds., American Society for Microbiology, Washington, D.C., 1974, 393.

166. Dowell, V. R., Jr., Thompson, F. S., Whaley, D. N., Alpern, R. J., Felner, J. M., Armfield, A. Y., McCroskey, L. M., and Wiggs, L. S., Differential Characteristics of Anaerobic Bacteria, Center for Disease Control, Department of Health, Education, and Welfare, Atlanta, 1970.

167. Ajello, L., Georg, L. K., Kaplan, W., and Kaufman, L.. Laboratory Manual for Medical Mycology, PHS Publ. 994, U.S. Government Printing Office, Washington, D.C., 1963.

168. Ajello, L., Georg, L. K., Kaplan, W., and Kaufman, L., Mycotic infections, in *Diagnostic Procedures for Bacterial Mycotic and Parasitic Infections*, 5th ed., Bodily, A. L., Updyke, E. L., and Mason, J. O., Eds., American Public Health Association, New York, 1970, 662.

169. Beneke, E. S. and Rogers, A. L., *Medical Mycology Manual*, 3rd ed., Burgess, Minneapolis, 1971.

170. Moore, W. E. C. and Holdeman, L. V., Gram-positive nonsporeforming anaerobic bacilli, in *Manual of Clinical Microbiology*, Blair, J. E., Lennette, E. H., and Truant, J. P., Eds., American Society for Microbiology, Washington, D.C., 1970, 290.

171. Pine, L. and Georg, L., The classification and phylogenetic relationships of the Actinomycetales, *Int. Bull. Bacteriol. Nomencl. Taxon.*, 15, 143, 1965.

172. Gordon, M. A., Aerobic pathogenic *Actinomycetaceae*, in *Manual of Clinical Microbiology*, 2nd ed., Lennette, E. H., Spaulding, E. H., and Truant, J. P., Eds., American Society for Microbiology, Washington, D.C., 1974, 176.

173. Gordon, R. E., Some criteria for the recognition of *Nocardia madurae* (Vincent) Blanchard, *J. Gen. Microbiol.*, 45, 355, 1966.

173a. Mackinnon, J. E. and Artagaveytia-Allende, R. C., The main species of pathogenic aerobic actinomycetes causing mycetomas, *Trans. R. Soc. Trop. Med. Hyg.*, 50, 31, 1956.

174. Mariat, F., Activité uréasique des Actinomycètes aérobies pathogènes, *Ann. Inst. Pasteur* (Paris), 105, 795, 1963.

175. Becker, B., Lechevalier, M. P., and Lechevalier, H. A., Chemical composition of cell-wall preparations from strains of form-genera of aerobic actinomycetes, *Appl. Microbiol.*, 13, 236, 1965.

176. Gordon, R. E. and Mihm, J. M., Identification of *Nocardia caviae* (Erickson), Nov. comb., *Ann. N.Y. Acad. Sci.*, 98, 628, 1962.

177. Gonzales-Mendoza, A. and Mariat, F., Sur l'hydrolyze de la gélatine comme caractère différential entre *Nocardia asteroides* and *N. brasiliensis, Ann. Inst. Pasteur* (Paris), 107, 560, 1964.

178. Runyon, E. H., Pathogenic mycobacteria, *Adv. Tuberc. Res.*, 14, 235, 1965.

179. Runyon, E. H. and Committee Members of the American Society for Microbiology Taxonomy Subcommittee on Mycobacteria, *Mycobacterium tuberculosis, M. bovis* and *M. microti* species descriptions, *Zentralbl. Bakteriol. Parasitenkd. Infektionskr. Hyg. Abt. 1 Orig.*, 204, 405, 1967.

180. Harrington, R., Jr. and Karlson, A. G., Differentiation between *Mycobacterium tuberculosis* and *Mycobacterium bovis* by *in vitro* procedures, *Am. J. Vet. Res.*, 27, 1193, 1966.

181. Karlson, A. G. and Lessel, E. F., *Mycobacterium bovis*, nom. nov., *Int. J. Syst. Bacteriol.*, 20, 273, 1970.

182. Hepper, N. G. G., Karlson, A. G., Leary, F. J., and Soule, E. H., Genitourinary infection due to *Mycobacterium kansasii, Mayo Clin. Proc.*, 46, 387, 1971.

183. Winter, F. E. and Runyon, E. H., Prepatellar bursitis caused by *Mycobacterium marinum (balnei)*: Case report, classification, and review of the literature, *J. Bone Jt. Surg. Am. Vol.*, 47, 375, 1965.

184. Wayne, L. G., Doubek, J. R., and Deaz, G. A., Classification and identification of mycobacteria. IV. Some important scotochromogens, *Am. Rev. Respir. Dis.*, 96, 88, 1967.

185. Wayne, L. G., On the identity of *Mycobacterium gordonae* Bojalil and the so-called tap water scotochromogens, *Int. J. Syst. Bacteriol.*, 20, 149, 1970.

186. Engback, H. C., Vergmann, B., Baess, I., and Will, D. H., *M. xenopei*: A bacteriological study of *M. xenopei* including case reports of Danish patients, *Acta Pathol. Microbiol. Scand.*, 69, 576, 1967.

187. Schaefer, W. B., Incidence of the serotypes of *Mycobacterium avium* and atypical mycobacteria in human and animal diseases, *Am. Rev. Respir. Dis.*, 97, 18, 1968.

188. Wayne, L. G., Classification and identification of mycobacteria. III. Species within Group III, *Am. Rev. Respir. Dis.*, 93, 919, 1966.

189. Kubica, G. P., Silcox, V. A., Kilburn, J. O., Smithwick, R. W., Beam, R. E., Jones, W. D., Jr., and Stottmeier, K. D., Differential identification of mycobacteria. VI. *Mycobacterium triviale* Kubica sp. nov., *Int. J. Syst. Bacteriol.*, 20, 161, 1970.

190. Kubica, G. P., Differential identification of mycobacteria. VII. Key features for identification of clinically significant mycobacteria, *Am. Rev. Respir. Dis.*, 107, 9, 1973.

191. Kestle, D. G., Abbott, V. D., and Kubica, G. R., Differential identification of mycobacteria. II. Subgroups of Groups II and III (Runyon) with different clinical significance, *Am. Rev. Respir. Dis.*, 95, 1041, 1967.

192. Fogan, L., Atypical mycobacteria: Their clinical, laboratory, and epidemiologic significance, *Medicine* (Baltimore), 49, 243, 1970.

193. Stanford, J. L. and Beck, A., Bacteriological and serological studies of fast growing mycobacteria identified as *Mycobacterium friedmannii, J. Gen. Microbiol.*, 58, 99, 1969.

194. Borghans, J. G. A. and Stanford, J. L., *Mycobacterium chelonei* in abscesses after injection of diphtheria-pertussis-tetanus-polio vaccine, *Am. Rev. Respir. Dis.*, 107, 1, 1973.

195. Carpenter, C. M. and Miller, J. N., The bacteriology of leprosy, in *Leprosy in Theory and Practice*, 2nd ed., Cochrane, R. G. and Davey, T. F., Eds., John Wright and Sons, Bristol, Engl., 1964, 13.

196. MacCallum, P., Talhurst, J. C., Buckle, G., and Sissons, H. A., A new mycobacterial infection in man, *J. Pathol. Bacteriol.*, 60, 93, 1948.

197. Harris, A., Rosenberg, A. A., and Riedel, L. M., A microflocculation test for syphilis using cardiolipin antigen, *J. Vener. Dis. Inform.*, 27, 169, 1946.

198. Hunter, E. F., Deacon, W. E., and Meyer, P. E., An improved FTA test for syphilis, the absorption procedure (FTA-ABS), *Public Health Rep.*, 79, 410, 1964.

199. Galton, M. M., Powers, D. K., Hall, A. D., and Cornell, R. G. A., Rapid macroscopic-slide screening test for the serodiagnosis of leptospirosis, *Am. J. Vet. Res.*, 19, 505, 1958.

200. Stoenner, H. G. and Davis, E., Further observations on leptospiral plate antigens, *Am. J. Vet. Res.*, 28, 259, 1967.

201. Alexander, A. D., Leptospira, in *Manual of Clinical Microbiology*, 2nd ed., Lennette, E. H., Spaulding, E. H., and Truant, J. P., Eds., American Society for Microbiology, Washington, D.C., 1974, 347.

202. Baker, L. A. and Cox, C. D., Quantitative assay for genus specific leptospiral antigen and antibody, *Appl. Microbiol.*, 25, 697, 1973.

203. Current problems in Leptospirosis, Report of a WHO Expert Group, *WHO Tech. Ser.*, No. 380, 1967.

204. Galton, M. M., Menges, R. W., Shotts, E. B., Jr., Nahmias, A. J., and Heath, E. W., Jr., Leptospirosis. Epidemiology, Clinical Manifestation in Man and Animals and Methods in Laboratory Diagnosis, Communicable Disease Center, Atlanta, 1962.

205. Crawford, Y. E., Mycoplasma, in *Manual of Clinical Microbiology*, Blair, J. E., Lennette, E. H., and Truant, J. P., Eds., American Society for Microbiology, Washington, D.C., 1970, 252.

206. Kenny, G. E., Serological comparison of ten glycolytic *Mycoplasma* species, *J. Bacteriol.*, 98, 1044, 1969.

207. Kenny, G. E., Immunogenicity of *Mycoplasma pneumoniae*, *Infect. Immun.*, 3, 510, 1971.

208. Pollack, J. D., Somerson, N. L., and Senterfit, L. B., Isolation, characterization, and immunogenicity of *Mycoplasma pneumoniae* membranes, *Infect. Immun.*, 2, 326, 1970.

209. Hayflick, L., *The Mycoplasmatales and the L-phase of Bacteria*, Appleton-Century-Crofts, New York, 1969.

210. Madoff, S., *Mycoplasma and the L-forms of Bacteria*, Gordon and Breach, New York, 1971.

211. Shepard, M. C., Differential methods for the identification of T mycoplasma based on demonstration of urease, *J. Infect. Dis.*, 127, 22, 1973.

212. Ormsbee, R. A., Rickettsiae, in *Manual of Clinical Microbiology*, 2nd ed., Lennette, E. H., Spaulding, E. H., and Truant, J. P., Eds., American Society for Microbiology, Washington, D.C., 1974, 812.

213. Burman, W. P., in *Progress in Microbiological Techniques*, Collins, C. H., Ed., Butterworths, London, 1967.

214. Skali, P., Air-borne Bacteria Studies in the Detroit, Michigan, Area, Final Report, Contract cd3-1633, Detroit City Health Department, Michigan Department of Health and Public Health Service, 1953.

215. Lee, R. E., Jr., Harris, K., and Akland, G., Relationship between viable bacteria and air pollutants in an urban atmosphere, *Am. Ind. Hyg. Assoc. J.*, 34, 164, 1973.

216. Andersen, A. A., New sampler for the collection, sizing, and enumeration of viable airborne particles, *J. Bacteriol.*, 76, 471, 1958.

217. Flesch, J. P., Norris, C., and Nugent, A., Calibrating particulate air samplers with monodispersed aerosols: application of the Andersen Cascade Impactor, *Am. Ind. Hyg. Assoc. J.*, 28, 507, 1967.

218. Jutze, G. A. and Tabor, E. C., The continuous air monitoring program, *J. Air Pollut. Control Assoc.*, 13, 278, 1963.

219. West, P. W. and Gaeke, G. C., Fixation of sulfur dioxide as disulfitomercurate (II), subsequent colorimetric estimation, *Anal. Chem.*, 28, 1916, 1956.

220. Littman, F. E. and Benioliel, R. W., Continuous oxidant recorder, *Anal. Chem.*, 25, 1480, 1953.

221. Morris, R. A. and Chapman, R. L., Flame ionization hydrocarbon analyzer, *J. Air Pollut. Control Assoc.*, 11, 467, 1961.

222. Saltzman, B. E., Colorimetric microdetermination of nitrogen dioxide in the atmosphere, *Anal. Chem.*, 26, 1949, 1954.

223. Standard Method of Test for Particulate Matter in the Atmosphere, Optical Density of Filter Deposit, Standards on Methods of Atmospheric Sampling and Analysis, American Society for Testing and Materials, Philadelphia, 1962, 497.

224. Lee, R. E., Jr., Caldwell, J. S., and Morgan, G. B., The evaluation of methods for measuring suspended particulates in air, *Atmos. Environ.*, 6, 593, 1972.

225. Treskunov, A. A., Principles of development of instruments for determining bacterial contamination of the air, *BioMed. Eng.*, 5, 36, 1971.

226. Ryan, W. J., Isolation of *Salmonella* from sewage by anaerobic methods, *Microbiology* (USSR), 5, 533, 1972.

227. Collins, V. G. and Willoughby, L. G., The distribution of bacteria and fungal spores in Blehnam Tarn with particular reference to an experimental overturn, *Arch. Mikrobiol.*, 43, 294, 1962.

228. Staples, D. G. and Fry, J. C., A medium for counting aquatic heterotrophic bacteria in polluted and unpolluted waters, *J. Appl. Bacteriol.*, 36, 179, 1973.

229. Windle Taylor, E., *The Examination of Waters and Water Supplies*, 7th ed., Churchill, London, 1958.

230. Ferrer, E. B., Stapert, E. M., and Sokolski, W. T., A medium for improved recovery of bacteria from water, *Can. J. Microbiol.*, 9, 420, 1963.

231. Hayes, F. R. and Anthony, E. H., Lake water and sediment. VI. The standing crop of bacteria in lake sediments and its place in the classification of lakes, *Limnol. Oceanogr.*, 4, 229, 1959.

232. Jones, J. G., Studies on freshwater bacteria: Effect of medium composition and method on estimates of bacterial population, *J. Appl. Bacteriol.*, 33, 679, 1970.
233. Standard methods for Enumeration of Water, Sewage, and Industrial Wastes, American Public Health Association, Washington, D.C., 1965.
234. Schmidt, E. L. and Bankole, R. O., Detection of *Aspergillus flavus* in soil by immunofluorescent staining, *Science*, 136, 776, 1962.
235. Eren, J. and Pramer, D., Application of immunofluorescent staining to studies of the ecology of soil microorganisms, *Soil Sci.*, 101, 39, 1966.
236. Hill, I. R. and Gray, T. R. G., Application of the fluorescent-antibody technique to an ecological study of bacteria in soil, *J. Bacteriol.*, 93, 1888, 1967.
237. Schmidt, E. L., Bankole, R., and Bohlool, B., Fluorescent antibody approach to study of rhizobia in soil, *J. Bacteriol.*, 95, 1987, 1968.
238. Schmidt, E. L., Fluorescent antibody techniques in the study of microbial ecology, in *Modern Methods in the Study of Microbial Ecology*, Uppsala, Sweden, 1973.
239. Gray, T. R. G., Baxby, P., Hill, I. R., and Goodfellow, M., Direct observation of bacteria in soil, in *The Ecology of Soil Bacteria*, Gray, T. R. G. and Parkinson, D., Eds., Liverpool University Press, Liverpool, 1968, 1971.
240. Casida, L. E., Jr., Infrared color photography: selective demonstration of bacteria, *Science*, 159, 199, 1968.
241. Nikitin, D. I., Lokhmacheva, R. A., and Vasilёva, L. V., Microorganisms growing on soil fulvic acids, *Proc. 9th Int. Congr. Microbiol. Moscow*, 1966, C 2/15, 269.
242. Nemec, P. and Bystricky, V., Peculiar morphology of some microorganisms accompanying Diatomaceae, Preliminary report, *J. Gen. Appl. Microbiol.*, 8, 121, 1962.
243. Orenski, S. W., Bystricky, V., and Maramorosch, K., The occurrence of microbial forms of unusual morphology in European and Asian soils, *Can. J. Microbiol.*, 12, 1291, 1966.
244. Alexander, F. E. S. and Jackson, R. M., Examination of soil microorganisms in their natural environment, *Nature*, 174, 750, 1954.
245. Kubiena, W. L., *Micropedology*, Collegiate Press, Ames, Iowa, 1938.
246. Hepple, S. and Gurges, A., Sectioning of soil, *Nature*, 117, 1186, 1956.
247. Burges, A. and Nicholas, D. P., Use of soil section in studying amounts of fungal hyphae in soil, *Soil Sci.*, 92, 25, 1961.
248. Jones, D. and Griffiths, E., The use of soil sections for the study of soil microorganisms, *Plant Soil*, 20, 232, 1964.
249. Perfil'ev, B. V. and Gabe, D. R. The capillary microbial-landscape method in geomicrobiology, and the use of the microbial-landscape method to investigate bacteria which concentrate manganese and iron in bottom deposits, in *Applied Capillary Microscopy: The Role of Microorganisms in the Formation of Iron-Manganese Deposits*, Gurevich, M. S., Ed., Sinclair, F. L., translation, Consultants Bureau, New York, 1965, 1, 9.
250. Perfil'ev, B. V. and Gabe, D. R., *Capillary Methods of Studying Microorganisms*, Akad. Nauk SSSR, Moscow, 1969; Engl. trans., Oliver and Boyd, Edinburgh.
251. Stotzky, G., Microbial respiration, in *Methods of Soil Analysis*, Vol. 2, Black, C. A. et al., Eds., American Society of Agronomy, Madison, 1965, 1550.
252. Gray, T. R. G. and Williams, S. T., Microbial productivity of soil, in *Microbes and Biological Activity*, Hughes, D. E. and Rose, A. H., Eds., Cambridge University Press, Cambridge, 1971, 225.
253. Domsch, K. H., Bodenatmung, Sammelbericht über Methoden und Ergebnisse, *Zentralbl. Bakteriol. Parasitenkd. Infektionskr. Hyg. Abt. 2*, 116, 33, 1962.
254. Stotzky, G., Activity, ecology and population dynamics of microorganisms in soil, *CRC Crit. Rev. Microbiol.*, 2, 59, 1972.
255. Davis, J. G., Microbial aspects of pollution. Some general observations, *Soc. Appl. Bacterial Symp. Ser.*, 1, 1, 1971.
256. Sharpe, A. N. and Jackson, A. K., Comparison of poured and surface inoculated plates of Baird-Parker's medium for enumerating *Staphylococcus aureus* in foods, *J. Appl. Bacteriol.*, 35, 681, 1972.
257. Baird-Parker, A. C., The use of Baird-Parker's medium for the isolation and enumeration of *Staphylococcus aureus*, in *Isolation Methods for Microbiologists*, Shapton, D. A. and Gould, G. W., Eds., Academic Press, London, 1969.
258. Sharpe, A. N., Biggs, D. R., and Oliver, R. J., Machine for automatic bacteriological pour plate preparation, *Appl. Microbiol.*, 24, 70, 1972.
259. Keller, P., Sklan, D., and Gordin, S., Effect of diluent on bacterial counts in milk and milk products, *J. Dairy Sci.*, 57, 127, 1974.
260. *Standard Methods for the Examination of Dairy Products*, 12th ed., American Public Health Association, Washington, D.C., 1967, 34.
261. Bhadsavle, C. H., Shehata, T. E., and Collins, E. B., Isolation and identification of psychrophilic species of *Clostridium* from milk, *Appl. Microbiol.*, 24, 699, 1972.
262. Shehata, T. E. and Collins, E. B., Isolation and identification of psychrophilic species of *Bacillus* from milk, *Appl. Microbiol.*, 21, 466, 1971.
263. Juffs, H. S., Identification of *Pseudomonas* spp. isolated from milk produced in South Eastern Queensland, *J. Appl. Bacteriol.*, 36, 585, 1973.
264. Jones, L. M. and Morgan, W. J. B., A preliminary report on a selective medium for the culture of *Brucella*, including fastidious types, *Bull. WHO*, 19, 200, 1958.

265. Morgan, W. J. B., Comparison of various media for the growth of *Brucella, Res. Vet. Sci.,* 1, 47, 1960.
266. Painter, G. M., Deyoe, B. L., and Lambert, G., Comparison of several media for the isolation of *Brucella, Can. J. Comp. Med. Vet. Sci.,* 30, 218, 1966.
267. Ryan, W. J., A selective medium for the isolation of *Brucella abortus* from milk, *Mon. Bull. Minist. Health Public Health Lab.,* 26, 33, 1967.
268. Farrell, I. D., The Use of Antibiotics and Antibacterial Agents for the Selective Isolation of *Brucella abortus,* Ph.D. thesis, University of Liverpool, 1969.
269. Farrell, I. D. and Robertson, L., A comparison of various selective media, including a new selective medium for the isolation of brucellae from milk, *J. Appl. Bacteriol.,* 35, 625, 1972.
270. Mair, N. S., A selective medium for the isolation of *Brucella abortus* from herd samples of milk, *Mon. Bull. Minist. Health Public Health Lab.,* 14, 184, 1955.
271. Alton, G. G. and Jones, L. M., Laboratory techniques in brucellosis, *WHO Monogr. Ser.,* No. 55, 1967.
272. Thatchero, F. S. and Clark, D. S., Eds., *Microorganisms in Foods: Their Significance and Methods of Enumeration,* University of Toronto Press, Toronto, 1968, 94.
273. Erdman, I. E., ICMSF methods studies. IV. International collaborative assay for the detection of *Salmonella* in raw meat, *Can. J. Microbiol.,* 20, 715, 1974.
274. Mossel, D. A. A., The presumptive enumeration of lactose negative as well as lactose positive *Enterobacteriaceae* in foods, *Appl. Microbiol.,* 5, 379, 1957.
275. Mossel, D. A. A., Current Methods of Examination of Foods for Coliforms, Tech. Circ. No. 526, Food Research Association, Leatherhead, Surrey, 1973.
276. Mossel, D. A. A., Bacteriological safety of foods, *Lancet,* 1, 193, 1974.
277. Mossel, D. A. A., Visser, M., and Cornelissen, A. M. R., The examination of foods for *Enterobacteriaceae* using a test of the type generally adopted for the detection of Salmonellae, *J. Appl. Bacteriol.,* 26, 444, 1963.
278. Henriksen, S. D., A study of the causes of discordant results of the presumptive and completed coliform tests on Norwegian waters, *Acta Pathol. Microbiol. Scand.,* 36, 87, 1955.
279. Drion, E. F. and Mossel, D. A. A., Mathematical-ecological aspects of the examination for enterobacteriaceae of foods processed for safety, *J. Appl. Bacteriol.,* 35, 233, 1972.
280. Committee on *Salmonella,* An Evaluation of the *Salmonella* Problem, Publ. 1683, National Research Council, National Academy of Sciences, Washington, D.C., 1969.
281. Silliker, J. H. and Gabis, D. A., ICMSF methods studies. I. Comparison of analytical schemes for detection of *Salmonella* in dried foods, *Can. J. Microbiol.,* 19, 475, 1973.
282. Goresline, H. E., Ed., *Recommended Methods for the Microbiological Examination of Foods,* American Public Health Association, New York, 1958, 155.
283. Bacteriological Analytical Manual, 2nd ed., U.S. Department of Health, Education, and Welfare, Public Health Service, Food and Drug Administration, Division of Microbiology, Washington, D.C., 1972.
284. Thatcher, F. S. and Clark, D. S., Eds., *Microorganisms in Foods: Their Significance and Methods of Enumeration,* University of Toronto, Toronto, 1968, 90, 94.
285. Compliance Program Guidance Manual. Salmonellae Sampling Plans, United States Food and Drug Administration, Washington, D.C., 1972.
286. Edel, W. and Kampelmacher, E. H., Comparative studies on the isolation of "sublethally injured" salmonellae in nine European laboratories, *Bull. WHO,* 48, 167, 1973.
287. Gabis, D. A. and Silliker, J. H., ICMSF methods studies. II. Comparison of analytical schemes for detection of *Salmonella* in high-moisture foods, *Can. J. Microbiol.,* 20, 663, 1974.
288. Idziak, E. S., Airth, J. M. A., and Erdman, I. E., ICMSF methods studies. III. An appraisal of 16 contemporary methods used for detection of *Salmonella* in meringue powder, *Can. J. Microbiol.,* 20, 703, 1974.
289. Palumbo, S. A., Huhtanen, C. N., and Smith, J. L., Microbiology of the frankfurter process: *Salmonella* and natural aerobic flora, *Appl. Microbiol.,* 27, 724, 1974.
290. Ng, H., Bayne, H. G., and Garibaldi, J. A., Heat resistance of *Salmonella:* the uniqueness of *Salmonella senftenberg* 775W, *Appl. Microbiol.,* 17, 78, 1969.
291. Winter, A. R., Stewart, G. F., McFarlane, V. H., and Solowey, M., Pasteurization of liquid egg products. III. Destruction of salmonella in liquid whole egg, *Am. J. Public Health,* 36, 451, 1946.
292. Casman, E. P., Staphylococcal food poisoning, *Health Lab. Sci.,* 4, 199, 1967.
293. Gilbert, R. J., Wieneke, A. A., Lanser, J., and Simkovičová, M., Serological detection of enterotoxin foods implicated in staphylococcal food poisoning, *J. Hyg.,* 70, 755, 1972.
294. Simkovičová, M. and Gilbert, R. J., Serological detection of enterotoxin from food-poisoning strains of *Staphylococcus aureus, J. Med. Microbiol.,* 4, 19, 1971.
295. McClung, L. S., Human food poisoning due to growth of *Clostridium perfringens (C. welchii)* in freshly cooked chicken: preliminary note, *J. Bacteriol.,* 50, 229, 1945.
296. Hobbs, B. C., Smith, M. E., Oakley, C. L., Warrack, G. H., and Cruickshank, J. C., *Clostridium welchii* food poisoning, *J. Hyg.,* 51, 75, 1953.
297. Sutton, R. G. A. and Hobbs, B. C., Food poisoning caused by heat-sensitive *Clostridium welchii.* A report of five recent outbreaks, *J. Hyg.,* 66, 135, 1968.

298. **Hall, H. E., Angelotti, R., Lewis, K. H., and Foter, M. J.,** Characteristics of *Clostridium perfringens* strains associated with food and food-borne disease, *J. Bacteriol.,* 85, 1094, 1963.
299. Foodborne Outbreaks, January—June 1970, Center for Disease Control, Atlanta, 1970.
300. Foodborne Outbreaks, Annual Summary, Center for Disease Control, Atlanta, 1970.
301. **Bryan, F. L.,** What the sanitarian should know about *Clostridium perfringens* foodborne illness, *J. Milk Food Technol.,* 32, 381, 1969.
302. **Strong, D. H., Canada, J. C., and Griffiths, B. B.,** Incidence of *Clostridium perfringens* in American foods, *Appl. Microbiol.,* 11, 42, 1963.
303. **Smith, L. D. and Gardner, M. V.,** The occurrence of vegetative cells of *Clostridium perfringens* in soil, *J. Bacteriol.,* 58, 407, 1949.
304. **Hauschild, A. H. W. and Hilsheimer, R.,** Enumeration of food-borne *Clostridium perfringens* in egg yolk-free tryptose-sulfite-cycloserine agar, *Appl. Microbiol.,* 27, 521, 1974.
305. **Hauschild, A. H. W. and Hilsheimer, R.,** Evaluation and modifications of media for enumeration of *Clostridium perfringens, Appl. Microbiol.,* 27, 78, 1973.
306. **Shahidi, S. A. and Ferguson, A. R.,** New quantitative, qualitative, and confirmatory media for rapid analysis of food for *Clostridium perfringens, Appl. Microbiol.,* 21, 500, 1971.
307. **Harmon, S. M., Kautter, D. A., and Peeler, J. T.,** Improved medium for enumeration of *Clostridium perfringens, Appl. Microbiol.,* 22, 688, 1971.
308. **Handford, P. M. and Cavett, J. J.,** A medium for the detection and enumeration of *Clostridium perfringens* (*welchii*) in foods, *J. Sci. Food Agric.,* 24, 487, 1973.
309. **Mossel, D. A. A. and Pouw, H.,** Studies on the suitability of sulphite cycloserine agar for the enumeration of *Clostridium perfringens* in food and water, *Zentralbl. Bakteriol. Parasitenkd. Infektionskr. Hyg. Abt. 1 Orig. Reihe A,* 223, 559, 1973.
310. **Marshall, R. S., Steenbergen, J. F., and McClung, L. S.,** Rapid technique for the enumeration of *Clostridium perfringens, Appl. Microbiol.,* 13, 559, 1965.

Chapter 3

NEW METHODS OF DETECTION AND IDENTIFICATION OF BACTERIA

INTRODUCTION

Procedures currently used to identify bacteria yield the correct answer in most instances, but at least a 24-hr period is required for identification. In recent years many new methods of detection and identification of bacteria have been developed for the following purposes:

1. To expedite the identification procedures which may save time as well as the life of the infected patient.

2. To increase the sensitivity of detection of a few organisms in a specimen.

3. To increase the reliability and accuracy of bacterial identification methods.

4. To decrease the cost due to lengthy procedures needed to perform a battery of tests.

5. To cope with the increasing work load in bacteriology laboratories. In diagnostic bacteriology alone, it is estimated that during the past decade the work load in many laboratories has increased at a rate of about 10% per year.

6. To utilize the knowledge of advanced instrument technology for practical application in the bacteriology laboratory.

Many clinicians do not see any particular need for development of rapid methods in bacteriology because roughly 80% of the repetitive exclusive bacteriology which passes through all routine laboratories can wait until the following day for an answer. Also, many of the tests performed by the so-called newer rapid techniques have not stood the test of time; often they have proven to be complicated and expensive. Nevertheless, it can be argued from many points of view that a rapid identification of the causative agent is desirable; these reasons can range from factors of direct concern to the patient and his subsequent treatment to problems concerning cross-infection in the ward. Detailed investigations of bacterial isolates are of immense importance for research purposes as well as for providing valuable training for future bacteriologists.

Various approaches have been used in recent years to rapidly detect and identify the bacteria in biological fluids and other environments. Most of these methods utilize the known chemical, physical, and physiological characteristics of bacteria. This chapter describes recent developments in the methodology of bacterial identification with a view toward their applicability and usefulness in the diagnostic, public health, or food bacteriology laboratory. In general, many of the rapid methods reported for bacterial identification are far less precise than the slower conventional methods (see Chapter 2). However, a few are established methods based on proven principles (for example, immunofluorescence and gas chromatography) for which adequate published methodology, sensitivity, and performance data are available. Because of increasing interest in the practical application of gas-chromatographic methods for the purpose of bacterial identification, Chapter 4 deals solely with this topic. Other tests, which are either in the experimental stage or based on principles of questionable validity, are discussed briefly in this chapter.

METHODS BASED ON MORPHOLOGY AND GROWTH CHARACTERISTIC OF BACTERIA

Many newer methods have been developed based on the principles of bacterial morphology (size, shape, and arrangement), staining properties, and cultural characteristics on solid medium or in liquid culture. While some methods have been modified to increase the efficiency of detection and enumeration of the organisms, others are aimed at improving the identification and characterization procedures.

Cell Size

The size ranges of the different taxonomic groups of bacteria vary, but most individual bacterial cells are 1 to 3 μm. Values in terms of equivalent dry weight are extremely variable, but 10^{-13} to 10^{-12} g is representative of most individual bacterial cells. The shape, dimensions, and dry weight equivalent of bacteria vary in different species as well as within the same bacterial species or strain, depending on growth conditions. In spite of the large variations, ranges of values for bacterial dimensions and mass are sufficiently characteristic so as to be used as the basis of differentiation. Turbidity determinations have long been used to estimate bacterial concentrations. However, particle counting methods by various instruments have recently been applied for the determination of the number of bacteria present in a sample.

Coulter Counter[®]

Particle counting by the Coulter counter is based on the simultaneous passsage of a conducting fluid and current through a small aperture which serves as a resistance element in a measuring circuit. When a suspended particle passes through the aperture, the resistance is momentarily changed; the magnitude of the resistance pulse is a measure of the volume of the particle.[1] Threshold dials may be set for counting only those pulses of a certain magnitude. The concentration, size distribution, and volumes of small particles in liquid suspension can be rapidly determined with the Coulter counter.[2] Sensitivity of detection depends on the cell volumes of the particles. In bacteria with cell volumes greater than 1 to 2 μm^3, concentrations down to several hundred cells per milliliter can be determined. However, with small cell volumes, the background noise levels are relatively larger, and the necessity to increase the sensitivity of the instrument increases the detection of foreign particles that are undetected at lower sensitivity settings. Anderson and Whitehead[3] describe a method which directly uses a master sample for both the Coulter counts and the plate counts; this allows the suspensions at the lowest cell concentration to be plated without further dilutions, thereby eliminating any error due to the dilution factor. They used a Model B Coulter counter fitted with a 30-μm orifice tube, which was calibrated against 0.796- and 1.3-μm diameter particles. Ringer solution (used as the electrolyte) was refiltered through a stack of three Oxoid[®] membranes (grade 0.45) until it gave an acceptably low mean background count. The threshold dial settings were related to the mean maximum and minimum dimensions of the organisms. A good correlation was obtained between the counts by nutrient agar plates and the Coulter counter (Table 3.1). Truant et al.[4] favorably evaluate the Model B Coulter counter for enumerating organisms. Kniseley and Throop[5]

TABLE 3.1

Comparative Plate and Coulter Counts of Different Bacteria

Organism (size range)	Mean colony count	Mean Coulter count
Aerobacter sp. (1.5−1.9 μm)	620	440
	6,020	5,275
	6,740	5,530
	13,600	13,100
Escherichia coli I (1.25−1.5 μm)	3,500	3,207
	3,500	3,207
	3,500	2,857
	32,840	32,690
	35,200	35,100
Salmonella typhimurium (0.6−1.4 μm)	3,360	1,265
	7,360	8,477
	22,140	23,200
	26,820	25,960

Note: Correlation coefficient (r) = 0.95.

From Anderson, G. E. and Whitehead, J. A., *J. Appl. Bacteriol.*, 36, 353, 1973. With permission.

used the Model A Coulter counter to determine the effect of lysostaphin on bacterial lysis by measuring the reduction in cell counts as lysis occurs. Garret and Wright[6] have also used the Coulter counter to study the kinetics and mechanisms of action of drugs on microorganisms.

The Coulter counter may be used to count mixed cultures if certain precautions are followed.[7] A mixed culture is a series of overlapping volume distributions; if they do not overlap appreciably, they can be counted with a conventional Coulter counter. With overlapping of size distributions, use of a multichannel pulse-height analyzer with special apertures,[1] hydrodynamic focusing,[8] or electron shaping of peaks[9] can give volume distributions of bacteria with very good resolution.

Application of the Coulter counter to mixed cultures is limited by biological considerations (rather than instrument resolution) which may involve changes in mean size and shape of bacterial volume distribution with environmental changes. Previously studied organisms include *Escherichia coli* B/r and *Dictyostelium discoedeum*,[10] *Azotobacter vinelandii* OP and *Tetrahymena pyriformis* W,[11] and *Lactobacillus casei* and *Saccharomyces cerevisiae*.[12] Microorganisms studied fall into three categories:

1. *E. coli, L. casei,* and *Bacillus licheniformis* with cell volumes of 0.3 to 2 μm^3.

2. Large bacteria and yeasts (*Azotobacter, S. cerevisiae*) with cell volumes of 2 to 100 μm^3.

3. Protozoa (*Dictyostelium, Tetrahymena,* and *Colpoda*) with volumes of 300 to 20,000 μm^3.

Rigid cell walls of bacteria and yeast minimize volume changes or lysis after dilution into fluids with osmotic properties different from culture medium. Chains, cell clumps, involution forms, or budding yeast may appear as counts in an interval assigned to a larger microorganism. Frequent microscopic examination of mixed cultures is necessary to avoid errors due to these causes. However, with the Coulter counter (in comparison with other methods of estimation of bacterial population), the greatest discrepancies are due to low population densities of the organisms.

Royco® Particle Counter*

This instrument measures quantity and diameter of micrometer-size particles present in the air or other gases. It can be programmed to count all particles within one or more size ranges, giving a separate total for each range or all particles larger than any selected range. One of fifteen individual channels covering a band 0.3 μm and above can be selected to measure particles within a particular size; 30,000 particles per minute can be counted with a coincidence loss of <10% using a sample flow rate of 0.01 ft^3/min. The air sample is passed through a light beam where the measured particles scatter light onto a phototube. The pulses from the phototube are analyzed, sorted, and counted electronically according to the particle size selected for the count. The instrument is calibrated with uniformly sized polystyrene latex particles** (grades from 0.5 to 5 μm) that are disseminated with an aerosol generator consisting of an atomizer that forms a fine mist of distilled water which carries the particles. The mist is passed down a dryer tube to give a fine dispersion of the calibration particles that then go to the counter. The instrument may be potentially useful for the monitoring of a "clean room" and the detection of biological aerosols resulting from leaks in culture apparatus.

πMC®System***

The πMC system consists of a microscopic television camera linked to a small digital computer. It detects particles in the microscopic field, projects them onto a viewing screen, and performs a variety of counting and measuring functions.

Velocimeter®†

Model 6100 laboratory Velocimeter is a sound velocity probe. The U.S. Air Force has used this sonar principle in determining bacterial contamination of jet fuels.

Soloway and Louder[13] described an instrument for sizing particles in flowing liquids which uses forward lobe scattering in a dark field microscope system. Continuous flow centrifugation coupled with density flotation in Ludox® or polyvinyl alcohol and dextran combinations are used to separate microorganisms from other materials such as soil.

A major problem with the new methods of bacterial detection and enumeration is interference by extraneous material in the sample. An assay may be accurate and reproducible when it is applied to a small number of purified or isolated bacteria suspended in a bland medium. Alternatively, bacterial colony-forming units on the surface of agar medium may be counted. The use of a highly specific method does not necessarily resolve the problem of interference; it may be necessary to fractionate the sample before analysis. Particles collected from air samples are fractionated on a size basis with a liquid pre-impinger that removes particles >4 μm in diameter; a multistage liquid impinger that separates particles into size ranges of >6, 3 to 6, and <3 μm in diameter; or through filters of graded pore size.[14] Liquid samples are fractionated by membrane filtration, differential or gradient centrifugation, and a liquid two-phase polymer system.[15-18] If membrane filtration is adopted for the fractionation purposes, it is important to realize that bacteria may be retained even though they have smaller diameters than the mean pore size of the filter.

---

*Rayco Instruments, Inc., Menlo Park, Cal.
**Dow Chemical Company, Physical Research Laboratory, Midland, Mich.
***Millipore Corporation, Bedford, Mass.
†NUC Corporation, Electronic Systems Division, Paramus, N.J.

Other Physical Properties of Cells

Electrophoretic Mobility

Electrophoretic mobility is a measure of movement of a particle in a solution when it is subjected to an externally applied electric field. The direction and rate of the particle depend on polarity and density of the surface charges.[19] Most particles acquire an electric charge in aqueous suspensions due to ionization of their surface groups and absorption of ions. Electrophoretic mobility studies may yield information about composition of surfaces and physical behavior of particles.

Microelectrophoresis, involves direct observation under the microscope of visible particles as they migrate in an electric field. It consists of a cell which can be focused under a microscope, electrodes, and an arrangement for filling and emptying the cell with provision for temperature control. Mobility may be influenced by diffusion of ions through the cell membrane;[20] by presence of capsules, mucilage, or fimbriae;[21] and by motile flagellates.

Surface groups have been identified by comparing pH-mobility curves of untreated cells with curves of cells altered by specific chemical or enzymic treatments. Cohen,[22] the first to do so, treated cells of *Bacillus proteus* with benzenesulfonyl chloride, resulting in the treated cells having a higher negative charge which in turn suggested that imidazole and amino groups had been substituted. Surface amino groups can be detected by treatment with an ethanolic solution of fluoro-2,4-dinitrobenzene, and carboxyl groups can be detected by treatment of acid-washed cells with ethanolic diazomethane.[23] Some C-terminal groups at the bacterial surface are detected by treatment with specific amino acid decarboxylases followed by electrophoresis.[24] Surface phosphate groups may be identified by the reduction of mobility produced in the presence of UO_2[25] or Ca[26] or by pretreatment with alkaline phosphatase.[27]

Bacteria of clinical importance studied by the above methods include: *Escherichia coli* (Davies et al.[28] suggested a polysaccharide covering, James and List[21] studied the effect of fimbriae on mobility, and Gittins and James[23] further studied surface structure); *Aerobacter aerogenes* (Barry and James[29,30] found the need for controlling ionic strength, Plummer and James[31] correlated changes in mobility with changes in capsular size, and Gittens and James[23]

studied surface features by chemical treatment); and *Bacillus subtilis* and *B. megaterium* (studied by Douglas[32,33] and Douglas and Parker[34]). *Streptococcus pyogenes* and *S. faecalis* were studied by Plummer et al.[35] and Hill et al.[24,27,36] after treatment with hyaluronidase. Lipid content of *S. pyogenes* was studied by Hill et al.[37] Tetracycline-sensitive and -resistant strains were examined by Norrington and James.[38] Antibody bound to cells of *S. pyogenes* was detected by Hill et al.[39] Mobility of *S. faecalis* was studied by Schott and Young.[40] *Micrococcus lysodeikticus* was studied by Few et al.[41] Conductivity studies on an unknown species of *Micrococcus* were conducted by Einolf and Carstensen.[42] *Staphylococcus aureus* pH-mobility curve studies were done by James and Brewer,[43] who also identified the teichoic acid and protein overlying the glycopeptide layer.[44] Two strains of *Mycobacterium phlei* were studied by Adams and Rideal;[45] one had a phospholipid and protein surface, and the other was almost entirely covered by phospholipid.

Bacterial polyribosomes, ribosomes, and ribosomal subunits possess characteristic electrophoretic mobilities in polyacrylamide-agarose composite gels, containing 2 to 3% acrylamide.[46] Such a procedure is valuable in the research laboratory. The type of RNA contained in ribosomal subunits, ribosomes, and polyribosomes can be identified by a second electrophoresis. Similarly, the subunit composition of ribosomes and polyribosomes is established. The principal subunits observed were a faster (30S) subunit containing 16S RNA and a slower (50S) subunit containing 23S RNA. Polyribosomes resembled each other in their proportional content of these subunits.

Electrophoretic mobility studies are valuable in the agricultural industry. Marshall[47] studied fast- and slow-growing *Rhizobium* species and compared the mobilities of normal strains of *Rhizobium trifolii* with those of mutants unable to form nodules on clover roots. Protective layers of colloidal clay of soil bacteria were studied by Marshall.[48-50]

Electron Microscopy

Examination of biological material at higher resolution and greater depth of field is possible with scanning electron microscopy (EM), as opposed to conventional microscopy.[51,52]

Application of EM to studies of whole colonies was not possible until methods of preparing dried colonies with minimum distortion were devised by Whittaker and Drucker.[53] They examined colonies of *Streptococcus mutans, Streptococcus* sp. D182, *Staphylococcus aureus,* and *Candida albicans.* Roth[54] examined colonies of *Bacillus, Brucella, Diplococcus, Mycobacterium, Myxococcus, Neisseria, Pseudomonas, Rhizobium, Staphylococcus,* and *Streptococcus* species. Drucker and Whittaker[55] studied colonies of *Aerococcus* sp., *Corynebacterium xerosis, Spirillum rubrum, Streptomyces scabies,* and *Vibrio metschnikovii* by electron microscopy.

Drucker and Whittaker[55] freeze-dried well-separated colonies on plates and glued the dried colonies to alloy stubs and metal coated with gold-palladium alloy. The original shape of the colonies was preserved. Colonies of *Aerococcus* sp. consisted of densely packed sheets of cells covered by extracellular material. The layers cavities were seen beneath the extracellular material. Colonies of *Corynebacterium xerosis* showed a microscopically irregular surface and were densely and haphazardly packed. No extracellular material was seen. Colonies of *Spirillum rubrum* had typical spirella and small amounts of associated extracellular material. Cells appeared joined by gum or "cellular bridges," and localized orientation of cells was apparent. Colonies of *Streptomyces scabies* appeared to have "tunnels" surrounded by cells. A surface film was present at the center of the colonies with no covering at the colony edges. *Vibrio metschnikovii* colonies had densely packed cells with a flat surface; their outline was indistinct because of the extracellular material. Correlations were noted between those colonies presenting a glossy macroscopic appearance and production of extracellular material and those colonies with a rough form and haphazard internal colonial "micro-structure."

Freeze-dried colonies of *Clostridium butyricum, C. sporogenes, Dermatophilus* sp., *Lactobacillus casei, L. fermenti, Mycobacterium phlei, Neisseria catarrhalis,* and *N. pharyngis* were metal coated and examined in the scanning electron microscope.[56] The *Neisseria* colonies had a distinctive covering film; extracellular material was noted in colonies of *Clostridium, Dermatophilus,* and *Mycobacterium.* The two *Lactobacillus* species had differing colonial architecture and belonged to different metabolic subgroups. *C.*

butyricum colonies were seen as flat sheets of rod-shaped cells arranged in parallel bundles. Some cells were surrounded by extracellular material. *C. sporogenes* colonies were similar, with extracellular material abundant at the edge of the colonies. *Dermatophilus* colonies had irregular arrangement of cells, and the outlines were obscured by extracellular material through which tunnels passed. Colonies of *L. casei* had no extracellular material or surface covering and were densely packed with irregularly arranged coccobacilli. In contrast, colonies of *L. fermenti* were densely packed with parallel bundles of chaining rods which had a banded appearance in some colonies. It must be remembered that *L. casei* is homofermentative, whereas *L. fermenti* is heterofermentative. Drucker and Whittaker[55] demonstrated that it was possible to examine the arrangement of cells in colonies, their packing, the presence of intercellular material, and surface covering film.

Banding was described for *Streptococcus mutans*[53] and *S. faecalis;*[57] this was attributed to sequential cell wall synthesis. Since this feature was not observed in other genera examined, Drucker[56] suggested that bands of synthesized cell wall may be characteristic of lactic acid bacteria. Colonies of *Mycobacterium phlei* had thick sheets of cells bound together by extracellular material which may be mycobacterial lipid. Colonies of *Neisseria pharyngis* and *N. catarrhalis* were comprised of irregularly arranged cocci with a regularly patterned surface membranous covering. Membranous covering found on *Neisseria* colonies had previously been demonstrated on colonies of *Candida albicans.*[53] Species of *Clostridium* had a colonial microstructure very similar to that of *Neisseria* species, presumably a reflection of the similarity of colony form and chemistry of the organisms.

Ultraviolet Spectroscopy

Since bacterial species vary both chemically and morphologically, Torten and Schneider[58] performed studies using circular dichroism and absorption spectra. Circular dichroism of washed cells suspended in 0.1 mM phosphate buffer (pH 7.2, 10^8 to 10^9 cells/ml) was measured from 310 to 185 nm in 1.0-mm path cells with a Cary® Model 60 spectropolarimeter with CD attachment. Absorption spectra were recorded on a Cary Model 15 spectrophotometer from 360 to 185 nm using cells with a 1.0-mm light path.

Each bacterial species tested gave a distinctive set of spectral patterns (Figures 3.1A to C). The following species were tested: (1) a Gram-negative rod (*Escherichia coli*), (2) a Gram-positive coccus (*Staphylococcus albus*), (3) a Gram-positive rod (*Bacillus cereus,* vegetative form), (4) a spirochete (*Leptospira grippotyphosa*), and (5) an acid-fast bacterium (*Mycobacterium smegmatis*).

Growth Characteristics of Bacteria

The most common methods for detection and enumeration of bacteria utilize some aspect of viability because expression of the ability to reproduce is one of the most reliable indicators of the presence of bacteria. Attempts have been made to modify classical microbiological methods (plating, culturing, etc.) by reducing growth time needed to yield results. If the bacteria in the sample are viable and results are not required in less than a few hours, growth methods of detection and identification of bacteria are preferred. These methods include identification of characteristic metabolic products, increase or change in turbidity, colony counting, gas production, respiration, and calorimetry.

Sharpe and Kilsby[59] described a rapid and inexpensive technique for enumeration of bacteria. A sample suspension is diluted in molten agar and plated out in standard petri dishes as a series of 0.1-ml agar droplets. This method has several advantages over conventional pour plate counting methods, particularly in quality control work in food factories. Sharpe et al.[60] modified the above method by using a foot-operated diluter/dispenser and projection viewer. With a foot-operated diluter/dispenser, decimal or centimal dilutions are made and the droplets are plated out rapidly with a minimum of effort. The projection viewer allows easy enumeration of colonies by throwing a bright image of a droplet, magnified to the size of a standard petri dish, toward the technician as he sits at the bench. The instrument is useful for rapid bacterial counting and saves cost and labor with reductions in incubation time, incubator space, and preparative work. Sharpe et al.[60] compared 525 standard pour or spread plate counts with droplet counts using the diluter/dispenser and the projection viewer; there were no significant differences for most of the samples tested.

Many conventional bacteriological methods have been modified by miniaturizing techniques.[61] The usefulness of these miniaturized methods in the bacteriology laboratory is that they offer substantial savings of space, material, labor, and time. Multiple inoculation devices (such as bolts, needles, pins, plastic stamps, velveteen, syringes, Pasteur pipettes, capillary tubes, and a "piggy-back"method) have been used with these microtechniques by various workers in the field. Several miniaturized microbiological techniques have been described by Fung and Hartman.[62] These techniques were combined with multiple-inoculation methods for rapid characterization of *Staphylococcus aureus* strains isolated from turkey products. It was found that the miniaturized procedures require only 5% of the materials and 10% of the time required by conventional procedures in studies where large numbers of isolates are examined. The authors suggested that these methods may be adapted to similar conditions where identification of large numbers of isolates is desired, such as the rapid identification of bacteria isolated from foods, soils, and other environments.[63,64]

In clinical specimens such as blood or urine, the concentration of bacteria is usually determined by either the pour plate or the spread plate technique. These conventional methods are recognized as the most reliable for the diagnosis of urinary tract infections. A faster and simpler screening test for quantitative urine culture was reported by Nabbut and Hakim.[65] An ordinary microscope slide coated with McConkey agar on one side and blood agar on the other side is inoculated by dipping it in freshly voided urine. A number of urine specimens were cultured simultaneously by the dip-slide and viable count methods. Of these specimens, 27.7% had more than 10^5 bacteria per milliliter urine, 8.4% had between 10^4 and 10^5, 15.4 to 16.9% had less than 10^4, and about 50% showed no growth (Table 3.2). Other simple methods similar to the dip-slide methods have been reported previously.[66-68] These methods may also be used in bacteriology as screening tests to determine bacterial counts in milk at collection stations and in a community water source.

Various growth media have been utilized in an effort to isolate and identify bacteria from clinical specimens and other biological samples. In some instances the modified media improved the isolation procedures by facilitating the growth conditions. Mara[69] found lactose tryptone ricinoleate broth (0.3% lactose, 2% Oxoid tryp-

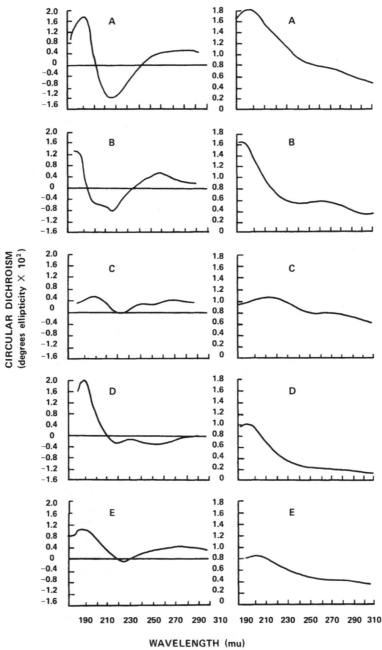

CIRCULAR DICHROISM (degrees ellipticity × 10²)

WAVELENGTH (mu)

A. *E. coli*
B. *B. cereus*
C. *Staph. albus*
D. *L. grippotyphosa*
E. *M. smegmatis*

FIGURE 3.1. Circular dichroism and ultraviolet absorption spectra of five bacterial species: (A) *Escherichia coli*; (B) *Bacillus cereus*; (C) *Staphylococcus aureus*; (D) *Leptospira grippotyphosa*; (E) *Mycobacterium smegmatis*. (From Torten, M. and Schneider, A. S., *J. Infect. Dis.*, 127, 319, 1973. With permission. Copyright by the University of Chicago Press.)

TABLE 3.2

Dip-slide and Spread-plate Results of 130 Urine Samples

	Dip slide		Spread plate	
Bacterial count	Blood agar	MacConkey agar	Blood agar	MacConkey agar
No growth (negative)	63	68	61	69
$<10^4$/ml (negative)	20	19	22	17
10^4-10^5/ml (doubtful)	11	10	11	11
$>10^5$/ml (positive)	36	33	36	33

From Nabbut, N. H. and Hakim, A., *Leban. Med. J.*, 26, 41, 1973. With permission.

tone, 0.1% sodium ricinoleate — pH 7.1) to be a more convenient medium than the currently recommended combination of lactose ricinoleate broth and tryptone water for the detection of *Escherichia coli* from various sources such as sewage effluent, polluted water, etc. A significantly higher number of coliform and fecal coliform organisms were isolated using lactose-glutamic acid medium than by using lauryl sulfate tryptone, lactose, brilliant green bile, or EE (Difco) broths.[70] Coliforms/Enterobacteriaceae were isolated less often with brilliant green bile and lauryl than with the other media (Table 3.3). For the detection of fecal coliforms, EE broth proved less useful than the other media. All five media showed more coliforms/Enterobacteriaceae at 30°C after 48 hr than after 24 hr. Few false positive results were obtained using any of these media.

Enumeration of bacteria in milk to evaluate the fitness of pasteurized milk for human consumption is a routine procedure in dairy bacteriology laboratories. Many conventional procedures and their simplified versions have been developed in recent years, offering savings in time and money if used in place of the standard plate count procedure. The oval tube count and the plate loop count are among these newer techniques using a 0.001-ml calibrated loop to deliver samples.[71,72]

Microcalorimetry

Boling et al.[73] obtained characteristic profiles for 17 species from 10 genera for different members of the family Enterobacteriaceae by observing the heat produced during their growth in liquid media (Figure 3.2a to f). A sample volume of 4 ml was used, and the growth of the organisms in brain-heart infusion broth (BHI) over periods of 8 to 14 hr was observed, yielding curves of heat production against time. Forrest[74] showed that heat production ceases abruptly following exhaustion of available glucose by *Streptococcus faecalis*. This was also observed in BHI. With some organisms (*Klebsiella* and *Enterobacter cloacae*) sudden changes in heat production occur at many successive points during growth in BHI. *Enterobacter aerogenes* and *Klebsiella,* which exhibit similar behavior in biochemical tests, are clearly different with respect to heat production. Some strains of *E. cloacae* can be separated from each other, although each possesses characteristics which permit species identification.

Staining Methods

Many varieties of dyes have been reported to inhibit growth of bacteria. Therefore, certain dyes can be used as bacteriostatic or bactericidal agents. Alternatively, the differentiating properties of dyes can be utilized for the identification and differentiation of bacteria when they are incorporated into a growth medium. Fung and Miller[75] described a rapid screening procedure in which they tested the effects of 42 dyes on growth of 30 species of bacteria on solid media. It was found that Gram-negative organisms showed greater resistance to dyes than Gram-positive organisms and that basic dyes were more inhibitory than acidic or neutral dyes at the same concentration. The authors concluded that many dyes not commonly used could be utilized for development of new selective and differential media. Dyes are commonly used in the bacteriology laboratory for the demonstration and tentative identification of organisms. Many staining methods (such as Gram stain, methylene blue, and acid-fast staining) are performed on a routine basis in bacteriology. These methods have also been modified to some extent for better differentiation of bacteria in clinical material and biological samples. Engbaek et al.[76] compared various staining methods for demonstration of tubercle bacilli in sputum, including the Tan-Tiam-Hok method modified by Devulder with heat coloration, the Tan-Tiam-Hok method modified by Devulder with cold coloration, the Armand method, the Ziehl-Neelsen method, and fluor-

TABLE 3.3

Comparison of Five Media for the Determinations of Coliforms/Enterobacteriaceae

Ranking score for products
(no. of samples)

Medium	Dehydrated (265)	Deep frozen (278)					
		Fish (55)	Meat (38)	Poultry (39)	Vegetable (86)	Prep. dish (60)	Total (278)
BGB	1014.15−	196.0−	116.5	121.0	271.5	201.5	906.5−
LST	908.0−	184.0	129.0	122.5	271.0	193.5	900.0
LB	725.5+	168.0	127.5	100.0	257.0	169.0	821.5
EE	808.0	154.5	107.5	145.0−	292.5−	193.0	892.5
LGA	519.0+	122.5+	89.5+	96.5	198.0+	143.0+	649.5+
5% limits for ranking scores	738−855	139−192	92−136	95−138	224−240	153−207	775−894

Note: +, Significantly low score value (medium better than the others); −, significantly high score value (medium not as good as the others) (see Youden).[3a] The 5% limits mentioned indicate the rank scores which are still just acceptable.

From Moussa, R. S., Keller, N., Curiat, G., and DeMan, J. C., *J. Appl. Bacteriol.*, 36, 619, 1973. With permission.

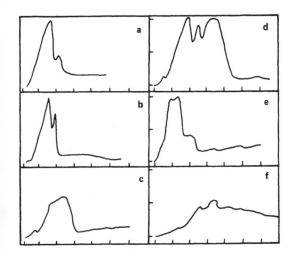

FIGURE 3.2. Six named heat profiles. For each, the abscissa represents time, with 2-hr intervals indicated by scale marks, and the ordinate represents heat production, with scale marks at 0, 20, and 40 μcal 5^{-1} ml^{-1}. (a) *Enterobacter aerogenes*; (b) *Klebsiella*; (c) *Proteus vulgaris*; (d) *Enterobacter cloacae*; (e) *Escherichia coli*; (f) *Proteus rettgeri*. (From Boling, E. A., Blanchard, G. C., and Russell, W. J., *Nature*, 241, 472, 1973. With permission.)

escent microscopy. A large number of slides containing varying numbers of tubercle bacilli were examined for the reproducibility of the staining method. Based on their study, the Ziehl-Neelsen method was recommended for a small laboratory and fluorescence microscopy was recommended for larger laboratories.

Convit and Pinardi[77] describe a simple laboratory method for the differentiation of *Mycobacterium leprae* from BCG, *M. smegmatis*, an atypical mycobacteria, a hamster lesion mycobacterium, and *M. lepraemurium*. The differentiation was accomplished by the use of the Ziehl-Neelsen, Truant, and Baker staining methods combined with treatment with pyridine. Smears and cryostat sections were treated for 2 hr with pyridine at room temperature. *M. leprae* lost its acid fastness while the other mycobacteria did not. The method required a suspension of at least 10^7 bacilli per milliliter. The stain-negative pyridine-extracted *M. leprae* slides revealed abundant mycobacteria when stained by Gram's method.

Pital et al.[78] report the use of fluorescein isothiocyanate in a direct staining method for bacterial detection. The method provides a broad-spectrum detection capability in contrast to the high specificity of immunofluorescent staining. Heat-fixed smears of bacteria and proteins (either alone or in the presence of atmospheric debris and soil) were stained with the dye, washed with alkaline buffer (pH 9.6), and examined micro-scopically with ultraviolet radiation. A stable and apparently specific linkage formed with protein and nonprotein substances was readily destained. A number of bacterial species, hamster kidney cells, wheat germ, and egg albumin were detected. However, the potential sources of error in this method are the reaction of fluorescein isothio-cyanate with nonprotein substances of unusual chemical configuration and the emission of green autofluorescence by nonprotein substances.

A rapid membrane filtration staining method for determining the number of viable bacteria in food and food processing equipment was reported by Winter et al.[79] Bacteria are rinsed from food or swab samples with sterile diluent and concentrated by filtration onto membrane filters. The filters are incubated on suitable media for 4 hr at 30°C, heated at 105°C for 5 min, and stained. Counts on dried membranes (rendered transparent) compared with those of the standard plate count method show a correlation coefficient of 0.906 (Table 3.4).

Filtration Membranes

Current methods to estimate the extent of microbial contamination on foods and equipment include standard plate count, spread plates, shake tubes, most probable number, and dilution to extinction. The disadvantage shared by these methods is the 24 to 72 hr required for microbial cells to grow to visible concentrations. Methods introduced to shorten the incubation period include the Frost "little" plate,[80] agar strips, roll, and oval tube methods. However, Winter et al.[79] doubt their reliability. The microscopic methods of Breed[81] and Brew[82,83] are applicable to milk but not vegetables, since there is a problem in distinguishing food particles from bacteria and living from dead cells. Frazier and Gneiser[84] used a membrane filter and 8 to 18 hr of incubation, but found difficulty in applying this technique to fresh and frozen vegetables. Winter et al.[79] believe that the method they propose is superior in that it overcomes prior difficulties. With their method, the microbial cells rinsed from food or swab samples are concentrated on 0.45-μm membrane filters and incubated on a sterile absorbent pad. Then, they are treated with 2 ml of medium (glucose-tryptone-yeast extract) for 4 hr at 30°C, heated at 105°C for 5 min, and stained with Janus B green. Membranes are dried at 60°C for 15 min,

TABLE 3.4

Comparison of Quick Counting (QC) and Standard Plate Count (SPC) Methods[a]

Product	Sample	SPC	QC
Broccoli	1	3.4×10^6	2.3×10^6
	2	4.2×10^6	2.4×10^6
	3	3.5×10^6	2.2×10^6
	4	3.2×10^6	2.2×10^6
	5	$1.5 \text{ N } 10^6$	TNC[b]
	6	2.1×10^6	TNC[b]
	7	2.6×10^6	1.1×10^6
	8	7.3×10^6	1.7×10^6
	9	4.4×10^6	1.2×10^6
	10	3.3×10^6	1.2×10^6
	11	2.8×10^7	4.1×10^7
	12	2.3×10^7	4.1×10^7
	13	1.7×10^7	4.0×10^7
	14	1.6×10^7	4.0×10^7
Brussel sprouts	1	8.2×10^3	TLC[c]
	2	3.1×10^3	TLC[c]
	3	6.7×10^4	6.6×10^4
	4	2.4×10^4	2.3×10^4
	5	1.1×10^5	2.0×10^4
Carrots	1	2.3×10^4	4.1×10^4
	2	6.1×10^5	6.9×10^5
	3	6.7×10^5	6.8×10^5
	4	6.3×10^5	6.5×10^5
	5	2.5×10^4	2.2×10^4
	6	5.1×10^4	5.2×10^4
	7	7.3×10^7	6.7×10^7
	8	8.0×10^7	6.9×10^7
Cauliflower	1	2.1×10^4	2.2×10^4
	2	3.0×10^4	2.5×10^4
	3	2.5×10^4	2.6×10^4
	4	2.8×10^6	3.0×10^6
Green beans	1	4.2×10^4	4.1×10^4
	2	6.1×10^4	5.8×10^4
	3	2.0×10^4	1.8×10^4
	4	1.5×10^5	1.3×10^5
	5	4.7×10^5	6.7×10^5
	6	4.8×10^5	7.3×10^5
	7	3.9×10^7	2.3×10^7
	8	4.3×10^7	2.9×10^7
Lima beans	1	2.8×10^4	3.0×10^4
	2	2.5×10^4	2.7×10^4
	3	6.8×10^6	6.0×10^6
	4	5.9×10^6	7.3×10^6
	5	3.1×10^4	3.0×10^4
	6	3.4×10^4	3.1×10^4
	7	6.9×10^7	5.1×10^7
	8	6.8×10^7	7.4×10^7

TABLE 3.4 (continued)

Comparison of Quick Counting (QC) and Standard Plate Count (SPC) Methods[a]

Product	Sample	SPC	QC
Mixed vegetables	1	7.0×10^6	4.3×10^6
	2	6.7×10^6	4.5×10^6
	3	7.1×10^6	5.1×10^6
	4	7.2×10^6	5.2×10^6
Peas	1	2.3×10^4	1.5×10^4
	2	3.4×10^4	2.2×10^4
	3	6.8×10^6	7.1×10^6
	4	7.1×10^6	5.0×10^6
	5	8.0×10^5	7.1×10^5
	6	1.0×10^2	TLC[c]
	7	3.5×10^3	3.9×10^3
	8	5.1×10^4	5.5×10^5
Peas and carrots	1	3.0×10^6	1.9×10^6
	2	4.0×10^4	2.9×10^4
	3	3.1×10^4	3.5×10^4
	4	5.2×10^6	4.9×10^6
Spinach	1	2.3×10^4	1.8×10^4
	2	2.2×10^4	2.0×10^4
	3	1.6×10^4	9.1×10^4
	4	2.0×10^4	7.7×10^4

[a]Statistical analysis: $x = 9.666 \times 10^6$; $y = 9.019 \times 10^6$; $\Sigma x = 5.993 \times 10^8$; $\Sigma y = 5.592 \times 10^8$; $n = 62$ (TLC and TNC disregarded); $r = \Sigma xy\ nxy/[(\Sigma x^2 - nx^2)\ (\Sigma y^2 - ny^2)]^{1/2} = 0.906$.

[b]Too numerous to count.

[c]Too low to count.

From Winter, F. H., York, G. K., and El-Nakhal, H., *Appl. Microbiol.*, 22, 89, 1971. With permission.

rendered transparent with immersion oil, and examined microscopically. Over 60 comparisons with the same samples determined by the standard plate count method resulted in a correlation coefficient of 0.906.

Benefits of this method include rapid results (within 4 hr) within ranges of "low," "moderate," or "excessive" contamination, i.e., 50,000, 50,000 to 100,000, and 100,000 cells per gram. This method may be used as a rapid monitoring system, but it does not replace the standard plate count method. It does not yield the same degree of accuracy within a 4-hr period.

Membrane techniques are of use in evaluating water supplies. For this purpose, membrane techniques were compared with MacConkey broth and glutamate medium using multiple tube methods.[85] The membrane filtration technique involved a preliminary incubation of two membranes on a 0.4% enriched Teepol® medium (0.4 ET) for 4 hr at 30°C, followed by 14 hr at 35°C for the total coliform count and 44°C for the *Escherichia coli* count. Attempts were repeated in order to incorporate the advantages of glutamate into the membrane filtration technique. In addition, both membrane techniques were evaluated by incubating at 25°C on 0.4 ET for 6 hr and also at 35 and 44°C for 18 hr. The glutamate medium was prepared as described in Reports on Public Health and Medical Subjects.[86]

The multiple tube technique using MacConkey broth and improved formate lactose glutamate medium (IFLG) was used as described in Report No. 71.[85] Presumptive positive results at 18 hr were recorded as coliform organisms; those at 24 and 48 hr were confirmed by subculture to lactose ricinoleate broth (LRB) incubated at 37°C. Presence of *Escherichia coli* was confirmed by subculturing all presumptive positive tubes to LRB for gas formation and to peptone water for indole production with incubation at 44°C.

Unchlorinated water samples were used as available. Samples of chlorinated water were prepared from raw waters by the marginal chlorination method,[87] which is based on chlorination in the presence of excess ammonia at very low temperatures.

For unexplained reasons, laboratories differed in comparisons between membranes and multiple tube methods. Therefore, it was recommended[87] that before adopting membrane filtration as a routine procedure, an adequate parallel series of tests should be run to compare membranes with multiple tubes in order to establish the equivalence or superiority of one over the other. Results from all laboratories showed significant differences in the results obtained with different media.[85] The standard and extended membrane methods gave significantly higher results than the glutamate tube method for coliform organisms in both chlorinated and unchlorinated waters, significantly lower results for *Escherichia coli* with chlorinated waters, and equivocal results with unchlorinated waters. Transport membranes did not perform as well as standard membrane methods; however, results were usually in agreement with glutamate tubes, except for *E. coli* in chlorinated waters. The glutamate membranes were found to be unsatisfactory. Preliminary incubation of glutamate at 30°C was not required.

Industries have applied membrane techniques in the sterility testing of creams and ointments. Factors influencing reisolation of each of three organisms from white soft paraffin and ceto-macrogol cream base were examined using a membrane filtration technique.[88] The major difficulty in isolation of viable contaminants from creams and ointments is the separation of viable cells from the lipid fractions of the formulation. There are two approaches. One uses various oil solvents, such as isopropyl myristate,[89] hexane,[90] or light liquid paraffin. Following homogenization, the preparation is filtered through a membrane of 0.45-μm pore size to retain bacterial cells; the membrane is placed on a suitable recovery medium. The drawback to this procedure is the toxicity of the solvent to living cells; this is enhanced by the elevated temperatures used. In the second approach, the preparation is homogenized in an aqueous solution containing a surface-active agent (such as Tween® 80 or Lubrol® W) which aids dispersion of the oily material and release of the viable cells into the aqueous phase.[91-93]

Sampling from the aqueous phase, viable counts are obtained on nutrient agar plates or by a membrane filtration method. Antibacterial substances are more readily separated using the first method. *Escherichia coli, Pseudomonas aeruginosa,* and *Staphylococcus aureus* were the test organisms.

The ointment and cream bases were contaminated with about 10^4 viable cells per gram from fresh overnight cultures of bacteria. Ointments are less sensitive to variations and show a more consistent recovery rate. Creams require more rigid conditions. Low levels of contamination remain undetected because of the difficulty in increasing sample size. The methods discussed are used routinely to check for bacterial contamination of ophthalmic ointments, steroid and emollient creams, and topical preparations containing antibacterial substances. Emulsified formulations have proven most difficult to examine for the presence of viable microorganisms.

METHODS BASED ON BIOCHEMICAL ACTIVITY OF BACTERIA

Newer methods for the identification of bacteria which utilize biochemical properties of the organisms include chemical composition of the organisms, growth and metabolic characteristics, enzyme activities, multiple test systems containing biochemical reactants, and the use of sugar fermentation capabilities. Many of these methods are useful in the detection and identification of a broad range of organisms, whereas other methods are more or less specific for a certain group of bacteria. The following section describes some of the newer methods which have been tested in many laboratories and are currently being used in addition to the conventional biochemical methods of bacterial identification.

Analysis of Chemical Constituents and Metabolic Products of Bacterial Cells

Many methods have been reported that depend on analysis for a particular constituent common to bacteria in general. Bacteria are composed of the basic elements C, H, O, N, P, and S which are similar to those found in other microbial or mammalian cells. However, the molecules formed by these elements may vary not only in bacterial and mammalian cells, but also within the bacterial species and strains. In utilizing these differences of bacterial constituents, the organisms should be isolated before analysis in order to eliminate the interfering materials present in the biological sample. It is known that the chemical composition, particularly carbohydrate, lipid, and protein contents, of bacteria varies markedly in different species or with different growth conditions. RNA and DNA contents in bacteria also vary but not as widely as other chemical components. Therefore, the methods must be standardized with regard to growth medium, purification of bacterial cell preparation, and growth cycle of the organism before the analysis of the organism is carried out. Among the recent techniques that have been used to precisely measure the chemical constituents of bacteria for detection and identification purposes are chromatograph techniques, mass fragmentography, and radiochemical techniques.

Chromatography

Paper (PC), thin-layer (TLC), and gas-liquid (GLC) chromatography have been adapted to identify bacteria.

Paper Chromatography

The identification of lower-chain fatty acids

produced by anaerobic bacteria aids in the classification of the bacteria.[94] GLC has been used for the detection of volatile, short-chain fatty acids produced by anaerobes in culture media.[95] Because of its simplicity, small cost, and small space requirement, Slifkin and Hercher[96] favor the feasibility of paper chromatography (PC) to determine characteristic fatty acid metabolites of anaerobic bacteria.[97-99]

Modifications of two procedures were used to evaluate PC as an alternative to GLC in identification of anaerobic bacteria in the clinical laboratory. In the first procedure, volatile fatty acids (including lactic acid) are converted to esters and then to hydroxamic acids and are detected as colored iron complexes.[100-101] This procedure was modified by Seligman and Doy.[102] In the second procedure, succinic acid and C_3-C_6 volatile fatty acids are separated as their ethylamine salts. The identification of succinic acid was via isooctane and a methyl alcohol solvent system modification. Both ethylamine and hydroxylamine derivatives were required for PC. The former was used for identification of propionic, butyric, valeric, caproic, and succinic acids. The hydroxylamine procedure was used to identify acetic acid; lactic acid could not be distinguished from formic acid.

PC analysis correlated with results obtained by GLC. Both acetic and propionic acids from *Propionibacterium* and *Veillonella* species were detected by PC. Determination of butyric acid assisted in distinction of *Fusobacterium* from *Bacteroides*. A large hydroxamic acid spot in the area of lactic-formic acids, along with other diagnostic criteria, assisted in identification of *Lactobacillus*.

Isomers of fatty acids and alcohols of fermentation were not detected by the PC procedures.[95] A "best-fit" analysis was applied with some bacteria. Slifkin and Hercher[106] found PC profiles as useful as fatty acid profiles obtained by GLC for the identification of the genera and speciation of the clinical and reference organisms. *Clostridium pasteurianum* and *Clostridium sticklandii* with essentially identical biochemical characteristics, were differentiated by PC analysis according to the VPI schema of fatty acid profiles (Table 3.5).

PC development time is approximately 5 hr (about 8 to 16 specimens per sheet) vs. 15 to 40 min per organism by GLC determination.

Thin-layer Chromatography

Since many groups of nonfermenting bacteria produce arginine dehydrolase, a rapid method for detecting this enzyme might identify and differentiate nonfermenters in the clinical laboratory. Current methods (method of Thornby as used by Taylor and Whitby)[107] require 2 to 7 days. Williams et al.[108] developed a faster procedure by using the principles of the Soru method[109] which requires 24 hr and detects the enzyme in β-hemolytic streptococci by using circular paper chromatography. The test is positive for fluorescent pseudomonads as well as *Pseudomonas maltophilia*. Other *Pseudomonas* sp., *Mima*, and *Herellea* are negative. With the Thornby method, tubes were inoculated by stabbing the medium, which was then covered with a 5-mm layer of sterile liquid paraffin and incubated at 35°C for 7 days. A positive test was indicated by a deep pink to red color, whereas a negative test was demonstrated by an absence of color change in the medium or a yellow color.

TLC method — Two microliters of cell suspensions (which were originally grown for 18 to 24 hr at 35°C on triple-sugar iron or Trypticase® soy agar slants) followed by a loopful of growth inoculated into 0.25 ml of 0.01 M L-arginine hydrochloride (pH 6.4) and incubated for 2.5 hr at 37°C were applied on a cellulose precoated sheet.* Controls were arginine, citrulline, and ornithine. The chromatogram was developed in n-butanol-acetone-acetic acid-water (35:35:10:20) for approximately 45 min. After drying, the chromatogram was sprayed to saturation with a 0.4 M solution of ninhydrin in n-butanol-acetone (1:1). Development of purple spots occurred in 15 to 20 min. *Escherichia coli*, with aqueous solutions of 0.01 M urea and agmatine as controls, was used to see the effect an arginine decarboxylase system would have on the arginine substrate.[110] Rf values were as follows: ornithine, 0.18; arginine, 0.22; citrulline, 0.24; agmatine, 0.42. All organisms which gave a positive Thornby reaction showed conversion of arginine to ornithine on TLC. Negative Thornby organisms showed no conversion on TLC, except *Pseudomonas maltophilia* which gave identical trailing and spotting as *E. coli*. Williams et al.[108] believe the TLC method is more accurate and sensitive than the methods of Pickett and Pedersen[111] and Gilardi[112] (which rely on pH change) or the method of Stanier et

*MN Polygram Cell 300, Brinkman Instruments, Inc., Westbury, N.Y.

TABLE 3.5

Comparison of Fatty Acid Profiles of Various Anaerobic Bacteria by GLC and PC

Bacteria	No. of strains tested	VPI manual GLC profile[a]	PC profile[b]	GLC profile[c]
Propionibacterium acidi-propionici	1	Pas (lfiv)[d]	PAs	PAs
Peptococcus asaccharolyticus	3	Ab (slf)	Ab	Ab
P. magnus	4	A (lsf)	A (LF)[e]	Alf
P. prevotii	3	Ab (Lspf)	Ab	Ab
P. variabilis	4	A (lspb)	A	A
Peptostreptococcus intermedius (VPI)	1	L (asḟ)	(LF)	Lfas
P. productus (VPI)	1	As (lp)	A (lp)s	Als
Veillonella alcalescens	9	ap (l)	ap	ap
V. parvula	4	ap (l)	ap	ap
Fusobacterium russii	1	BA (Lfs)	BA	BA
F. mortiferum	1	BAp (LF iv s)	BAp	BAp
F. naviforme	1	BLap (fs)	BLap	BLap
F. necrophorum	1	Bap (Lsf)	Baps	Baps
F. varium	4	SA (Lf ib iv)	SA (lf)	SAl
Bacteroides fragilis subsp. *fragilis*	8	SAp (lib iv f)	SAp	Sapl
B. oralis	1	SA (Lf ib iv)	SA (lf)	SAl
B. praeacutus	1	AB iV iBp (slf)	ABV (fl)	AB iV iB pl
Clostridium barkeri ATCC 25849	1	BL (sṗ)	B (LF)	BL
C. bifermentans	3	AFp ic ib iv c (bsl)	A (lf) pcbv	AFp ic ib iv c
C. haemolyticum ATCC 9650	1	PBA (ls)	PBA	PBA
C. lentoputrescens	1	Bap (l)	Bap (lf)s	Bap ls
C. mangenotii ATCC 25761	1	iv ic ib a p s l (f)	v c b a p s	(f) APIBB IV VIC es
C. pasteurianum	1	BA (lsf)	BA	BA
C. scatologenes ATCC 25775	1	ABC iv v pib (fsl)	BAC vpb	ABC iv v pibl
C. sporogenes	1	Ab iv ib (ic spc)	Abv	Ab iv ib s
C. sticklandii	1	AB iV ib lps	ABV (lf) ps	Bap lb iv ls
Lactobacillus sp.	14	L (afs)	La	La

[a]Fatty acid profiles obtained directly from VPI Anaerobe Laboratory manual.
[b]Fatty acid profile derived from combination of ethylamine and hydroxamate paper chromatography.
[c]Gas-liquid chromatography analysis obtained in this investigation.
[d]Letter denotes fatty acid: A, acetic; B, butyric; C, caproic; F, formic; L, lactic; S, succinic; V, valeric. Size of letter denotes relative major and minor fatty acid concentrations. Letter in parentheses denotes fatty acids produced by occasional strains in the species. The letter "i" preceding a letter denotes its isomer.
[e](LF) denotes that lactic acid is not distinguishable from formic acid by PC analysis.

From Slifkin, M. and Hercher, H. J., *Appl. Microbiol.*, 27, 500, 1974. With permission.

al.[113] (in which residual arginine is measured after incubation). TLC requires 4 hr if fresh cultures are available from slants.

Jenkins et al.[114] used the procedure of Marks et al.[115] to extract lipids and examine them by TLC. Solvent systems were

1. *N*-propanol:water:ammonia, 75:22:3 or *N*-propanol:*n*-butanol:water:ammonia, 57:20:20:3
2. *N*-Butanol:acetic acid:water, 60:20:20
3. *N*-Propanol:water:diisopropyl ether, 45:10:45 or *N* propanol:water:ammonia: diisopropyl ether, 45:9.5:0.5:45

Mycobacterium scrofulaceum was identified by chromatographs sprayed lightly with orcinol-sulfuric acid. Heated golden-brown spots were seen when developed in Solvents 1 and 2, and yellow spots were seen when developed in Solvent 3. Chromatographs developed in two dimensions by Solvents 1 and 2 are characteristic for each of the serotypes scrofulaceum, Gause, and Lunning. Of 131 strains classified as *M. scrofulaceum* on general cultural and biochemical characters, 127 provided patterns of the type described and 4 gave a nonspecific picture.

M. xenopi strains gave a constant picture of a pale yellow spot of Rf 0.95 in Solvent 1 and Rf 0.85 in Solvent 3. The change of Rf is much less than that met with the AIS (*avium-intracellulare-scrofulaceum*) complex in which some microbiologists assimilate *M. xenopi*. *M. gordonae* strains gave a highly distinctive lipid pattern. The specific spots straddle or lie close to the phenol red marker from the medium when developed in Solvent 1 and are dense and have a characteristic green or greenish-brown color. In two-dimensional analysis each spot split into a stationary and mobile component with Solvent 2. *M. flavescens* appears to be heterogeneous. With 32 strains, 11 different lipid patterns were observed. *M. aquae* spots are green, brown, or intermediate. They resemble those of *M. kansasii* in distribution but not in color. *M. kansasii* spots are bluish except for a frequent tinge of brown.[116] In two-dimensional analysis the green or brown color of *M. aquae* spots is muted or lost, and this feature emphasizes the resemblance to the *M. kansasii* pattern.

Jenkins et al.[114] showed that *Mycobacterium avium*, *M. intracellulare*, and organisms classed as *M. scrofulaceum* have a lipid structure of the same general character. This supports the contention of Tsukamura et al.[117] based on the numerical analysis of many properties, that the three species belong to a single complex.

Following the VPI publication[118] which showed that many genera and species of anaerobic bacteria produce a characteristic array of fermentation acids, Altenbern[119] sought to differentiate enteric bacteria by analysis of fermentation acids. Cells from a single colony of bacteria were allowed to grow in a small (0.25 ml) amount of liquid medium. After incubation, 0.05 ml of 5% sulfuric acid was added to stop further growth and metabolism and to permit extraction of the organic acids arising from fermentation. The organic acids (succinic, fumaric, and pyruvic) arising from the fermentation permitted differentiation of the genera *Escherichia*, *Enterobacter*, and *Klebsiella* (Table 3.6).

The fermentation acids were extracted into diethyl ether. A 30-μl solution of ether extract was spotted on Gelman TLC sheets, type SG, 8 × 8 in. The solvent system was chloroform-methanol-formic acid in an 80:1:0.75 ratio. After develop-

TABLE 3.6

Fermentation Acid Production by Wild Type and Authentic Strains of *Escherichia*, *Enterobacter*, and *Klebsiella* spp.[a]

| Organism | No. of strains | Organic acid in μg/30 μl ether extract | | |
		Succinic	Fumaric	Pyruvic
Escherichia coli	12	5.7–8.2	1.8–2.7	0
Enterobacter spp.	5	1.7–2.1	1.0–1.2	0
Klebsiella spp.	8	1.4–2.0	1.0–2.2	0.7–1.3
Enterobacter cloacae, Bact-Chek	1	2.1	1.2	0
E. aerogenes	1	3.7	1.1	0
ATCC #13048	1	3.7	1.1	0
E. liquefaciens ATCC #14460	1	1.4	0.2	0
E. hafnia ATCC #11604	1	2.9	1.1	0
K. pneumoniae ATCC #13883	1	3.1	1.4	0.9

[a]Single colonies incubated in hydrolyzed casein medium for 3.5 hr at 38°C, acidified, and ether extracted; 30 μl of ether extract is analyzed by TLC. Data for wild type strains are presented as the full range of values.

From Altenbern, R. A., *Proc. Soc. Exp. Biol. Med.*, 144, 551, 1973. With permission.

ment, drying, and rehumidification, the sheets were sprayed with an aqueous-indicator solution of bromcresol purple. Organic acids appeared as yellow spots on a bluish-purple background. The complete chromatography procedure is described by Bleiweis et al.[120]

Escherichia coli strains produced three times as much succinic acid and twice as much fumaric acid as *Klebsiella* and *Enterobacter*. *Klebsiella* showed distinct and measurable amounts of pyruvic acid and thus could be differentiated from *Enterobacter*. No other detectable acids are produced by these three genera during fermentation; trace amounts of propionic acid are common to all three. Altenbern[119] favored TLC over GLC because: (1) minor cost of TLC; (2) TLC simultaneously analyzes many samples; (3) GLC first needs to convert the organic acids to volatile methyl esters which adds to the time required for analysis. TLC takes 4.5 hr (3.5 hr for inoculation and 1 hr for chromatographic analysis). This can be cut to 3 hr with a larger inoculum (four colonies rather than one).

Gas Chromatography and Mass Spectrometry

Gas chromatography has been applied to the analysis of the pyrolysis products of bacteria, resulting in its suggestion as an acceptable method for rapidly and specifically identifying bacterial species. Oyama[121] and Reiner[122] were among the first to suggest the use of gas chromatography in the identification of bacteria. Typical bacterial constituents from the cell wall, cell membrane, and cytoplasm may be chromatographed, and the patterns may be compared for the identification of the organism. The chromatographic patterns of the bacterial cells are prepared after pyrolysis or after hydrolysis and extraction of certain classes of compounds such as carbohydrates, lipids, amino acids, etc. Because of the widespread use of gas chromatography for the identification of bacteria, this topic will be discussed separately in Chapter 4. For rapid biological purposes, Mitz[123] proposed the specific identification of small numbers of bacteria involving pyrolysis and product analysis with gas chromatography and mass spectroscopy. In the identification of unknown organisms or compounds, standard chromatographic or mass spectroscopic patterns of known organisms or products must be used as a basis of comparison.

However, one disadvantage is the variation of patterns concomitant with variations in retention time, temperature, and type of individual columns. Many laboratories use mass spectroscopy for the identification of bacteria based on the fragmentation patterns of chemical constituents. Mass spectroscopy is an expensive technique to use on a routine basis, and there are many technical problems at present which prevent its use as a routine technique in the bacteriology laboratory.

Luminol® Chemiluminescence

Other specific constituents of bacteria may be analyzed by sensitive methods which provide information for the characterization of the organisms. For example, the catalytic effect of hematin on the chemiluminescence of Luminol can be used to determine extremely small amounts (10^{-11} g) of various hematin compounds.[124] In the presence of alkali hydrogen peroxide or sodium perborate and an activating agent (such as sodium hypochloride, potassium ferricyanide, or a transition metal) Luminol (5-amino-2,3-dihydro-1,4-phthalazinedione) produces photons. If the reaction is allowed to take place in a dark chamber, the emitted light can be measured photometrically.

Determination of ATP

Adenosine triphosphate is apparently present in all living organisms including bacteria. Levin et al.[125] adapted the firefly bioluminescent reaction for the detection and counting of bacteria based on the presence of ATP in cells. D'Eustachio et al.[126] and Chappelle and Levin[127] correlated ATP concentration with bacterial numbers. Light emission occurs when the enzyme luciferase reacts with luciferin (in firefly tails), Mg^{++}, and ATP and is directly proportional to the amount of ATP present. The method involves adding a crude buffered extract of firefly tails containing magnesium ions to a sample contained in a light-proof chamber adjacent to a phototube and measuring the peak light emission with a photometer. Certain extraneous reactions (e.g., involving ATP production in the reaction mixture) are largely eliminated if purified luciferase and luciferin are used instead of crude extracts of firefly tails. An instrument (the Luminescence Biometer®*) using this principle has recently been

---

*E. I. duPont de Nemours and Company, Instruments Products Division, Wilmington, Del.

marketed.[128] It is claimed that this instrument detects as little as 10^{-13} g ATP per 10-μl sample. Extraction of bacteria with butan-1-ol is recommended and is introduced as an aqueous phase to partition the hydrophilic ATP. A 10-μl sample of the aqueous phase is injected into the luciferin-luciferase reaction mixture, and the resulting light flash is automatically converted to concentration of ATP or microorganisms. Assuming that bacterial ATP falls within a range of 2.2 to 10.3 × 10^{-16} g per cell, detection of ATP from 100 to 300 bacteria is feasible. However, because an extraction process has to be used, a large number of bacteria is required, and about 2×10^4 is probably a realistic estimate of the minimum number of bacteria detectable. Because the bacterial ATP contact varies in different as well as in the same species with the physiological state and environment of the bacteria, the minimum number of bacteria that is required to give a detectable response in the firefly luminescence assay may be considerably higher than in certain environmental conditions with freshly grown bacteria in the laboratory. Also, the measurement of ATP may not necessarily be indicative of viability of the bacterial cells.[129] The ATP content of cells apparently depends on the metabolic state of the cells at the time of measurement of their mode of death. Thus, results obtained from organisms exposed to heat, antibiotics, or other deleterious agents might not be valid in terms of projected viable progeny. The method is not applicable to the detection of bacteria in the direct analysis of clinical specimens.

Mitz[123] reported an assay based on the colorimetric determination of another coenzyme, diphosphopyridine nucleotide (NAD), in bacteria. However, the method has to be tested further before its application to the detection and differentiation of bacteria can be realized.

Radiochemical Techniques

The methods of the radioactive antibody technique (uptake of ^{125}I or ^{131}I by bacterial cells) are discussed in a subsequent section. The determination of a few bacteria by this method is possible after separation of the labeled immune complex; the variation in the level of non-specifically attached radioactivity is less than the radioactivity of the immune complex.

Another approach utilizes the detection of radioactive CO_2 and other gases formed as end products during the metabolism of labeled substrate by microorganisms. Levin et al.[130] were the first to report on the use of radioisotopes to detect the presence of microorganisms of medical significance. A membrane filter was used to collect bacteria; it was subsequently immersed in medium containing ^{14}C-labeled substrates. Metabolically produced $^{14}CO_2$ was collected. Improvements were made on the basic test, including use of ^{14}C-lactose for a one-step presumptive coliform test and use of ^{14}C-formate in an inhibitory broth for a one-step confirmatory fecal coliform test. Levin found a quantitative relationship in his studies between evolved radioactivity and number of organisms. He has continued to develop the technique to include a means of life detection on other planets.[131-136a]

Macleod et al.[136b,c] reported a membrane filtration method for the rapid detection of small numbers of viable bacteria based on their ability to take up ^{32}P as orthophosphate. A medium containing potassium chloride, magnesium sulfate, glucose, and $^{32}PO_4{}^{3-}$ was inoculated with a small number of bacteria and incubated with shaking at 37°C; parallel controls did not contain bacteria. After incubation (usually 1 hr), 1-ml samples were filtered and washed on sterile Millipore® membrane filters (HA; 0.45 μm). The membrane filters were placed on planchets and air-dried; retained radioactivity was determined with a thin end window Geiger counter attached to a Picker® scaler. Several thousand bacteria of different species can be detected with this method. However, the extent of interference due to many substances in a biological mixture or natural sample (e.g., water) may cause the results to vary.

By measuring ^{14}C glucose, Previte[137] detected aerobic and anaerobic species of significance in foods. *Salmonella typhimurium* and *Staphylococcus aureus* were inoculated into tryptic soy broth containing ^{14}C glucose. Detection times ranged from 10 to 3 hr for inocula of 10^0 to 10^{14} cells per milliliter of broth. Previte also incubated heat-shocked spores of *Clostridium sporogenes* or *C. botulinum* in tryptic soy broth supplemented with thioton and $NaHCO_3$. The medium was rendered anaerobic with N_2. The spores required an increased amount of ^{14}C glucose and 3 to 4 hr longer for detection than did comparable numbers of aerobic vegetative cells. His results demonstrated the importance of availability of sufficient label in the media and the potential of the application of this technique for sterility testing of foods.

Phosphorescent Decay

When proteins or aromatic amines are irradiated with ultraviolet radiation, they exhibit fluorescence and phosphorescence. Adelman et al.[138] irradiated intact living or killed cells of five bacterial strains and found that the general forms of the total and phosphorescent emission curves were different, if not unique, from one species or strain to another. They proposed analysis of phosphorescent decay curves as a means of identifying bacteria.

Use of Selective Media for the Detection And Identification of Bacteria

Many attempts have been made to modify classical microbiological methods based on growth or expression of the ability to reproduce in culture media. There is continuing investigation of test procedures to obtain results rapidly and accurately. New media or substrate-supplemented media are being developed; commercially available forms include reagent tablets,[139] multiple media in a single tube, prepackaged microtest packets, and reagent-impregnated paper strips. Selective and differential media were used by Smith et al.[140] for the quantitative determination of airborne staphylococci. Five percent human blood in Columbia agar base (BBL)* was used as a total count nonselective reference medium. Mannitol salt agar medium (BBL) was used to aid in identification of airborne *Staphylococcus aureus*. Methyl green DNAase agar (MGD) was used because bacterial nuclease activity can be directly observed without flooding plates with acid. Orth and Anderson[142] successfully used a similar medium to isolate food-borne staphylococci. A concentration of 30 μg/ml of lincomycin was added to blood and MGD medium as selective agent for endemic *S. aureus* resistant to this antibiotic. The airborne microflora were characterized by Gram stain, colonial morphology, and the catalase test. It was found that the MGD agar without inhibitory agents provided a means of directly selecting *S. aureus* on the basis of DNAase activity and enhanced pigmentation, yielding quantitative data which compared favorably with blood agar. Mannitol salt agar was least effective for quantitating *S. aureus*.

Numerous reports have appeared in the literature concerning methods and media for the isolation of the fecal streptococci. Sixty-eight media are known for the selective isolation or enumeration of these organisms. The more important inhibiting substances which have been used in selective media for these organisms include sodium azide, thallium salts, potassium tellurite, crystal violet, citrate, antibiotics, sodium taurocholate and bile salts, Tween 80, selenite, tetrathionate, sodium chloride, and phenethyl alcohol. Most of the selective media depend on the incorporation of either sodium azide, thallium salts, or potassium tellurite. To increase selectivity, these substances may be used in conjunction with dyes (particularly crystal violet), antibiotics, an incubation of 45°C, esculin, 2-,3-,5-triphenyltetrazolium chloride (TTC), Tween® 80, citrate, sodium taurocholate, sodium chloride, or others. Pavlova et al.[143] evaluated the selectivity of five media containing different classes of inhibitory agents for the isolation and enumeration of fecal streptococci from natural sources such as feces, sewage, and food. Their scheme for identification of fecal streptococci is presented in Figure 3.3. They found that the lowest level of fecal streptococcus recovery occurred on M-enterococcus and azide-sorbitol agars. However, thallus acetate agar, which yielded the highest counts, was the least selective since 58% of the isolates were nonfecal streptococci. The dehydrated media KF and PSE agars yielded the highest recovery of fecal streptococci while showing the lowest percent of nonfecal streptococci. In general, the PSE agar has a number of advantages over the other media. First, it provides a wide range of sources including food, feces, and sewage. Second, it exhibits a low percent of nonfecal streptococci. Third, the results are available after 18 to 24 hr incubation. The recently formulated PSE agar has been described as suitable for rapid isolation and enumeration of fecal streptococci.[144] Brooks and Goodman[145] devised a rapid test system employing modified inoculation procedures, modified media, and reagents which provided a highly effective and efficient procedure for routine differentiation of Enterobacteriaceae. Known species of Enterobacteriaceae were tested using standard media and reagents modified so as to allow for simple inoculation with test results available within 4 hr. For rapid tests, 0.5 ml of conventional medium, prepared at twice the normal concentration, was

*Baltimore Biological Laboratory, Division of Becton, Dickinson and Company, Cockeysville, Md.

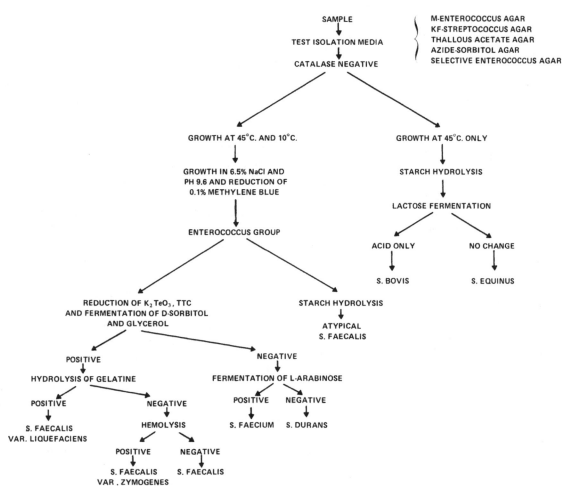

FIGURE 3.3. Scheme for identification of enterococci. (From Pavlova, M. T., Brezenski, F. T., and Litsky, W., *Health Lab. Sci.,* 9, 289, 1972. With permission.)

inoculated with 0.5 ml of a heavy water suspension of the bacterial cells. Lactose fermentation was found to be unreliable in the rapid tests. The media modified were lysine decarboxylase, ornithine decarboxylase, urease, tryptophan broth for indole production, and glucose and lactose broth. A solution of DL-phenylalanine in saline was used for the phenylalanine diaminase test. This method provided a small-scale system for rapid and effective differential testing of Enterobacteriaceae.

A rapid procedure for the detection of *Salmonella* in food and food ingredients was developed by Hoben et al.[146] A lysine-iron-cystine neutral red (LICNR) broth medium[147] was modified and used to detect the presence of viable *Salmonella* organisms in a variety of food, food ingredients, and feed materials by using a two-step enrichment technique. Tetrathionate broth was used to enrich samples with incubation at 41°C for 20 hr, followed by transfer to LICNR broth and incubation at 37°C for 24 hr for further enrichment and for the detection of *Salmonella* organisms by color change. Of the 110 samples tested, 94% were found to be presumptively positive by this method and were confirmed as positive by a culture method. Using this new medium (LICNR broth), it was possible to detect *Salmonella* in food samples within 3 days. The procedure does not require any special equipment or antisera. The authors concluded that the LICNR broth is quite useful for rapidly screening the presence of viable *Salmonella* organisms. Naguib[148] reported the suitability of different types of Bacto® strips, in comparison to the conventional methods using the corresponding selective media, for the estimation of coliforms in pasteurized milk. Among the selective media tested were desoxycholate agar (Difco), brilliant green bile broth (Difco), and violet red-bile agar (Difco). Three types of Suisse Bacto strips were tested; desoxycholate (cowein and coweiten), brilliant green bile (cobrien and cobrieten), and violet red bile (cored) were used. It was concluded that the Bacto strip technique offers a satisfactory and simple method for routine and qualitative estimation of coliforms. Therefore, it is applicable to the detection of coliform in milk, as it gives more rapid results than the conventional methods, using selective media. The selective media are preferable only when accurate coliform counts are needed.

Multiple Test Systems

Several multiple test systems have recently been developed which are intended to simplify the identification of bacteria, particularly enteric bacteria. These methods, used for differential identification of enteric and related Gram-negative bacteria, have been dependent mostly on biochemical tests. Goldin[149] compared five different multiple test systems to a conventional system for the identification of members of Enterobacteriaceae from clinical specimens. The multiple test media included the following.

Kahn's medium (Oxoid) — This consists of two tubes, the first containing glucose, mannitol, and urea in agar slants and the second containing sucrose and salicin in a semisolid medium. Lead acetate and modified Kovac's reagent paper strips are suspended from the plug of the second tube for H_2S and indole production, respectively.

Gillies composite medium — This is similar to the medium used by Gillies.[150] It is essentially similar to Kahn's medium except in the type of pH indicators used.

Hemmes-7-in medium — This is a single-tube semisolid medium in which seven different reactions (including dextrose, lactose and sucrose fermentation; production of H_2S; urease detection; motility; and indole production) are read.

R-B system[®]* — This consists of a commercially prepared two-tube system. The first tube is intended for observation of lactose and dextrose fermentation and gas production, H_2S production, lysine decarboxylation, and phenylalanine

determination. The small second tube is used for determination of motility, production of indole, and ornithine decarboxylation. Indole is read after the addition of a modified Kovac's reagent furnished with the tube.

Enterotube[®]** — This is a plastic tube divided into eight compartments, each containing a slant of a test medium. A wire which serves as an inoculating medium needle extends through the center of the tube. Dextrose fermentation, H_2S and indole production, phenylalanine deaminase, urease activity, dulcitol and lactose fermentation, lysine decarboxylase, and citrate utilization can be determined. Indole production and phenylalanine deaminase are read by injecting Kovac's reagent and a 10% ferric chloride solution through the specific plastic windows with the Enterotube in an upright position to prevent the flow of these reagents into other compartments. This system will be described further elsewhere in this chapter. Goldin[149] used 200 consecutive Gram-negative enteric bacilli isolated from urines, throats, sputa, bronchial lavages, drainages, blood cultures, and miscellaneous sources were used to test the above-listed media. Of these, the Enterotube system gave the best correlation (92%) with the conventional system of identification of bacteria as described by Edwards and Ewing[151] and Shaffer and Goldin.[152] Reactions were generally clear-cut and easy to interpret. However, additional biochemical tests were occasionally required, particularly in the *Klebsiella-Enterobacter-Serratia* group. Goldin[149] concluded that a multiple test system is advantageous, particularly in smaller laboratories with limited facilities for enteric bacteriology. However, a clinical bacteriologist should be aware of the limitations of these multiple test systems and be prepared to do adequate supplementary tests when results are questionable. Some of the advantages and disadvantages of multiple test systems have been discussed in a recent article by Lapage.[153] Various multiple test systems are commercially available. Some of these systems are briefly discussed in the following section.

Enterotubes

Elston et al.[154] also evaluated the Enterotube system in parallel with conventional tests to determine its value in the identification of enteric and certain Gram-negative bacilli. The Enterotubes

*Diagnostic Research, Inc., Roslyn, N.Y.
**Roche Diagnostics Division, Hoffman-LaRoche, Inc., Nutley, N.J.

are plastic tubes with one side round and the adjacent side flat. The flat side is covered with a thin plastic seal. Each tube has eight compartments, each containing a given biochemical test medium. The type of medium in the compartments and the reactions were as follows.

Dextrose agar — The fermentation of dextrose changed the indicator, phenol red, from pink to yellow.

Hydrogen sulfide-indole agar — Hydrogen sulfide was demonstrated by the formation of a black precipitate along the inoculation line, whereas indole production was tested by injecting 0.3 ml of Kovac's reagent through the plastic film with a needle and syringe onto the culture growth in the compartment. A red color indicated the presence of indole.

Phenylalanine deaminase agar — Deamination was tested by injecting 0.3 ml of a 10% ferric chloride solution through the plastic film with a needle and syringe onto the culture growth in the compartment. The formation of a dark green color on the surface of the culture slant indicated a positive test. Kovac's and ferric chloride solutions used in the indole and phenylalanine deaminase tests should be injected with the Enterotubes in an upright position to prevent the flow of these reagents into other compartments via the inoculating canal.

Urea agar — The production of ammonia from urea was shown by a change in the phenol red indicator from yellow to pink.

Dulcitol and lactose agar — The fermentation of these carbohydrates was indicated by a change in the phenol red indicator from pink to yellow in their respective compartments.

Lysine-lactose agar — A positive lysine decarboxylase test was noted by a change in the phenol red indicator from yellow to pink or orange.

Simmon's citrate agar — A positive test was shown by a change in the bromthymol blue indicator from green to blue. All reactions in the compartments were read after incubation at 37°C from 18 to 24 hr and again at 48 hr.

Among the conventional biochemical tests used were carbohydrate test medium, hydrogen sulfide (Klinger iron agar), tryptone broth, phenylalanine deaminase agar, urea agar, lysine medium, lysine-lactose agar, lysine broth, and Simmon's citrate agar. By using these two test systems, Elston et al.[154] found no discrepancies in the indole test

and one discrepancy for dextrose (Table 3.7). However, in 7 of 242 hydrogen sulfide tests, 3 of 242 phenylalanine tests, 22 of 242 urease tests, 15 of 242 dulcitol tests, 12 of 242 lactose tests, 27 of 217 lysine decarboxylase tests, and 5 of 242 citrate tests, the Enterotube results were contrary to those obtained with conventional methods. It was concluded that although there were obvious deficiencies in the system, the Enterotube method is simple and convenient, with the added advantage that all media are inoculated at once from a single colony.

Two-tube R-B System

This system (described before) has been modified to include a constriction at the base of both tubes. The constriction in the first tube separates the lysine decarboxylase test from the remainder of the tests in that tube. In the second tube the constriction confines the indole-ornithine motility medium which contains less agar than used previously. This modified R-B system was evaluated by McIlroy et al.[155] who found the following test reactions in the R-B system comparable to those in the conventional method: pigmentation of glucose, hydrogen sulfide production, and lysine and ornithine decarboxylase activities (Table 3.8). The production of gas from glucose was positive in the R-B system more often than in the conventional method; however, the motility test and the production of indole were positive less often in the R-B system. The authors reported that the preliminary identification of Enterobacteriaceae with the R-B system is enhanced if Simmon's citrate and Christian's urea tests are used concomitantly.

Four-tube R-B System

This system was evaluated by Brown[156] (14 tests) for identification of *Serratia marcescens* and *S. liquefaciens*. The tests contained in the first two tubes have been described previously.[157,158] Tube three (Cit/Rham) contained a slant of Simmon's citrate and a base of medium for rhamnose fermentation; tube four (Soranase) contained medium for testing for extracellular DNAase production and raffinose, sorbitol, and arabinose fermentation. The author reported that the 14 biochemical tests of the expanded R-B system increased the reliability of Enterobacteriaceae identification. All 417 cultures of *S. marcescens* and *S. liquefaciens* were identically

TABLE 3.7

Comparison of a Number of Biochemical Tests by Conventional Tests and the Enterotube® System

Organism	No. of strains	Dextrose C[a]	Dextrose E	H₂S C	H₂S E	Indole C	Indole E	Phenyl-alanine C	Phenyl-alanine E	Urease C	Urease E	Dulcitol C	Dulcitol E	Lactose C	Lactose E	Lysine C	Lysine E	Citrate C	Citrate E
Shigella																			
Group A	2	2	2	0	0	1	1	0	0	0	0	0	0	0	0	0	1	0	0
Group B	2	2	2	0	0	2	2	0	0	0	0	0	0	0	0	0	0	0	0
Group D	1	1	1	0	0	0	0	0	0	0	0	0	0	0	0	0	0	0	0
Alkalescens-Dispar	2	2	2	0	0	2	2	0	0	0	0	1	0	0	0	2	2	2	2
Escherichia[b]	60	60	60	0	0	59	59	0	0	0	0	30	26[c]	59	56	0	3	2	2
Klebsiella pneumoniae	60	60	60	0	0	2	2	0	0	51	46	7	6	60	58	3	2	60	60
Enterobacter aerogenes	6	6	6	0	0	0	0	0	0	6	3	1	0	6	6	1	0	6	6
E. cloacae	4	4	4	0	0	0	0	0	0	2	2	0	1	4	4	0	0	4	4
E. hafnia	12	12	12	0	0	0	0	0	0	7	0	2	3	6	4	0	5	12	12
Serratia marcescens	4	4	4	0	0	0	0	0	0	1	1	0	0	0	0	4	4	4	4
Proteus mirabilis	27	27	27	23	19	0	0	26	26	27	27	0	0	0	0	0	12	20	19
P. rettgeri	10	10	10	0	0	10	10	8	9	10	10	0	0	0	0	0	0	7	4
P. morganii	4	4	4	2	0	4	4	4	3	4	4	0	0	0	0	0	0	0	0
P. vulgaris	1	1	1	1	1	1	1	1	1	1	1	0	0	0	0	0	0	0	0
Providence group	3	3	3	0	0	3	3	3	2	0	0	0	0	0	0	0	1	3	3
Salmonella																			
S. paratyphi B	2	2	2	2	2	0	0	0	0	0	0	2	2	0	0	2	2	2	2
S. typhi	2	2	2	2	2	0	0	0	0	0	0	2	2	0	0	2	1	2	2
S. gallinarum	1	1	1	0	0	0	0	0	0	0	0	1	1	0	0	1	0	1	1
Salmonella group E	1	1	1	1	1	0	0	0	0	0	0	1	1	0	0	1	1	1	1
Citrobacter group	11	11	11	2	2	2	2	0	0	4	1	9	9	6	3	3	3	11	11
Arizona arizonae	2	2	2	2	2	0	0	0	0	0	0	0	0	2	0	2	2	2	2
Pseudomonas aeruginosa	20	1	0	0	0	0	0	0	0	7	5	0	0	0	0			20	20
Herellea	5	5	5	0	0	0	0	0	0	2	2	0	0	0	0			5	5

[a] C, conventional; E, Enterotube

[b] Includes *Escherichia coli* and atypical *Escherichia* (*Paracolon coliforme*, *E. intermedia*).

[c] With *Escherichia*, 23 strains were positive by both methods.

From Elston, H. R., Baudo, J. A., Stanek, J. P., and Schaab, M., *Appl. Microbiol.*, 22, 408, 1971. With permission.

TABLE 3.8

Positive Tests Observed Compared to Prediction by Ewing's Classification[a]

	% positive		
Test	R-B	Ewing	Conventional
Glucose	88[b]	c	100
Lactose	26	c	53
Gas	73	c	63
H$_2$S	28	30	27
Indole	23[d]	36	33
Ornithine decarboxylase	62	62	63
Motility	55	76[e]	60
Lysine decarboxylase	50	50	52
Urea		40	40
Citrate		60	63

[a]See Ewing, W. H., Differentiation of Enterobacteriaceae by Biochemical Reactions, National Communicable Disease Center, Atlanta, 1968.
[b]100% if adjusted to positive when H$_2$S obscured the test result.
[c]Not comparable because of difference in methods.
[d]Significantly lower than predicted.
[e]Plain agar medium.

From McIlroy, G. T., Yu, P. K. W., Martin, W. J., and Washington, J. A., II, *Appl. Microbiol.*, 24, 358, 1972. With permission.

identified by the R-B system and corresponding conventional tests.

API® System*

One of the several devices available commercially for identification of Enterobacteriaceae is the API system, a plastic strip holding 20 miniaturized compartments or capsules, each containing a dehydrated substrate for a different test.[159] The technique is basically a modification of one of the many "little tube" methods, as enumerated by Hartman.[160] The API system has been used for the identification of *Lactobacillus,* enteric *Moraxella* and *Pseudomonas* cultures, *Arizona,* and other members of family Enterobacteriaceae from clinical sources.[161] Washington et al.[161] used the API system of 20 biochemical tests to identify 128 Enterobacteriaceae, 5 *Aeromonas,* and 1 *Yersinia enterocolitica.* The results of tests for H$_2$S and indole production; citrate utilization; lysine and ornithine decarboxylase; arginine dehydrolase; nitrate reduction; β-galactosidase; and fermentation of arabinose,

*Analytab Products, Inc., New York, N.Y.
**Warner Chilcott, Diagnostic Division, Morris Plains, N.J.

rhamnose, mannitol, and glucose showed almost complete agreement between the two systems (Table 3.9). The authors found that the Analytab system (API) was the most complete commercially available test for the identification of the family Enterobacteriaceae, with an accuracy of nearly 90%. However, the principal disadvantages of the test are the time required to prepare and inoculate the 20 tests (approximately 3 min) and the care required in the tedious task of filling each tube.

Smith et al.[162] tested 366 cultures from different sources by the API system and conventional system for the identification of *Edwardsiella, Klebsiella, Providencia, Salmonella, Shigella, Proteus, Enterobacter, Arizona, Citrobacter, Escherichia,* and *Serratia* species. Overall accuracy of identification with the API system was reported to be 96.4%;[162] of the 13 cultures misidentified, 7 were atypical strains. Their data are presented in Tables 3.10 and 3.11.

The API system was used by Starr et al.[163] to identify anaerobic bacteria. This included tests for neutral red and nitrate reduction; H$_2$S, urease, and indole production; hydrolysis of gelatin and esculin; and fermentation of glucose, mannose, fructose, galactose, mannitol, lactose, sucrose, maltose, salicin, glycerol, xylose, arabinose, and starch. All procedures were conducted in an anaerobic glove box.[164] A total of 104 cultures, including 18 reference strains and 86 diagnostic cultures (*Bacteroides, Clostridium, Fusobacterium, Propionibacterium,* and *Lactobacillus* species), were examined in this study. The authors reported that 91% of the total tests were performed with the two systems in agreement. Greater than 90% agreement between the two systems (conventional and API systems) was obtained with 12 of 17 differential tests compared (Table 3.12). The tests for nitrate reduction and H$_2$S gave the poorest agreement, 77.8 and 80.8% respectively; only 66% of the 86 diagnostic cultures could be presumptively identified with the micromethod system supplemented only with microscopy and colonial characteristics. However, the agreement was increased to 93% by using supplementary tests such as gas chromatography and Ehrlich's reagent instead of Kovac's reagent.

Patho-Tec® Systems**

The Path-Tec system consists of ten reagent-impregnated strips designed to determine bio-

TABLE 3.9

Comparison of Positive Tests in Analytab® and Conventional Systems[a] for the Identification of *Enterobacteriaceae* (128 Strains), *Acromonas* (5 Strains), and *Yersinia enterocolitica* (1 Strain)

Test	No. positive by Analytab			No. positive by conventional		
	24 hr	48 hr	>48 hr	24 hr	48 hr	>48 hr
H₂S	22	2		24	1	1
Citrate	68	8	4	70	8	3
Urea	37			51	3	4
Indole	52			51	1	
Voges-Proskauer	34			43		
Lysine decarboxylase	47	16	5	61	2	1
Arginine dihydrolase	9	17	5	6	19	4
Ornithine decarboxylase	60	4		63	2	1
Deaminase	27			31		
Gelatin	14	12	1	7	5	10
Arabinose	75			79	2	
Rhamnose	73			74	2	
Sucrose	58			58	9	1
Inositol	59			36	14	
Sorbitol	75			83		
Glucose	124			124		
Mannitol	93			96		
o-Nitrophenyl-β-D-galactopyranoside	77	6		83		
Nitrate reduction[b]	102			106	1	

[a]Fermentation tests by conventional means were not performed on melibiose and amygdaline.
[b]111 tests performed in parallel.

From Washington, J. A., II, Yu, P. K. W., and Martin, W. J., *Appl. Microbiol.*, 22, 267, 1971. With permission.

chemical characteristics of the Enterobacteriaceae. This system differs from other kits in that it is designed to enable one to obtain answers within 4 hr and thus identify or generate significant data from organisms on the same day they are isolated. Also, it is available as separate biochemical tests and provides a flexible system which allows the user to select individual tests as indicated. In addition, all substrate and detection reagents (except KOH for the Voges-Proskauer test) are dry on the strips, resulting in prolonged stability and allowing test results to be read immediately after the incubation period is over.

Jeans[165] compared the Patho-Tec strips impregnated with four of the common biochemical reactions with those tests routinely employed in a hospital bacteriology department. All of the work was carried out on Gram-negative bacteria. The cytochrome oxidase and urease

Patho-Tec strips compared favorably with the well-established methods for testing biochemical activity of bacteria. However, Jeans did not find all of the Patho-Tec reactions to be clearly readable. A small number of equivocal results were obtained with the lysine decarboxylase test, and the phenylalanine deaminase test was particularly difficult to assess with weakly reacting strains of *Salmonella*. The author suggests the addition of a few extra reactions to the four tests to supply more complete identification.

Blazevic et al.[166] evaluated ten biochemical test strips in the Patho-Tec system as compared to standard tests for accuracy and efficacy in identification of 193 Gram-negative bacilli (Table 3.13). They reported that the test agreement was 100% for oxidase; and phenylalanine deaminase; 99% for indole, nitrate, and Voges-Proskauer; 98% for malonate; 97% for lysine decarboxylase; 90%

TABLE 3.10

TABLE 3.11

Agreement Between API® System and Conventional Identification

Organism	C/T[a]	% correct
Enterobacter cloacae	25/25	100
Enterobacter hafnia	19/19	100
Edwardsiella	18/18	100
Klebsiella	21/21	100
Proteus mirabilis	16/16	100
Proteus morganii	20/20	100
Proteus vulgaris	11/11	100
Providencia	28/28	100
Salmonella	28/28	100
Shigella	12/12	100
Enterobacter aerogenes	21/22	95.5
Proteus rettgeri	18/19	94.7
Arizona	27/29	93.1
Escherichia coli	26/28	92.9
Citrobacter	21/23	91.3
Enterobacter liquefaciens	19/21	90.5
Serratia	23/26	88.5
Average		96.4

[a]Number correct per number tested.

From Smith, P. B., Tomfohrde, K. M., Rhoden, D. L., and Balows, A., *Appl. Microbiol.*, 24, 449, 1972. With permission.

Comparative Test Results with the API® and Conventional Systems

Test	A/T[a]	Agreement (%)
Glucose	366/366	100
Sorbitol	365/366	99.7
Phenylalanine	365/366	99.7
Sucrose	364/366	99.5
Ornithine	363/366	99.2
Mannitol	362/366	98.9
Arginine	361/366	98.6
Rhamnose	361/366	98.6
Lysine	358/366	97.8
Indole	357/366	97.5
Arabinose	355/366	97.0
H_2S	350/366	95.6
Gelatin	345/366	94.3
Inositol	344/366	93.8
Melibiose	338/366	92.3
Acetoin	338/366	92.3
Citrate	337/366	91.2
Urea	331/366	90.4
Average		96.5

[a]Number of results in agreement per number tested.

From Smith, P. B., Tomfohrde, K. M., Rhoden, D. L., and Balows, A., *Appl. Microbiol.*, 24, 449, 1972. With permission.

for urease; 84% for H_2S; and 75% for esculin hydrolysis. Most of the commonly isolated Enterobacteriaceae were identified correctly within 4 hr. Although the ten Patho-Tec tests do not allow for identification of all organisms and additional tests were indicated for some of the strains tested (e.g., *Proteus morganii* and *P. rettgeri*), the decreased time necessary for identifying common isolates is definitely an advantage in the laboratory.

Auxotab®*

The enteric 1 card consists of ten capillary units, each containing a specific biochemical test, a viability control (which is resagurin reduction), and tests for malonate utilization, phenylalanine deaminase, hydrogen sulfide production, sucrose fermentation, *O*-nitrophenyl-β-D-galactopyranoside, lysine decarboxylase, ornithine decarboxylase, urease, and tryptophan. Washington et al.[167] evaluated the Auxotab Enteric 1 system with 160 freshly isolated and

stock cultures of Enterobacteriaceae. They reported that the use of this technique was laborious and a potential hazard to those working with it. The percentage of correct identification was 83.8% at the species level and 90% at the generic level. Rhoden et al.[168] tested the validity of the system with 417 stock cultures instead of fresh isolates. In double-blind studies with the Auxotab, 87% of the strains tested were correctly identified (Table 3.14). The authors concluded that there is a need for modification of the Auxotab system with regard to ease of handling, time required for use, and accuracy of identification of enteric bacteria.

Fermentation of Sugars

Bacteriologic oxidations may be either aerobic or anaerobic. If the hydrogen acceptor is molecular oxygen, the reaction is aerobic. Anaerobic oxidation-reduction reactions are called fermentations. Many organic compounds can be fermented.

*Colab Laboratories, Inc., division of Wilson Pharmaceutical and Chemical Corporation, Glenwood, Ill.

TABLE 3.12

Results of Conventional and Micromethod System Tests Performed on 104 Strains of Anaerobic Bacteria[a]

Substrate or test	No. of tests				No. of tests in agreement	Tests in agreement (%)
	Micro + Conv −	Micro − Conv +	Micro + Conv +	Micro − Conv −		
Glucose	1	1	91	5	96	92.3
Mannose	3	5	63	33	96	92.3
Mannitol	1	1	13	89	102	98.1
Lactose	2	5	73	24	97	93.3
Sucrose	1	1	33	63	96	92.3
Maltose	2	4	69	29	98	94.2
Salicin	3	5	27	69	96	92.3
Glycerol	5	5	16	78	94	90.3
Xylose	1	1	36	66	102	98.1
Arabinose	1	0	14	89	103	99.0
Starch	17	1	39	47	86	82.7
H₂S	9	11	25	59	84	80.8
Esculin	5	9	57	33	90	86.6
Urea	1	0	1	102	103	99.0
Indole	5	7	14	78	92	88.4
Nitrate	22	1	28	53	81	77.8
Gelatin	3	5	25	71	96	92.3

[a] A total of 1,768 tests were performed; of these, 1,612 (91.2%) were in agreement.

From Starr, S. E., Thompson, F. S., Dowell, V. R., Jr., and Balows, A., *Appl. Microbiol.*, 25, 713, 1973. With permission.

TABLE 3.13

Accuracy of Identification Using the Patho-Tec® "Rapid I-D System"

Organism	No. correct/ no. tested	Correct (%)
Arizona	9/9	100
Citrobacter diversus	10/10	100
Edwardsiella	7/7	100
Enterobacter aerogenes	10/10	100
Escherichia coli	10/10	100
Klebsiella	10/10	100
Proteus mirabilis	10/10	100
Shigella	10/10	100
Yersinia enterocolitica	2/2	100
Proteus vulgaris	9/10	90
Citrobacter freundii	7/10[a]	70
Salmonella	6/10[a]	60
Providencia	4/8[a]	50
Enterobacter cloacae	4/10[a]	40
Proteus rettgeri	1/10	10
Proteus morganii	0/10	0
Enterobacter[b]	?/7	Additional tests indicated
Serratia	?/10	Additional tests indicated
Acinetobacter	?/10	Additional tests indicated
Pseudomonas aeruginosa	?/10	Additional tests indicated
Pseudomonas maltophilia	?/10	Additional tests indicated

[a] Additional tests indicated for some strains.
[b] Four *E. liquefaciens*, one *E. hafnia*, and two *E. agglomerans*.

From Blazevic, D. J., Schreckenberger, P. C., and Matsen, J. M., *Appl. Microbiol.*, 26, 889, 1973. With permission.

TABLE 3.14

Accuracy of Identification with the Auxotab® System

Organism	No. correct/ no. tested	Correct (%)
Citrobacter	23/23	100
Klebsiella	29/29	100
Providencia	28/28	100
P. vulgaris	11/11	100
Salmonella	28/28	100
Shigella	19/19	100
Escherichia coli	27/28	96.4
Proteus morganii	19/20	95.0
Enterobacter hafnia	26/29	89.7
Proteus mirabilis	23/26	88.5
Edwardsiella tarda	14/17	82.3
Proteus rettgeri	18/22	81.8
Arizona	23/29	79.3
Enterobacter cloacae	21/29	72.4
Serratia, Enterobacter liquefaciens, E. aerogenes	54/79	68.4
Total	363/417	Average 87.1%

From Rhoden, D. L., Tomfohrde, K. M., Smith, P. B., and Balows, A., *Appl. Microbiol.*, 25, 284, 1973. With permission.

Since carbohydrates are important nutrients for most bacteria, simple sugars can serve as the main source of energy for many kinds of microorganisms. The value of fermentation of carbohydrates by bacteria lies in the fact that the energy-rich bonds (e.g., ATP) generated in the process are utilized by the organisms for their biosynthetic processes. The end product of fermentation depends on the substrate, the enzyme present, and the conditions under which the reaction proceeds. The most common products of bacterial fermentations are lactic acid, formic acid, acetic acid, butyric acid, butyl alcohol, acetone, ethyl alcohol, and gases such as carbon dioxide and hydrogen. Fermentation reactions, while varying among species of bacteria, are of great constancy and therefore of extreme value in differentiating species of bacteria from their specific and characteristic action on a given sugar. Bacteria are classified as (1) those that do not ferment carbohydrates, (2) those that ferment carbohydrates with production of acid only, and (3) those that ferment carbohydrates with production of both acid and gas. In conventional methods, many carbohydrates are used in routine fermentation studies. Important sugars are glucose, lactose, maltose, sucrose, xylose, mannitol, and salicin (see Chapter 2). Using properties of bacterial fermentations, newer methods of differentiation have been reported. Many investigators have studied the action of plague and allied organisms on sugars and other carbohydrate substances and have utilized the fermentation reaction for differentiation. Several modifications in the preparation of medium have been suggested. Seal[169] reported a simplified new medium containing 1% peptone, 1% sugar, and 0.2 mg/ml of dried proteolysed liver at pH 7.4. For fermentation tests with plague and pseudotuberculosis organisms, it gave an efficiency similar to that of a peptone-water-sugar medium containing 2% rabbit serum. A solid glycerol agar medium was used to obtain a differential fermentation reaction between plague and pseudotuberculosis organisms. The new medium was found to be equally efficient in testing the fermentation reaction with diphtheria and diphtheroid organisms. The simplified medium could be used in the field work under limited laboratory facilities, and the same medium is equally suitable for the fermentation testing and differentiation of diphtheria organisms.

Baird-Parker[170] described a medium of glucose fermentation for testing Gram-positive cocci. On the basis of the ability of the organism to metabolize glucose under aerobic or anaerobic condition, classified cocci were classified under the genera *Staphylococcus* and *Micrococcus*. However, the clinical isolates cannot be classified on the basis of the fermentation properties of the bacteria alone. Chalmers[171] tested the media of Baird-Parker and the Subcommittee on Taxonomy of Staphylococci and Micrococci[172] and found that bromcresol purple had a marked inhibitory effect on the colony size and that the recommended concentration of 0.004% affected both the colony size and acid production. It was also noted that the fermentation tubes sealed with liquid paraffin did not remain anaerobic; therefore, the liquid paraffin was replaced with a plug of solid paraffin. When these precautions were taken, unequivocal results were obtained (*Micrococcus* produced acid in the oxidation tube only while *Staphylococcus albus* produced acid throughout the length of both tubes, exhibiting oxidation and fermentation of glucose).

Kellogg and Turner[173] developed a rapid sugar fermentation procedure for the confirmation of *Neiserria gonorrhoeae* from either primary isolation media or purification media. A lightly buffered salt solution with pH indicators and sugars was heavily inoculated with presumptively positive growth. By using the cells taken either directly from presumptively positive growth on TM medium or from secondary isolation plates, a rapid (1 to 2 hr) fermentation procedure was applied to the confirmation of *N. gonorrhoeae*. Proper fermentation patterns were obtained in 1 to 4 hr with no interference from inhibited contaminants or variation in results due to differing growth requirements of gonococcal strains (Table 3.15). Taylor and Keys[174] reported a simplified sugar fermentation technique incorporating glucose, maltose, lactose, and sucrose plates for confirming large numbers of *N. gonorrhoeae* cultures. Two fermentation plates were used, a triple sugar agar plate containing 0.5% each of maltose, lactose, and sugar and a single sugar agar plate containing 1% glucose. The medium CTA (BBL) was modified to contain a total of 1.78% agar and 1.0% vancomycin-colistimetrate-nystatin (VCN) inhibitor (BBL). A phenol red indicator was included in the CTA base medium. By using this technique for 279 of 288 presumptive positive cultures, the authors found that *Neisseria* showed typical fermentation patterns in both the glucose and combined maltose, lactose, and sucrose plates after incubation at 35°C in a candle jar. However, a heavy inoculum from a 24- to 48-hr actively growing primary culture was necessary to give satisfactory results on the sugar plates in 18 to 24 hr.

Enzymes

Many enzymes are apparently present in all bacteria, although there may be some differences in the physical and chemical structure of the enzyme molecules in different bacterial species. Measurement of enzyme activities is routine, used in bacteriology to characterize various species of organisms. The specific enzymes used for differentiation of bacteria include phenylalanine deaminase, cytochrome oxidase, lysine decarboxylase, catalase, coagulase, ornithine decarboxylase, and certain phosphatases. Reagent systems for identification of diagnostic enzymes are commercially available in the form of impregnated easy-to-use strips of paper suitable for application to bacterial cultures. Many combinations of enzyme tests packaged together (e.g., phenylalanine deaminase, cytochrome oxidase, and lysine decarboxylase) are also available.

A new indigogenic method for detecting the production of bacterial acid phosphatase to differentiate bacteria was described by Wolf et al.[175] Bacterial phosphatase enzyme hydrolyzes an indolyl substrate, with the formation of a highly insoluble blue-green indigo which is deposited on the bacterial growth. The test differentiates pathogenic from nonpathogenic staphylococci by demonstrating that pathogenic staphylococci produce acid phosphatase. However, differentiation of *Serratia* from the other members in the *Klebsiella-Enterobacter-Serratia* group of organisms is difficult because of cultural similarities on blood and eosine methylene. Wolf et al.[175] reported a simple, rapid, direct new indigogenic method for detecting production of bacterial acid phosphatase formation; *Serratia* produces the blue-green indigo color within 5 min, whereas *Enterobacter* produces the blue-green color in ½ hr. A deoxyribonuclease test may also be used for identification of *Serratia marcescens*.[176] Kedzia[177] proposed the acid phosphatase test for the detection of pathogenic staphylococci. However, at that time few data were available linking the presence of acid phosphatase with potential pathogenic staphylococci. Kocka et al.[178] evaluated the enzyme test with coagulase-positive and coagulase-negative staphylococci (Table 3.16). In order to elucidate the possible relationship of acid phosphatase ac-

TABLE 3.15

Rapid Fermentation Patterns of *Neisseria gonorrhoeae* Isolates on Thayer-Martin Medium

Organisms	Isolates tested (no.)	Glucose	Maltose	Fructose
N. gonorrhoeae	220	+	–	–
N. meningitidis	8	+	+	–
Group B	5	+	+	–
Group C	3	+	+	–
Slaterus X	5	+	+	–
Slaterus Y	1	+	+	–
Slaterus Z	7	+	+	–
Untypable	7	+	+	–
Rough	7	+	+	–
N. lactamicus	3	+	+	+
N. flava	5	+	+	+
N. perflava	3	+	+	–
N. subflava	3	+	+	+
N. sicca	3	–	–	–
N. flavascens	3	–	–	–
N. catarrhalis	4	–	–	–

Note: Tubes were incubated a maximum of 4 hr in a 35 to 36°C water bath.

From Kellogg, D. S., Jr. and Turner, E. M., *Appl. Microbiol.*, 25, 550, 1973. With permission.

TABLE 3.16

Acid Phosphatase Activity of Staphylococci

Optical density at 404 nm	Number of cultures	
	Coagulase positive	Coagulase negative
>1.0	10	0
0.9	6	0
0.8	7	0
0.7	9	0
0.6	5	0
0.5	9[a]	1[b]
0.4	6	2[b]
0.3	5[a]	6
0.2	0	8
0.1	0	10
<0.1	0	20[b]
Total	57	47

[a] Contains mannitol-negative cultures with optical density of 0.34, 0.32, 0.52, and 0.55.

[b] Contains mannitol-positive cultures with optical density of 0.51, 0.43, 0.03, and 0.06.

From Kocka, F. E., Magoc, T., and Searcy, R. L., *Am. J. Med. Technol.*, 39, 269, 1973. With permission. Copyright by the American Society for Medical Technology.

tivity with pathogenicity, it was compared with other tests such as coagulase, mannitol fermentation, hemolysis, DNAse, and pigment production. All coagulase-positive staphylococci isolated from 104 clinical specimens produced acid phosphatase. However, a few coagulase-negative isolates also elaborated this enzyme but to a lesser degree. Therefore, the acid phosphatase test may serve as an adjunct for the identification of *Staphylococcus aureus*.

Mitz[179] utilized the properties of enzyme activities in bacteria for rapid microbial detection purposes and selected two assay systems that are rapid and have high "turnover numbers" (molecules substrate decomposed/bacterium/minute) with their respective systems. The phosphatase assay system was selected (the activity is used as an index of bacterial action in milk). The assay depends on hydrolysis of *p*-nitrophenyl phosphate, with release of *p*-nitrophenol measured spectrophotometrically; alternatively, a fluorescent organic phosphate can be used, releasing a highly fluorescent product that can be measured fluorimetrically. Use of the latter substrate allowed detection of 10^7 bacteria within a few minutes. The second useful enzyme is bacterial esterase(s), found in many bacterial species. The assay of this enzyme depends on hydrolysis of phenyl acetate, with release of acetic acid which can be measured by automatic titration with standard potassium hydroxide to maintain the pH value of the reaction mixture at 7.4. The minimum number of bacteria detectable with the esterase assay varied from 10^3 to 10^6.

Rapid identification of *Proteus* and *Providencia* species by the examination of phenylalanine deaminase and β-D-galactosidase activities was reported by Closs and Digranes.[180] They studied 169 strains of *Proteus* and *Providencia* by using a simplified two-step identification procedure. In the first step, strains of the *Proteus-Providencia* group were identified on the basis of 1-hr tests for phenylalanine deaminase and β-D-galactosidase activity.[181,182] In the second step, differentiation between the four *Proteus* species and *Providencia* was based on H₂S production, urease activity, fermentation of maltose, and ornithine decarboxylase activity (HUMO) (Table 3.17). Identification based on the HUMO scheme accorded well with the final identification based on 19 characters. The HUMO scheme was recommended by the authors for use in routine diagnostic work.

TABLE 3.17

Reaction Pattern of 109 Strains of *Proteus* and *Providencia* in the HUMO Scheme

H	U	M	O	Number of strains
+	+	−	+	62 *Proteus mirabilis*
−	+	−	+	34 *Proteus morganii*, 2 *Proteus mirabilis*
+	+	+	−	13 *Proteus vulgaris*
−	+	+	−	15 *Proteus vulgaris*
−	+	−	−	23 *Proteus rettgeri*
−	−	−	−	20 *Providencia*

Note: +, positive reaction; −, negative reaction; H, hydrogen sulfide production; U, urease activity; M, fermentation of maltose; O, ornithine decarboxylase activity.

From Closs, O. and Digranes, A., *Acta Pathol. Microbiol. Scand. Sect. B*, 81, 684, 1973. With permission.

Levin et al.[183,184] described the specific gelation of amoebocyte lysate from *Linulus polyphemus* (horseshoe crab) in the presence of minute amounts of endotoxin elaborated by Gramnegative bacteria. This test was found to be clinically valuable in the detection of Gramnegative bacterial infection in various body fluids. Nachum et al.[185] evaluated the use of the *Linulus* lysate test in 112 patients as an aid in the early diagnosis and treatment of Gram-negative bacterial meningitis. Positive assays were obtained on all initial cerebrospinal fluid specimens from 38 patients with culture proven Gram-negative bacterial meningitis, whereas Gram stain demonstrated organisms in 25 of these specimens. Also, negative *Linulus* assays were obtained on all samples from 74 patients with Gram-positive bacterial meningitis, tuberculous meningitis, aseptic meningitis, or meningeal leukemia as well as patients without meningitis. The authors concluded that the *Linulus* assay is a rapid, sensitive test for the diagnosis of untreated Gram-negative bacterial meningitis and may be useful for the detection of persistent bacterial growth or residual endotoxin in the absence of other spinal fluid abnormalities.

Zolg and Ottow[186] used thin-layer chromatography (TLC) for detecting hippurate hydrolase activity for the differentiation of *Pseudomonas*, *Bacillus*, and Enterobacteriaceae. An acidified fraction of culture medium was extracted with chloroform, and the solvent layer containing ben-

TABLE 3.18

Detection of Hippurate Hydrolase by Identification of Benzoic Acid

Organisms tested	Benzoic acid shown by UV viewing (Rf = 0.15)	Spray number 1, 2, 3, 4, or 5
Pseudomonas mendocina ATCC 25411	–	–
P. stutzeri ATCC 17588	–	–
P. stanieri ATCC 17591	–	–
P. testosteroni ATCC 11996	–	–
P. fluorescens CCM 2115	–	–
P. aeruginosa CCM 1960	–	–
P. saccharophila CCM 1980	+	+
Pseudomonas sp. (No. 15, 16, 17, 19)	+	+
Enterobacter aerogenes	–	–
Serratia marcescens	–	–
Proteus vulgaris	+	+
Escherichia coli B	–	–
Bacillus brevis SMG 298	+	+
B. macerans ATCC 8244	–	–
B. psychrosaccharolyticus ATCC 23296	–	–
B. cereus var. *mycoides* CCM 48	–	–
B. firmus CCM 37	+	+
B. lentus CCM 35	–	–
B. pumilus ATCC 7061	+	+

Note: ATCC, American Type Culture Collection; SMG, Sammlung für Mikroorganismen, Göttingen; CCM, Czechoslovak Collection of Microorganisms, Brno.

From Zolg, W. and Ottow, J. C. G., *Experientia*, 29, 1573, 1973. With permission.

zoic acid was dried and separated by cellulose TLC aluminum sheets without fluorescence sheets or silica gel sheets with fluorescence indicator. The solvent system used was *n*-hexane-acetic acid (96:4). The fluorescent-treated plates were observed under UV light for deep purple benzoic acid spots. The cellulose sheets were sprayed either with *p*-diaminobenzaldehyde-acetic anhydride (red-orange spots with hippuric acid) or other color reagents such as $FeSo_4/H_2O_2/MnSO_4$ reagent. A number of bacterial strains in *Pseudomonas, Enterobacter, Serratia, Proteus, Escherichia*, and *Bacillus* were detected based on the identification of benzoic acid produced by the hippurate hydrolase enzyme in the organism (Table 3.18).

METHODS BASED ON IMMUNOLOGICAL AND SEROLOGICAL TESTS

Fluorescent Techniques

Since the introduction of fluorescent antibody techniques by Coons et al.,[187] the greatest appli-cation has been toward the detection and identifi-cation of pathogens in infection.

Direct or Indirect Fluorescent Antibody Staining Method

The simplest use of fluorescent techniques would be to apply a battery of fluorescent techniques directly to films of clinical material. Various methods of direct examination by the fluorescent antibody method have been described by Goldman,[188] Ellis and Harrington,[189] and Cherry and Thomason.[190]

The fluorescent antibody (FA) test uses an FA reagent which is a polyvalent conjugate of bac-terial antigenic groups (e.g., fluorescein-labeled *Salmonella* polyvalent O antiserum containing antibodies to O antigens A to G). An appropriate dilution of the conjugate in a buffered saline solution is used to stain all smears. The FA reagents are available commercially. Slides should be cleaned with 95% ethanol and scrubbed dry with a lintless towel. The cleaned slides are spotted with glycerol by using a syringe fitted with an 18-gauge needle which has the head cut off. The slides are sprayed with fluoroglide* (film bonding grade), allowed to dry for a few minutes, and then rinsed well in running tap water to remove all traces of glycerol. They are blotted gently to remove most of the water and allowed to air-dry. Each slide should be flamed well before using. The direct FA test is performed on smears from cultures grown in specific medium (e.g., from 24-hr tetrathionate with brilliant green (TET) enrichment broth cultures for *Salmonella* in foods and feed samples). A 2-mm loopful of each culture is placed in a well on the coated slides. After air-drying, the slides are fixed for 2 min in a fixative containing 60 parts of absolute ethanol, 30 parts of chloroform, and 10 parts of formaldehyde solution. The fixed slides are then rinsed briefly in 95% ethanol and allowed to drain dry. A small amount of conjugate is placed on each smear, and the slides are covered for 30 min with a large petri dish fitted with a moist piece of filter paper. Excess conjugate should be drained off on a paper towel. The slides are rinsed briefly in a bath of phosphate-buffered saline (PBS) solution and placed in fresh PBS for 10 min. Then they are dipped in distilled water to remove the salt, and cover glasses are mounted with buffered glycerol. Smears are examined under the 95X oil immersion objective of a microscope (e.g., Leitz Ortholus, Nikon, etc.) with fluorescence assembly. Appropriate light source and filters should be used in the microscopic examinations.

For diagnostic purposes, Garner and Robson[191] applied a variant of the fluorescent treponemal antibody (ABS) test to smears from genital sores. The direct fluorescent staining techniques may work well in the reference laboratory if the investigator is considering sonnei dysentery or coli serotypes and has a good idea what he is seeking. However, there is a problem of cross reactivity in many instances, e.g., a meningiococcus can be stained with a monospecific antigonococcal serum. The same serum stains contain gonococci equally well. However, it is not unreasonable that gonococci and meningiococci should show cross fluorescence even with carefully prepared sera. Another example of cross fluorescence is the reaction of neisserial antisera with staphylococci.

In diagnostic bacteriology the use of the fluorescent staining technique is not widely used at the present time because of its cross reactivity, lack of sensitivity, and lack of distinct advantages over the conventional methods such as Gram staining. However, in food microbiology and public health microbiology the value of the direct fluorescent staining technique has been demonstrated by many workers in the field. Thomason[192] reported a microcolony FA staining procedure and the direct FA procedure for detecting *Salmonellae* in 304 environmental, food, and feed samples. The author reported that the microcolony FA test detected all of the specimens found positive by culture, whereas the direct FA test missed 3.1% of them (Table 3.19). The specificity of the direct FA test was found to be similar to that of the microcolony FA test. The environmental samples consisted of surface swabs and both fresh- and salt-water samples. The food specimens examined were fresh fish and pork. The feed specimens were fish meal, bone meal, and pelleted pet foods. By using FA tests, *Salmonella* cells were detected from 86 of the 134 environmental samples, 30 of the 80 food samples, and 44 of the 90 feed samples. Using the direct FA tests, Smyster and Snoeyenbos[193] examined a number of samples of meat, bone meal, and fish meal. *Salmonella* were detected from 454 of the 550 samples examined by the FA technique and were recovered from 342 samples.

The applications of FA staining in the detection of *Salmonella* have been reported in meat,[194,195] eggs,[196] and other foods[197-201] (Table 3.20). They have also been applied for the detection of organisms in animal feedstuffs,[202-204] and stool samples during outbreaks of typhoid.[205-207] They have been used successfully in the diagnosis of pathogenic serotypes of *Escherichia coli* which have been associated with outbreaks of infantile diarrhea.[208-210] Abshire and Guthrie[211] reported an FA method for detection of fecal coliform pollution in water. They compared the results from the rapid FA method for detection of fecal coliform pollution with the EC broth method and with standard IMVIC typing; they found that potential fecal types were detectable by the FA method, while a few may be missed by the EC methods (Table 3.21). The authors concluded that the FA method for water quality testing offers the advantages of rapidity and accuracy over conventional methods.

*Chemoplast, Inc., Wayne, N.J.

TABLE 3.19

Comparison of Results Obtained by Direct FA and Microcolony FA with Cultural Tests for *Salmonella*

Specimen	No. examined	Test	Results FA positive/ culture positive	FA positive/ culture negative
Environmental	134	Direct FA	85/86	3/48
		Microcolony FA	86/86	1/48
Food	80	Direct FA	27/30	6/50
		Microcolony FA	30/30	13/50
Feed	90	Direct FA	43/44	3/46
		Microcolony FA	44/44	3/46
Total	304	Direct FA	155/160 (96.9)[a]	12/144
		Microcolony FA	160/160 (100.0)	17/144

[a]Values in parentheses are expressed as percentages.

From Thomason, B. M., *Appl. Microbiol.*, 22, 1064, 1971. With permission.

The indirect FA staining technique uses fluorescein-labeled goat antirabbit antisera and poly O antigens of the organisms to be labeled. Gibbs et al.,[212] using *Salmonella* poly O and fluorescein-labeled goat antirabbit antisera for the indirect FA test, found heterologus fluorescence with six species of enterobacteria. This was eliminated by absorption with heat-killed cells. In contrast, they reported that the combination of *Salmonella* poly O and fluorescein-labeled sheep antirabbit antisera did not stain these six species unequivocally, suggesting that the goat antirabbit fluorescent serum contained antibodies to these organisms. Baird-Parker[213] reported that by using the sheep antirabbit serum, two strains of staphylococci were isolated which fluoresced strongly and could have given false-positive FA results. These were identified as belonging to *Staphylococcus* Groups II and III. This heterologous fluorescence was eliminated by absorption of both the *Salmonella* poly O antiserum and the fluorescein-labeled sheep antirabbit immunoglobulin with heat-killed cells.[214] Gibbs et al.[215] reported that the heterologous fluorescence could be eliminated by absorption of both the *Salmonella* poly O antiserum and the fluorescein-labeled sheep antirabbit immunoglobulin with heat-killed cells and that absorption with both strains was necessary; the *Salmonella* poly O antiserum contained the heterologous antibodies. The fluorescein-labeled *Salmonella* poly O antiserum did not stain either the six species of enterobacteria previously found to stain in the

indirect FA test or the two strains of staphylococci isolated in their studies. The authors concluded that no single combination of enrichment broth and FA test gives unequivocal results in the screening of animal fecal material for salmonellae, but the staining of smears from tetrathionate broth by FA (direct or indirect) gives rise to a high percentage of false negative results (Table 3.22).

The applications of the FA technique for the identification of mycobacteria have been reported.[216-219] Martins et al.[220] obtained type-specific antisera to 11 strains of mycobacteria by injecting rabbits with cells killed by UV light. The antisera were conjugated with fluorescein antibody. The resulting antisera produced extensive cross reaction (Table 3.23). However, the type-specific antisera were produced only with the mycobacteria killed with UV light before they were injected into rabbits. The authors concluded that the specific antisera provided a rapid identification method for mycobacteria within a given serogroup. By using similar techniques, Gales et al.[221] produced multivalent fluorescent antisera for identification of organisms in the *Mycobacterium axium-M. intracellulare* complex and found that these provided a rapid identification method for the mycobacteria. No cross-reactivity problem was found in their studies.

Immunofluorescence Technique

The immunofluorescence technique (IFT) involves the use of immunoglobulin (e.g., IgG)

115

TABLE 3.20

Comparison of *Salmonella* Detection Methods for Food Samples and Ingredients[a]

Sample type	No. tested	Modified LICNR broth presumptive results		Cultural[b] procedure results		FA results[c]	
		+	−	+	−	+	−
Butter mix	4	1	3	0	4	1	3
Biben meat	13	13	0	13	0	13	0
Cheddar cheese[d]	2	1	1	1	1	1	1
Chicken breasts	4	2	2	2	2	−	−[e]
Chicken drumsticks	2	2	0	2	0	−	−
Chicken gizzards	3	3	0	3	0	−	−
Chicken livers	10	10	0	10	0	10	0
	14	12	2	12	2	−	−
Cocoa	1	0	1	0	1	−	−
Corn	2	0	2	0	2	−	−
Feed materials	18	8	10	7	11	−	−
	7	5	2	3	4	5	2
Hydrolysate	2	0	2	0	2	−	−
NFDM[d]	2	1	1	1	1	1	1
Noodles	5	5	0	5	0	5	0
Pet food mix	2	0	2	0	2	−	−
Pork sausage	2	0	2	0	2	−	−
Slants	2	2	0	2	0	2	0
Swabs (peptone water)	5	3	2	3	2	3	2
Swabs (tet broth)	9	0	9	0	9	−	−
Turkey hearts	1	0	1	0	1	−	−

[a]Subtotals for samples tested: no. tested, 50; (presumptive method) positive, 41; negative, 9; percent positive, 82; (cultural method) positive, 38, negative, 12; percent positive, 76; (FA method) positive, 41; negative, 9; percent positive, 82. Totals for samples tested: no. tested, 110; (presumptive method) positive, 68; negative, 42; percent positive, 62%; (cultural method) positive, 64; negative, 46; percent positive, 58.
[b]Biochemical and serological tests.
[c]Positive or negative fluorescence.
[d]One "spiked" with two to five *Salmonella* per gram (*S. typhimurium*).
[e]Sample not tested.

From Schulte, S. J., Witzeman, J. S., and Hall, W. M., *J. Assoc. Off. Anal. Chem.*, 51, 1334, 1968. With permission.

TABLE 3.21

Staining Results of Isolates

IMVIC types	EC reactions[a]	FA titers pooled conjugate[b]
Wastewater isolates		
++--	+(30)	4+(23); 3+(7)
(33 isolates examined)	-(3)	3+(1); 2+(2)
-+-+	+(10)	4+(7); 3+(3)
(14 isolates examined)	-(4)	4+(1); 2+(2); 1+(1)
--++	+(2)	-(1); 2+(1)
(10 isolates examined)	-(8)	-(5); 2+(2); 3+(1)
Unpolluted water		
--+-	+(2)	-(1); 2+(1)
(12 isolates examined)	-(10)	-(7); 1+(1); 2+(1); 3+(1)
Unpolluted soil		
--++	+(2)	-(2)
(10 isolates examined)	-(8)	-(6); 1+(1); 3+(1)
---+	+(0)	
(4 isolates examined)	-(4)	-(4)

[a]Positive reactions were those that exhibited gas production in EC broth in 24 hr at 45°C. (-) No fluorescence demonstrated. Numbers in parentheses represent number of strains tested or reacting.

[b]Pooled conjugate consisted of equal volumes of the ten antisera containing a protein concentration of 2 mg/ml. Pool was adsorbed with *Enterobacter aerogenes* and *Klebsiella pneumoniae* before testing.

From Abshire, R. L. and Guthrie, R. K., *Can. J. Microbiol.*, 19, 201, 1973. Reproduced by permission of the National Research Council of Canada.

TABLE 3.22

Direct and Indirect FAT in the Detection of Salmonellae

Sample	Staining method	No. of samples tested	No. of results by FAT			No. of samples positive by cultural methods
			+	False +	False -	
Meat products	Indirect	7	0	0	0	0
(steakburgers, chicken patties and roll, cooked, frozen hen meat)	Direct	14	0	0	0	0
Feedingstuffs	Indirect	31	11	4	0	7
(meat and bone, blood and poultry offal meals)	Direct	105	15	9	1	7
Broiler swabs	Indirect	30	1	1	0	0
	Direct	67	3	3	0	0
Giblets	Indirect	64	15	11	4	8
	Direct	117	21	10	2	13
Poultry plant swabs	Indirect	95	6	4	3	5
	Direct	59	13	9	1	5

From Gibbs, P. A., Patterson, J. T., and Murray, J. G., *J. Appl. Bacteriol.*, 35, 405, 1972. With permission.

TABLE 3.23

Cross Reactivity of Mycobacterial Antiserum Conjugates

Conjugates

Cells	M. tuberculosis	M. kansasii (Forbes)	M. kansasii (Brownell)	M. marinum	M. scrofulaceum	M. gordonae	M. intracellulare (Boone)	M. intracellulare (Howell)	M. terrae	M. smegmatis	M. fortuitum
M. tuberculosis	+	−	−	−	−	−	−	−	−	−	−
M. kansasii (Forbes)	−	+	+	−	−	−	−	−	−	−	−
M. kansasii (Brownell)	−	+	+	−	−	−	−	−	−	−	−
M. marinum	−	−	−	+	−	−	−	−	−	−	−
M. scrofulaceum	−	−	−	−	+	−	−	−	−	−	−
M. gordonae	−	−	−	−	−	+	−	−	−	−	−
M. intracellulare (Boone)	−	−	−	−	−	−	+	+	−	−	−
M. intracellulare (Howell)	−	−	−	−	−	−	−	+	−	−	−
M. terrae	+	+	+	+	+	+	+	+	+	+	+
M. smegmatis	−	−	−	−	−	−	−	−	−	+	−
M. fortuitum	−	−	−	−	−	−	−	−	−	−	+

From Martins, R. R., Walker, W. E., Batayias, G. E., and Gales, P. W., *Am. Rev. Respir. Dis.*, 108, 979, 1973. With permission.

which is conjugated with fluorescein isothiocyanate compound. Experimental animals (rabbits) are immunized by injection of a particular organism suspended (about 6 cells/ml) in 0.15 M NaCl and 5% formalin (v/v) every third to fourth day for 4 weeks. The animals are bled 7 days after the last injection. At least two or three rabbits should be immunized for each strain of bacteria to be studied. Rabbit immunoglobulin G (IgG) is prepared from sera by precipitation with ammonium sulfate at 37% saturation. The precipitate is dissolved in distilled water and the procedure is repeated twice. Then the precipitated and redissolved material is dialyzed against 0.0175 M Na_2HPO_4 (pH 6.3), further purified by chromatographic methods (e.g., diethylaminoethyl cellulose column), and finally concentrated (e.g., using ultrafiltration cell, etc.) to a protein concentration of about 20 mg/ml. Rabbit IgG is then conjugated by adding fluorescein isothiocyanate (FITC) (30 μg of dye per milligram of protein) under vigorous stirring at room temperature for 1 hr (pH 9.5). Nonreactive FITC is removed by passing the solution through a Sephadex® G-25* column equilibrated previously with phosphate-buffered saline (PBS) solution. Methods of preparation of FITC-labeled specific antibody fragment have been described by Forsum[222] and Nisonoff.[223] Absorption of antisera is performed with an equal volume of conjugated antiserum and packed living cells from a 48-hr culture of the heterologous, cross-reacting organisms at 37°C for 2 hr.

Smears of clinical specimens for immunofluorescent staining are made from a suspension of organisms or tissue material, etc. in FTA-hemagglutination buffer (BBL) and prereduced salt solution by streaking 0.01 ml of the suspension on a slide. Tissue sections of 4-μm thickness cut in a cryostat chamber are transferred to slides and fixed in acetone for 10 min. Bacterial cultures (24 to 48 hr growth) and freshly drawn human blood are smeared on slides by methods similar to those used for other staining procedures. The smears are stained by placing one drop of conjugate on a smear followed by incubating the slide for 20 min at 37°C. The smear is then washed with PBS for 15 min and mounted under a cover glass with phosphate-buffered glycerin (pH 7.2). The preparations are read under a suitable microscope with specific field, filters, and fluorescent devices. Usually, the specific staining of microorganisms is recorded as 3+, intensely fluorescent margins, well-marked edges; 2+, faintly fluorescent margins, edges usually diffuse; or 1+, barely distinguishable fluorescent margins, diffuse edges.

McCracken and Mauney[224] applied the technique of immunofluorescence to specimens obtained from 310 patients, 77 of whom were clinically diagnosed as having diphtheria. Commercial antisera prepared against the somatic antigens of *Corynebacterium diphtheriae* were used in these studies. Immunofluorescence of slides prepared directly from swabs were unsatisfactory, but when the swabs were subjected to prior incubation in a growth medium, the results of immunofluorescence and bacterial culture agreed in 95% of the specimens (Table 3.24). The authors concluded that immunofluorescence methods were useful and economical in the rapid identification of *C. diphtheriae* in large numbers of clinical specimens during an epidemic. Holmberg and Forsum[225] standardized the fractionated fluorescein-isothiocyanate (FITC) conjugated immunoglobulin G (dye-to-protein ratio < 10), produced against whole cells of *Actinomyces, Arachnia, Bacterionema, Rothia,* and *Propionibacterium* species which gave species specific conjugates with controlled nonspecific staining reactions when appropriately diluted on the basis of their antibody content (10 mg/ml). Serotype-specific conjugates were also available after dilution for all serotypes of these organisms except *Actinomyces viscosus* type 2 and *Propionibacterium acnes* type 1. The authors found that adequately adsorbed conjugates could be used to differentiate these serotypes from *A. viscosus* type 1 and *P. acnes* type 2, respectively. A serological classification in defined immunofluorescence corresponded to species and serotype designation proposed on the basis of other serological analysis and biochemical characteristics. The detection of certain *Actinomyces* of the family *Actinomycetaceae* and *Propionibacterium* species by defined immunofluorescence in direct smears prepared from clinical specimens agreed to 88% with parallel culturing when including a prereduced (PRAS) medium technique for isolation.

Reamer and Hargrove[226] developed a one-step selective and enrichment medium of lysine iron for

*Pharmacia AB, Sweden.

TABLE 3.24

Comparison of Results of Bacterial Culture and Immunofluorescence Following Incubation of Swabs

Group no.	Diagnosis	Positive fluorescence positive culture	Negative fluorescence negative culture	Negative fluorescence positive culture	Positive fluorescence negative culture	Totals
1	Clinical diphtheria	53 (68.8%)	12 (15.6%)	8[1] (10.4%)	4 (5.2%)	77 (100%)
2	Diphtheria contacts	1 (0.8%)	114 (98.4%)	0	1 (0.8%)	116 (100%)
3	Treated diphtheria	1 (2.2%)	45 (97.8%)	0	0	46 (100%)
4	Pharyngitis	0	71 (100%)	0	0	71 (100%)
	Totals	55 (17.7%)	242 (78%)	8[a] (2.6%)	5 (1.7%)	310 (100%)

[a]Includes one nontoxinogenic strain.

From McCracken, A. W. and Mauney, C. U., *J. Clin. Pathol.*, 24, 641, 1971. With permission.

the detection of salmonellae in nonfat dry milk by the FA technique. One hundred grams of nonfat dry milk was mixed aseptically with 1500 ml of the selective medium and incubated at 30°C for 5 hr and at 39°C for 17 hr. A sample (0.05 ml) was put on a slide, air-dried, and fixed in a Haglund's solution (60:30:10 absolute alcohol:chloroform:formalin mixture) for 30 sec, touched off on absorbent paper, and transferred to absolute alcohol for 30 sec before air-drying. Each smear was stained with polyvalent antiserum at a 1:2 dilution and incubated for 30 min. The incubation period was followed by a rinse in normal saline, 10 min in a phosphate-buffered saline, and a rinse in distilled water. The air-dried slides were mounted in a buffered glycerol FA mounting fluid, covered with a cover glass, and examined under a microscope. Many strains of salmonellae were tested by this procedure, and it was found to be sensitive enough to detect one *Salmonella* bacterium in 100 g of nonfat dry milk (Table 3.25).

Immunofluorescence-Membrane Filtration Technique

In this technique samples are filtered through a nonfluorescent black Multipore® membrane filter, and circular pieces (12.5 mm diameter) are cut out with a metal die. The bacteria are stained with a FA for 45 to 60 min, washed, mounted, and examined with a fluorescence microscope. The method may be quantitiated by using the formula

$$x = NR^2/20\, r^2$$

where

N = number of bacteria per 20 fields;
R = 20 or 40 mm, depending on the diameter of the membrane surfaces used;
r = diameter of field of vision (0.32 mm), with 40-mm diameter filtering surface;
x = 781.25 N (with 20-mm diameter filtering surface x = 195.31 N).

By this calculation, the concentration of bacteria specifically reacting with antibody can be determined. The technique easily allows quantitative determination of bacteria present in a minimum concentration of 10^5 per liter; however, with 10^3 bacteria per liter, the method is time consum-

TABLE 3.25

Applicability of Procedure to 100-g Samples of Nonfat Dry Milk Contaminated with Various Bacteria[a]

Sample contamination with[b]	Medium color	Fluorescent antibody
Uncontaminated nonfat dry milk	Red	−
Escherichia coli	Red	−
Salmonella tennessee	Black	+
Citrobacter freundii	Red	−
Proteus sp.	Red	−
Salmonella new-brunswick	Black	+
S. seftenberg	Black	+
S. montevideo	Black	+
S. cubana	Black	+
S. pullorum stock	Red	+
S. pullorum 2083-66	Black	+
S. pullorum 17-64	Red orange	+
S. pullorum 64/64	Black	+
S. paratyphi A (WMN)	Red	+
S. paratyphi (ATCC 9283)	Red	+

[a]Six separate tests were performed with identical results.
[b]Level of salmonellae contamination in these samples ranged from 1 in 50 to 1 in 100 g.

From Reamer, R. H. and Hargrove, R. E., *Appl. Microbiol.*, 23, 78, 1972. With permission.

ing.[227a] Danielsson and Laurell[227b] described the application of this method to the detection of small numbers of bacteria in water. When bacteria separated from a sample by filtration through a membrane filter were eluted by a simple washing procedure with glass beads or a magnetic stirrer, the lower limit of sensitivity for direct fluorescence microscopy was about 5000 bacteria per liter of water. When this technique was combined with an enrichment procedure, it was possible to demonstrate 2 to 50 bacteria per liter of water within 4 to 6 hr.

Closs[228] reported modification of the above method to detect *Haemophilus influenzae*, *Proteus rettgeri*, and *Pasteurella haemolytica*. Bacteria were collected on membrane filters (black Sartorius, MF 50), and cultured on a filter soaked with growth medium containing FA conjugate. Specifically stained microcolonies developed within 2 to 4 hr and were detected with the fluorescence microscope. Guthrie and Reeder[229] reported a further modification of the membrane filter-fluorescence technique. Water was filtered through a Millipore® membrane filter (HABG 047) that was then cultured on Trypticase soy agar at 35°C. Colonies appeared after 5 hr but incubation was continued for 12 hr. The membrane was overlayed for 5 min at 20°C with 1 to 2 ml of pooled normal rabbit serum. The serum was removed by suction, and the membrane was overlayed for 5 to 20 min with fluorescent rabbit antiserum. The antiserum was sucked through the membrane and washed with 10 to 15 ml of phosphate-buffered saline. The stained membrane was overlayed with mounting fluid and examined with a dissecting microscope (magnification × 10) using visible light to determine the total count and uv radiation to determine fluorescent colonies. The authors found that the lower magnification, although time consuming, can count all colonies on the filter.

Danielsson and Zieminska[230] reported the use of immunofluorescence (IF) for microbiological analysis of water for sanitary-epidemiological purposes. Using tap water and river water contaminated by addition of varying numbers of enteropathogenic strains of *Escherichia coli* 026 and 0111, they tested a combined technique of membrane filters and immunofluorescence (FM-IF) with application of monovalent rabbit antisera to *E. coli* 026 and 0111 labeled with fluorescein isothiocyanate (FIIC). The experi-

ments included determining the recovery of bacteria using white membrane filters and investigating the feasibility of trapping fluorescent microcolonies on black, nonfluorescent membrane filters (BMF). They report that combination of IF with the BMF technique is useful for detection of *E. coli* in water even if there is heavy contamination with accompanying microflora. Stained microcolonies were produced with bacteria cultured on BMF for 5 to 7 hr in the presence of labeled specific immune serum. The method of choice depended on the number of sought bacteria. For bacteria in excess of 10^4, direct staining on black nonfluorescent membrane filters is recommended. With the BMF-IF technique, results are obtained after about 90 min with samples of highly contaminated ($+10^4$ cells), otherwise 5.5 to 8.5 hr is needed.

The rapid detection of a small number of airborne bacteria with a membrane filter-fluorescent antibody method has been described by Jost and Fey.[231] Calibrated air samples from aerosols of *Serratia marcescens* were drawn through a Millipore aerosol holder (10 l/min) fitted with a nonfluorescent membrane filter, and the organisms were rapidly identified by the methods described before,[227] using a high-power incident light uv microscope.

Technical difficulties in immunofluorescent tracing have greatly hindered its widespread application. Geck[232] developed new serodiagnostic method using an India ink immunoreaction for the rapid detection of enteric pathogens. Smears are treated with India ink and immune serum for 5 min and are then read under a light microscope. Homologous binding of agglutinins (positive reaction) is indicated by the presence of a definite black contour corresponding in shape and size to the organism examined; the inside of the cell is ash-grey in color. The black contour is analogous to the fluorescent halo of the cell seen in immunofluorescent staining. In the absence of agglutinin-binding (negative result), bacteria are not visible; only remains and aggregates of India ink particles are seen, but these are easy to distinguish from bacteria showing homologous binding. Diagnostic sera frequently contain natural antibodies against cocci and aerobic and anaerobic spore-bearing bacteria. Because of the differences in morphology and size, specific binding with these organisms can easily be distinguished from specific reactions of Gram-negative enteric bacteria. The method was found practically equivalent in sensitivity with cultivation and in specificity with immunofluorescent tracing. Table 3.26 shows the results obtained by Geck[232] with the India ink immunoreaction for the identification of *Escherichia coli, Shigella sonnei,* and *Salmonella* strains. Geck claimed that this method is especially useful for replacing immunofluorescent examinations, as it unites the advantages of direct, indirect, and modified indirect staining.

Radioactive Antibody Technique

In addition to the fluorescent dyes as labels for antibody proteins, radioactive isotopes of iodine have been used. A sensitive assay based on bacterial uptake of radioactively labeled antibody has been reported by Hales and Randle[233] and Miles and Hales.[234] Strange et al.[235] described an assay based on the use of ^{125}I-labeled antibodies for the rapid specific detection and determination of small numbers of bacteria in aqueous suspension. A sample was treated with ^{125}I-labeled purified homologous antibody and filtered and washed on a Millipore membrane filter; the radioactivity of the separated, labeled, immune complex was measured. Although the method was found to be highly sensitive for detecting bacteria (minimum number about 500 organisms) and rapid (completed within 8 to 10 min), the sensitivity of assay depended on the quality, reaction concentration, and specific radioactivity of the ^{125}I-labeled antibody and the level and reproducibility of the blank value in the absence of bacteria. Sensitivity and accuracy decreased when the assay was applied to samples that contained particulate matter which nonspecifically attached antibody and was retained by a membrane filter. By using the specific immunological assay described above, only one species of organism may be detected and estimated at a time. However, by using a radioactively labeled homologous mixture of several type-specific antibodies, many homologous bacteria may be determined simultaneously. Strange et al.[235] applied this method to the detection and identification of vegetative cells of *Escherichia coli* and cells and spores of *Bacillus subtilis*. However, the determination of a few bacteria may be possible by this method only if, after separation of the labeled immune complex, the variation in the level of nonspecifically attached radioactivity is less than the radioactivity of the immune complex.

Fluorescent Spectrophotometry

In attempts to improve the reproducibility of standard immunofluorescent techniques, emphasis has been placed on standardization of reagents, methodology, and instruments.[236-238] Because of the molecular similarities between different groups of bacteria, the application of instrumental techniques for their identification is difficult. Fluorescent spectrophotometry has been applied to the differentiation of the bacteria based on distinguishing the differences in bonding among the different morphological forms of the same protein mix in the bacterial cells. It is an extremely sensitive device allowing for detection at very low levels and simultaneously responding to minor

TABLE 3.26

Model Experiments for Comparing the Effectiveness of Cultural Method and India Ink Immunoreaction

| Test organism | Cultivation | India ink immunoreaction | | Total |
		Positive	Negative	
Pathogenic *Escherichia coli* strains,	Positive	555 (80.4%)	8 (1.2%)	563 (81.6%)
10^3 to 10^9 cells/ml	Negative	72 (10.5%)	55 (7.9%)	127 (18.4%)
Total		627 (90.8%)	63 (9.2%)	690 (100.0%)
Shigella flexneri and *S.*	Positive	219 (78.2%)	6 (2.2%)	225 (80.4%)
sonnei, 10^3 to 10^9 cells/ml	Negative	28 (10.0%)	27 (9.6%)	55 (19.6%)
Total		247 (88.2%)	33 (11.8%)	280 (100.0%)
Salmonella strains,	Positive	233 (63.7%)	18 (5.1%)	241 (68.8%)
10^3 to 10^9 cells/ml	Negative	16 (4.6%)	93 (26.6%)	109 (31.2%)
Total		239 (68.3%)	111 (31.7%)	350 (100.0%)

From Geck, P., *Acta Microbiol. Acad. Sci. Hung.,* 18, 191, 1971. With permission.

changes in molecular bonding. Ginell and Feuchtbaum[239] reported the application of fluorescent spectrometry to the identification of *Escherichia coli, Sarcina lutea, Salmonella binza,* and *Proteus mirabilis.* Using three parameters (emission wavelength, excitation wavelength and intensity in the fluorescent spectrometry technique), topographic maps characteristic of the fluorescence of the chemical bonds of the organisms were produced. Further improvements in the technique may be possible by using an oscilloscope screen and dyes such as acridine orange.[240]

Immunoelectrophoresis

Immunoelectrophoresis or counterimmunoelectrophoresis (CIE) techniques have been reported to have the speed, specificity, and sensitivity for detection of antigen in serum or other clinical specimens of patients infected with *Neisseria meningitidis, Diplococcus pneumoniae, Haemophilus influenzae, Streptococcus* sp., *Mycobacteria,* and certain fungi or viruses.[241-247] CIE may be performed on a microscope slide (25 X 75 mm) coated with 1% agarose in 0.015 *M* barbital buffer (pH 8.6). Two slides each containing 12 pairs of diffusion wells (3 mm in diameter and 2 mm apart) are placed in an electrophoresis apparatus containing suitable buffer (e.g., 0.075 M barbital buffer, pH 8.6) in each reservoir. Specific bacterial antiserum is placed in the anode well, and the patient's serum or other clinical specimen is placed in the cathode well of each pair. The electric circuit is completed with filter paper wicks. A constant voltage is then applied for a certain length of time, after which the slide may be examined for a precipitin line between the wells. The concentration of antigen in body fluids may be determined by comparison with known concentrations of purified antigens of the organisms.

Fossieck et al.[248] used counter immunoelectrophoresis to test cerebrospinal fluids from patients with meningitis due to *Diplococcus pneumoniae* for the presence of pneumococcal polysaccharide. They determined the sensitivity of the CIE method of antisera for detection of antigen by reacting omniserum (a polyvalent antiserum containing antibodies to 82 different pneumococcal polysaccharide types) and monospecific antiserum with purified Type 3 pneumococcal polysaccharide in concentrations ranging from 100 to 0.01 µg/ml. They also tested

CSF and sera from patients either known to be free of bacterial infection or with infection due to microorganisms other than pneumococci by reacting the sera with various pneumococcal antisera in order to confirm the specificity of the CIE procedure. They found that as little as 0.05 µg/ml of Type 3 pneumococcal polysaccharide could be detected by CIE with monospecific antiserum to Type 3 or omniserum. A marked variation was found in the concentration and persistence in the CSF of pneumococcal polysaccharide in different patients before and after institution of antimicrobial therapy. Variations in persistence of antigen in CSF, urine, and serum were determined by the CIE method. The authors concluded that CIE is very useful as an aid in the rapid and specific diagnosis of pneumococcal meningitis in patients.

Ingram et al.[249] examined the specificity and sensitivity of CIE in detecting the capsular antigen (polyribose phosphate) in the body fluids of patients with *Haemophilus influenzae* type b disease. The concentration of antigen in body fluids was determined by comparison with known concentrations of polyribose phosphate diluted in normal body fluids of the type being tested. Supernates of overnight broth cultures were used as the test fluid for testing cross reactions with various bacterial species. As little as 1.5×10^2 pg of polyribiphosphate could be detected using the specific antisera with the CIE method. Certain pneumococcal polysaccharides (Types 6, 15, 29, and 35) and the capsular polysaccharide of certain *Escherichia coli* showed cross reactivity with anti-*H. influenzae* b serum. The CIE technique can be used to detect the capsular antigen of *H. influenzae* in body fluids of patients with systemic disease caused by that organism, but the antiserum employed must be tested for specificity and sensitivity before it is used. Bradshaw et al.[250] demonstrated that the cross-reacting antibody with the capsular polysaccharide of *E. coli* (EPS) can be eliminated by EPS absorption.

A rapid procedure for grouping A, B, C, D, G, and H streptococci with countercurrent immunoelectrophoresis was described by Wadström et al.[251] The group-specific carbohydrates were extracted with acid or formadride and treated with pronase. Grouping was performed with commercial antisera for these six groups, and the specificity and sensitivity of the CIE method were compared with those of the gel immunodiffusion (ID) method. The immunoprecipitates developed in the CIE method were up to 20 times more sensitive than those developed in ID or capillary precipitin tests. Also, the CIE method was found to be suitable for handling many samples at one time. The method is sensitive enough for routine groupings of streptococcal isolates in a diagnostic clinical laboratory. However, preparation of more specific antisera with higher titers of antibody against the group antigen are required to make the test more accurate and simple for routine groupings of many samples.

Rapid identification of β-hemolytic streptococci as to specific group was achieved by Dajani[246] using the CIE method (Table 3.27). No cross reaction occurred among the various groups and no false positive results were noted in his studies. The author also concluded that the CIE method is objective, inexpensive, rapid, and specific for the identification of β hemolytic streptococci in throat cultures from infected patients.

Coonrod and Rytel[252] evaluated the efficacy of a rapid and sensitive immunoprecipitin CIE for the detection of pneumococcal capsular antigen (PCA) in serum and urine of patients with pneumococcal pneumonia. In the 30 cases of pneumococcal pneumonia tested by the CIE method, PCA was detected in serum, untreated urine, or an ethanol precipitated fraction of urine (20-fold concentration in 20, 30, and 47% of the cases, respectively [Table 3.28]). Blood cultures were positive in only one half of the cases having demonstrable PCA in concentrated urine samples. The CIE method was found to be useful for the specific typing of PCA of Types 1, 3, 6, 8, 19, 29, 31, and 36, and the PCA persisted for days in body fluids regardless of antibody therapy. The authors also found that PCA was absent by the CIE test in serum or urine samples from 12 patients with pulmonary infection due to organisms other than pneumococci and in samples from 10 hospitalized patients without pulmonary infection. This study shows that the detection of PCA by the CIE method is useful for establishing an etiologic diagnosis of pneumococcal pneumonia.

Wright and Roberts[253] modified a two-dimensional immunoelectrophoretic method for differentiation of mycobacterial species and strains. In the two-dimensional immunoelectrophoresis (IEP), a complex of bacterial antigens is subjected to electrophoresis first through an agarose matrix in one direction and then through an antiserum-agarose matrix at right angles to the first direction for better separation of antigens.[254] By this procedure, a clear pattern of mycobacterial antigens may be obtained by using a highly sensitive antiserum. The vaccine was prepared by mixing culture filtrate and cell extracts of various strains of mycobacteria with alum; this was

TABLE 3.27

Comparison of Various Methods to Identify Streptococcal Groups from Original Plates with Group A Antiserum (R 343) and Reaction of Various Streptococcal Groups with Homologous Antisera

Streptococcal group in original specimens[a]	No. of specimens	No. bacitracin positive	No. CIE positive			
			A antiserum	B antiserum	C antiserum	G antiserum
A	85	84	69			
B	31	1	0	17	0	0
C	9	1	0	0	9	0
G	14	3	0	0	0	14
None (normal throat flora)	20		0			

[a]Grouping as determined by the serologic precipitation reaction.

From Dajani, A. S., *J. Immunol.*, 110, 1702, 1973. With permission.

TABLE 3.28

Frequency of Detection of PCA by CIE as Related to Clinical and Bacteriologic
Criteria Applicable for Diagnosis of 30 Cases of Pneumococcal Pneumonia

Diagnostic criteria[a]	No. cases	No. with PCA[b]		% with PCA[c]
		Urine	Serum	
Positive blood culture	11	7	5	63.6
Positive sputum culture, clinical course	7	3	1	42.8
Positive smear of sputum, clinical course	10	3	0	30.0
Clinical course	2	1	0	50.0
Total	30	14	6	46.7

[a]Gram-stained smears of sputa and cultures of sputum and blood were obtained prior to antibiotic therapy for all cases. Cases were classified according to the most definitive criteria applicable.

[b]Serum and a 20-fold concentrate of urine were tested by CIE using antiserum pools A to I.

[c]All cases with PCA detected in serum also had PCA in the urine.

From Coonrod, J. D. and Rytel, M. W., *J. Lab. Clin. Med.*, 81, 778, 1973. With permission.

followed by mixing with centrifugation pellets produced during preparation of the cell extract. The precipitate was diluted with unprecipitated cell extract and culture and injected (i.m.) into the rabbit. A booster injection consisting of unprecipitated cell extract and culture filtrate was given subcutaneously 40 days after the first injection. A sample of blood was drawn from the rabbit 7 to 10 days later, and the serum was tested against culture filtrate and cell extract in a conventional gel-diffusion plate. The development of eight or more precipitate bands in 48 hr indicated that the serum contained antibodies against the antigens. By using microscale homologous mycobacterial cell extract antigen-antiserum reactions, it was possible to differentiate *Mycobacterium tuberculosis* strains H37Ra and H37Rv, *M. bovis* strains BCG, *M. scrofulaceum, M. phlei,* and *M. intracellulare.* On the basis of the precipitin patterns obtained by using a small amount (1.4 ml) of a single high-titer antimycobacterium antiserum, mycobacteria may be differentiated in about 3½ hr with good sensitivity (29 to 60 precipitin curves).

Mycobacterial antigens may also be separated and isolated by continuous flow electrophoresis

(CFE).[247] The separation may be conducted in a self-contained refrigerated CFE apparatus (e.g. Model CP),* using a barbital buffer at pH 8.6 (0.025 ionic strength and a constant current of 100 mA). A culture filtrate is delivered to a cathode sample wick on the filter paper curtain from a refrigerated reservoir at a rate of approximately 0.2 ml/hr; 30 to 50 ml of filtrate is processed continuously for a period of 6 to 10 days during a single separation. The buffer is pumped at a flow setting between 9 to 11 from a refrigerated supply to the upper electrolyte reservoir, and its flow on the side wicks is controlled by adjusting the right and left siphons to a setting of 5. The separated fractions are collected from drip tapes on the curtain directly into the numbered tubes. At the end of a separation, the contents of identical collecting tubes are pooled, dialyzed at 5°C against deionized water to remove the buffer, and lyophilized. The fractions may then be examined for the presence of antigens by reaction with the reference antiserum in an immunodiffusion test or by immunoelectrophoresis.[255] The authors suggest that continuous flow electrophoresis might be useful in the fractionation and isolation of the

*Spinco Division, Beckman Instruments, Inc., Palo Alto, Cal.

polysaccharide and protein components from mycobacterial culture filtrates for their subsequent use in immunoelectrophoretic differentiations.

Gel electrophoresis has been used for the epidemiologic analysis of an outbreak of diphtheria based on historical information from patients and their contacts. *Corynebacterium diphtheriae* is grouped into three types (mitis, gravis, and intermedius) on the basis of colonial form, morphological structure, starch fermentations, hemolytic activity, clinical severity of infection, and serological behavior. Differentiation of *C. diphtheriae* has also been made on the basis of phage type.[256,257] As stated by Gundersen,[258] the only commonly recognized serological types are those of Robinson and Peeney[259] which comprise gravis Types I to V of Hewitt[260] and Tarnowski.[261]

Razin and Rottem[262] and Sacks et al.[263] used polyacrylamide gel electrophoresis for strain identification using whole cells of *Mycoplasma* and Enterobacteriaceae organisms. Larsen et al.[264] evaluated polyacrylamide gel electrophoresis in an effort to distinguish differences within *C. diphtheriae* gravis stock strains, Robinson and Peeney Serotypes I to V (Figure 3.4). Samples of phenol-acetic acid water extracts of cells adjusted to contain 250 μg of protein were electrophoresed by the method of Rottem and Razin[265] as modified by Larsen et al.[264,266] The gravis, mitis, and intermedius strains failed to produce characteristic electrophoretic patterns on Elek's medium, and supplemented HIB-agar plates were finally used. While studying the complex of soluble proteins in the cells, Nikolaeva and Safonov[267] found that alterations in medium composition caused variation in the electrophoretic patterns of strains of *Clostridium perfringens*.

Serological Methods

Serology is the examination of serum for antibodies related to microbial disease; consequently, the presence of antibodies in human serum is often sought as an aid to diagnosis. Several types of antigen-antibody reactions (e.g., agglutination reactions, precipitation reactions, and complement-fixation reactions) have been conventionally used. In the agglutination reaction, visible aggregates of bacteria are found when serum containing antibody is added to a suspen-

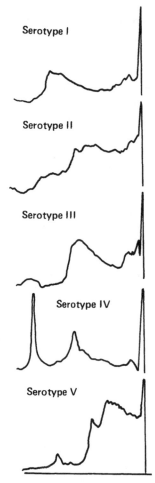

FIGURE 3.4. Densitometric tracings of polyarylamide gels as a means of characterizing *Corynebacterium diphtheriae*, gravis stock strains, Robinson and Peeney Serotypes I to V. (From Larsen, S. A., Bickham, S. T., Buchanan, T. M., and Jones, W. J., *Appl. Microbiol.*, 22, 885, 1971. With permission.)

sion of the organism. Precipitation (or flocculation) reactions show a ring of white precipitate formed when an antigen in solution is added to serum containing antibody. In complement-fixation reactions, complement (a group of substances present in fresh, normal serum) added to a serum-antigen mixture is removed if antibodies that can "fix" complement are present in the serum. Thus, if one can recognize complement in the mixture, then the presence of antibodies in the serum and thus the presence of infection can be determined. Complement is detected by the fact that appropriately sensitized red blood cells are lysed if complement is present. Conventionally, these tests are used as a routine for detection of bacterial infections. For example, the Wasserman reaction is commonly used as a complement-fixation test for the diagnosis of the infection with *Treponema pallidium* (syphilis); an agglutination reaction method is used to examine sera for the antibodies of salmonellae (typhoid and enteric fevers) and brucellae.

By using the principles of agglutination, precipitation, or complement-fixation reactions, newer techniques have been applied for the detection and identification of bacteria in clinical specimens and other biological samples. Kronvall[268] developed a new slide agglutination method for the serologic typing of pneumococci. In this method, when specific antipneumococcal antibodies were added to a stabilized suspension of staphylococci, they become bound to the protein A on the cell wall via the Fc structures of the γ-globulin.[269,270] The combining sites of the antibodies remained available to the corresponding type-specific antigen; when pneumococci were

mixed with a suspension of the stabilized staphylococci coated with the homologous antibody, strong agglutination occurred rapidly. By using this method, 89 strains of pneumococci were typed, which was in agreement with the capsular swelling reaction (Neufeld reaction) method. The main advantages of this new agglutination method are the ease of test, rapidity of results, and its potential wide application in serological work. Christensen et al.[271] applied a reagent staphylococci coated with antibodies against streptococci. Overnight cultures of streptococci were treated with trypsin and then directly tested for coagglutination of selected reagent staphylococci on glass slides. The results of 179 strains of streptococci tested by this method (Table 3.29) agreed with the grouping by Lancefield extracts in the CIE test. The Lancefield method of precipitation test is widely used for the grouping of streptococci; however, the coagglutination method was found to be an accurate, rapid, and simple procedure.

TABLE 3.29

Results of Testing a Panel of Laboratory Strains Against Selected Streptococcal Group-specific Reagent Staphylococci

Serological group of streptococcal strains tested	No. of strains	Coagglutination test using group-specific reagents				
		A	B	C	D	G
A[a]	9	++++	–	–	–	–
B[b]	6	–	++++	–	–	–
C[c]	8	–	–	++++	–	–
D[d]	1	–	–	–	++++	–
G[e]	3	–	–	–	–	++++
L[f]	7	–	–	–	–	–
N[g]	2	–	–	–	–	–

[a]SF 130, T-type 1; Richard, T-type 3; Symons, T-type 9; Blackmore, T-type 11; SF 42, T-type 12; Glover, T-type 13; SF 40, T-type 27; Hensson-Glossy, T-type 44; B 3264, T-type B 3264.
[b]*Streptococcus agalactiae:* S-B 2, S-B 8, S-B 17, S-B 20, S-B 37, S-B 39.
[c]*Streptococcus humanus:* S-CH 1, S-CH 2; *Streptococcus dysgalactiae:* S-C 4, S-C 18, S-C 19, S-C 29, S-C 31; *Streptococcus:* S-C 1.
[d]*Streptococcus:* S-D 1.
[e]*Streptococcus:* S-G 2, S-G 3, S-G 4.
[f]*Streptococcus:* S-L 2, S-L 3, S-L 4, S-L 7, S-L 8, S-L 9, S-L 10.
[g]*Streptococcus:* S-N 1, S-N 2.

From Christensen, P., Kahlmeter, G., Jonsson, S., and Kronvall, G., *Infect. Immun.,* 7, 881, 1973. With permission.

Aho and Sievers[272] screened the sera of 955 patients for neisserial antibodies with complement fixation (CFT) and the immunofluorescence test (IFT) for gonococcal arthritis. The CFT test was performed with one volume of a twofold dilution of the serum under test and one volume of each antigen (a mixture of 20 strains), complement (two full units) and hemolysin, and 2% sheep red cells in a total volume of 0.5 ml. An overnight fixation at 4°C was used before addition of the sensitized cells. Complete inhibition hemolysis in serum dilute 1:4 was considered to be a positive reaction. The results reported by Aho and Sievers[272] (Table 3.30) show that the screening for neisserial antibodies reveals a divergent category of patients in a hospital where the great majority are treated for chronic forms of arthritis. Both IF and CF tests do not properly differentiate the antibody response caused by the antigenically closely related meningococcus.

The most commonly used conventional method for identifying the serologic group of β-hemolytic streptococci is the precipitin test performed in capillary tubes.[273] Lancaster and Sherris[274] used the agar diffusion technique and found it reliable, economical, and easy to read. Rabinowitz[275] and Goldin and Glenn[276] used cellulose acetate membranes as a carrier for the reacting materials. Michael and Massell[277] used the gel double-diffusion technique in agar slides for grouping and typing of β-hemolytic streptococci with un-absorbed rabbit immune antisera. Prakash et al.[278] evaluated a microgel diffusion technique for grouping of β-hemolytic streptococci and compared the sensitivity and specificity of this method with that of the capillary tube method for the grouping of streptococci (Table 3.31). Agar gel diffusion was carried out with agar slides prepared by means of pipetting 3.5 to 4 ml of melted 1% agar (Difco agar) in 0.9% sodium chloride to give a 2- to 3-mm deep layer. Wells (2 mm in diameter were punched, and three sets were made on each slide. One set consisted of five equally spaced holes about 5 to 6 mm from the center well. The center well was filled with antiserum and the peripheral wells with the streptococcal extracts. Commercial antisera of Groups A, B, C, G, F, and O were routinely used in the tests. Slides were kept at 4°C in a moist chamber and read at various intervals. A clearly visible precipitating band

TABLE 3.30

Pertinent Data on Patients with Neisserial Antibodies

Case	Age	Sex	CF titre	IF test	RF test	Clinical data
1	42	F	8	+	−	Stage III RA, frequent urinary infection
2	60	F	8	+	−	Stage III RA, frequent urinary infection
3	42	F	4	+	+	Stage III RA, pyelonephritis
4	68	F	4	+	+	Stage III RA, frequent urinary infection
5	42	F	4	+	−	Stage III juvenile RA, frequent urinary infections
6	24	F	4	+	−	Sequela of an acute arthritis with later recurrence
7	45	F	32	+	−	Sequela of an acute arthritis
8	43	M	64	+	−	Acute arthritis, psoriasis, previous arthritis attacks
9	18	F	8	+	−	Acute arthritis, pyuria, later recurrence of arthritis
10	33	M	8	+	−	Acute arthritis following gonococcal urethritis
11	33	F	16	+	−	Subacute gonitis, possible sacroiliitis
12	48	F	16	+	−	Arthralgia, frequent urinary infections

Note: CF, complement fixation; IF, immunofluorescent; RF, rheumatoid factor; RA, rheumatoid arthritis.

From Aho, K. and Sievers, K., *Scand. J. Rheumatol.*, 1, 84, 1972. With permission.

TABLE 3.31

Time Intervals at which Precipitation Reaction in Capillary and Agar Gel Tests Gave Positive Results

Group of strains	No. of strains	Capillary method, time taken for precipitation			Agar gel diffusion method, time taken for precipitation line to develop		
		5 min	10 min	15–30 min	3–5 hr	16–24 hr	48 hr
A	88	76	9	3	14	64	10
G	28[a]	23	2	–	8	18	2
C	8	6	2	–	1	6	1
F	3	3	–	–	–	3	–
O	2	2	–	–	–	2	–
B	1	–	1	–	1	–	–
NG	8	–	–	–	–	–	–
Total	138	110	14	3	24	93	13
		84.6%	10.8%	2.3%	18.5%	71.5%	10%

[a]Three strains of Group G did not show precipitations by the capillary method.

From Prakash, K., Chawda, S., and Sharma, K. B., *Indian J. Med. Sci.*, 26, 578, 1972. With permission.

between the antigen and antibody wells, which remained readable for at least 5 days, was regarded as a positive result. Based on the tests of 138 strains of streptococci, the authors concluded that the agar gel method was more sensitive, reproducible, economical, reliable, and easier to read; no cross reaction was found. The results were obtained within 16 to 24 hr by this test procedure.

An accelerated procedure of enrichment serology for detecting salmonellae in foods has been described by Sperber and Deibel.[279] This technique involves preenrichment and selective enrichment of the food sample followed by elective enrichment in modified all-purpose Tween (APT) medium and serological detection of salmonellae growing in the latter medium by using precipitin reactions. The preenrichment of the dry food material and animal feed may be made by overnight incubation at 37°C in nutrient broth containing Evans peptone and sodium chloride at pH 7.2. The preenriched dry samples are then transferred to an enrichment medium such as selenite F and incubated at 41°C for 24 to 48 hr for selective enrichment. Alternatively, a commercial brand of tetrathionate broth may be used as the selective enrichment medium. The selective enrichment suppresses the growth of other organisms, for example, *Proteus* species. Boothroyd and Baird-Parker[280] compared the enrichment serology (ES) technique with the conventional method for detecting salmonellae using 2208 samples of raw food materials and products in order to determine the usefulness of the ES technique in the rapid detection of salmonellae. They found that by using the ES technique with a 24-hr elective enrichment step, 93 to 98% of the samples positive by the conventional procedure were also positive by the ES technique (Table 3.32). The ES technique seems to be a relatively simple, specific, and satisfactory method for screening large numbers of food samples for the presence of salmonellae, although a longer time is required by the ES technique compared to the FA technique. However, the ES technique, is similar in sensitivity to the FA test; it is also much simpler to perform and requires less costly equipment and technical expertise to operate.

TABLE 3.32

Comparison of Enrichment Serology (24-hr Elective Enrichment) with Conventional Plating Methods

	Enrichment (41°C) in broth							
	Selenite					Tetrathionate (Difco) and novobiocin		
	No. of salmonella detected					No. of samples with salmonella detected		
		Enrichment serology	Conventional				Enrichment serology	Conventional	
Food sample	No. of samples tested	24 hr[a]	24 hr[b]	48 hr	No. of samples tested	24 hr	24 hr	48 hr
Raw chicken	35	8	8	0	—[c]	—	—	—
Raw meats	39	0	0	0	88	37	38	0
Miscellaneous food products	—	—	—	—	41	0	0	0
Total	74	8	8	0	129	37	38	0

[a] Enrichment serology with 24-hr elective enrichment.
[b] Selective enrichment followed by plating on brilliant green agar after 24 or 48 hr.
[c] Not tested.

From Boothroyd, M. and Baird-Parker, A. C., *J. Appl. Bacteriol.*, 36, 165, 1973. With permission.

REFERENCES

1. Kubitschek, H. E., in *Methods in Microbiology*, Vol. 1, Norris, J. R. and Ribbons, D. W., Eds., Academic Press, London, 1969, 593.
2. Counting and Sizing Microorganisms with the Coulter Counter, Coulter Electronics, Inc., Hialeah, Fla.
3. Anderson, G. E. and Whitehead, J. A., Viable cell and electronic particle count, *J. Appl. Bacteriol.*, 36, 353, 1973.
3a. Youden, W. J., Ranking laboratories by Round-Robin tests, *Mater. Res. Stand.*, 3, 9, 1963.
4. Truant, J. P., Bret, W. A., and Merckel, K. E., Application of an electronic counter in the evaluation of significant bacteriuria, *He ıry Ford Hosp. Med. Bull.*, 10, 359, 1962.
5. Kniseley, S. H. and Throop, L. J., Electronic particle counting applied to the determination of lytic activity, *Anal. Biochem.*, 13, 417, 1965.
6. Garrett, E. R. and Wright, O. K., Kinetics and mechanisms of action of drugs on microorganisms, *J. Pharm. Sci.*, 56, 1576, 1967.
7. Drake, J. F. and Tsuchiya, H. M., Differential counting in mixed cultures with Coulter counter, *Appl. Microbiol.*, 26, 9, 1973.
8. Shuler, M. L., Aris, R., and Tsuchiya, H. M., Hydrodynamic focusing and electronic cell-sizing techniques, *Appl. Microbiol.*, 24, 384, 1972.
9. Harvey, R. J. and Marr, A. G., Measurement of size distributions of bacterial cells, *J. Bacteriol.*, 92, 805, 1966.
10. Tsuchiya, H. M., Drake, J. F., Jost, J. L., and Fredrickson, A. G., Predator-prey interactions of *Dictyostelium discoideum* and *Escherichia coli* in continuous culture, *J. Bacteriol.*, 110, 1147, 1972.
11. Jost, J. L., Drake, J. F., Fredrickson, A. G., and Tsuchiya, H. M., Interactions of *Tetrahymena pyriformis*, *Escherichia coli*, *Azotobacter vinelandii* and glucose in a minimal medium, *J. Bacteriol.*, 113, 834, 1973.
12. Megee, R. D., Drake, J. F., Fredrickson, A. G., and Tsuchiya, H. M., Studies in intermicrobial symbiosis, *Saccharomyces cerniciae* and *Lactobacillus casei*, *Can. J. Microbiol.*, 18, 1733, 1972.
13. Soloway, S. and Louder, N., An instrument for particle sizing and counting in liquids using the forward scattering lob, 18th Annu. Conf. on Analytical Chemistry and Applied Spectroscopy, Pittsburgh, 1967.
14. May, K. R., Multistage liquid impinger, *Bacteriol. Rev.*, 30, 559, 1966.
15. Schuval, H. I., Fattal, B., Cymbalista, S., and Goldblum, N., The phase separation method for the concentration and detection of viruses in water, *Water Res.*, 3, 225, 1969.
16. Schmidt, W. A. K., Die Verteilung von Adenovirus-komponenten im Zweiphasensystem wäBriger polymerer Lösungen, *Z. Naturforsch. Teil B*, 23, 90, 1968.
17. Tiselius A., Porath, J., and Albertsson, P. A., Separation and fractionation of macromolecules and particles, *Science*, 141, 13, 1963.
18. Albertsson, P. A., Particle fractionation in liquid two-phase systems. The composition of some phase systems and the behaviour of some model particles in them. Application to the isolation of cell walls from microorganisms, *Biochim. Biophys. Acta*, 27, 378, 1958.
19. Richmond, D. V. and Fisher, D. J., The electrophoretic mobility of micro-organisms, *Adv. Microb. Physiol.*, 9, 1, 1973.
20. James, A. M., Loveday, D. E. E., and Plummer, D. T., Some physical investigations of the behaviour of bacterial surfaces. XI. The effect of phenol and substituted phenols on the electrophoretic mobility of *Aerobacter aerogenes*, *Biochim. Biophys. Acta*, 79, 351, 1964.
21. James, A. M. and List, C. F., Some physical investigations of the behaviour of bacterial surfaces. XII. The effect of fimbriae on the electrophoretic mobility of some capsular and non-capsular bacteria of the *Coli-Aerogenes* group, *Biochim. Biophys. Acta*, 112, 307, 1966.
22. Cohen, S. S., The chemical alteration of a bacterial surface, with special reference to the agglutination of *Proteus* OX-19, *J. Exp. Med.*, 82, 133, 1945.
23. Gittens, G. J. and James, A. M., Some physical investigations on the behaviour of bacterial surfaces. VI. Chemical modification of surface components, *Biochim. Biophys. Acta*, 66, 237, 1963.
24. Hill, M. J., James, A. M., and Maxted, W. R., Some physical investigations of the behaviour of bacterial surfaces. VIII. Studies on the capsular material of *Streptococcus pyogenes*, *Biochim. Biophys. Acta*, 66, 264, 1963.
25. McQuillen, K., The bacterial surface. II. Effect of uranyl chloride on the electrophoretic mobility of bacteria, *Biochim. Biophys. Acta*, 6, 66, 1950.
26. Forrester, J. A., Dumonde, D. C., and Ambrose, E. J., The effects of antibodies on cells. II. Changes in the electrophoretic mobility of ascites tumour cells treated with antibodies and complement, *Immunology*, 8, 37, 1965.
27. Hill, M. J., James, A. M., and Maxted, W. R., Some physical investigations of the behaviour of bacterial surfaces. IX. Studies on the Streptococcal cell wall, *Biochim. Biophys. Acta*, 75, 402, 1963.
28. Davies, J. T., Haydon, D. A., and Rideal, E., Surface behaviour of *Bacterium coli*. I. The nature of the surface, *Proc. R. Soc. London Ser. B.*, 145, 375, 1956.
29. Barry, P. J. and James, A. M., Physical investigations of the behavior of bacterial surfaces. I. The electrophoretic mobility of *Aerobacter aerogenes*, *J. Chem. Soc.*, p. 3340, 1952.

30. Barry, P. J. and James, A. M., Physical investigations of the behavior of bacterial surfaces. II. The variation of the electrophoretic mobility of *Aerobacter aerogenes* with the age of the culture and nature of the culture medium, *J. Chem. Soc.*, p. 1264, 1953.

31. Plummer, D. T. and James, A. M., Some physical investigations of the behaviour of bacterial surfaces. III. The variation of the electrophoretic mobility and capsule size of *Aerobacter aerogenes* with age, *Biochim. Biophys. Acta*, 53, 453, 1961.

32. Douglas, H. W., Electrophoretic studies on spores and vegetative cells of certain strains of *Bacillus megaterium, Bacillus subtilis* and *Bacillus cereus, J. Appl. Bacteriol.*, 20, 390, 1957.

33. Douglas, H. W., Electrophoretic studies on bacteria. V. Interpretation of the effects of pH and ionic strength on the surface charge borne by *B. subtilis* spores, with some observations on other organisms, *Trans. Faraday Soc.*, 55, 850, 1959.

34. Douglas, H. W. and Parker, F., Electrophoretic studies on bacteria. II. The effect of enzymes on resting spores of *Bacillus megaterium, B. subtilis* and *B. cereus, Biochem. J.*, 68, 94, 1958.

35. Plummer, D. T., James, A. M., Gooder, H., and Maxted, W. R., Some physical investigations of the behaviour of bacterial surfaces. V. The variation of the surface structure of *Streptococcus pyogenes* during growth, *Biochim. Biophys. Acta*, 60, 595, 1962.

36. Hill, M. J., James, A. M., and Maxted, W. R., Some physical investigations of the behaviour of bacterial surfaces. X. The occurrence of lipid in the streptococcal cell wall, *Biochim. Biophys. Acta*, 75, 414, 1963.

37. Hill, M. J., James, A. M., and Maxted, W. R., The use of specific amino acid decarboxylases for the identification of C-terminal groups, *Biochim. Biophys. Acta*, 71, 740, 1963.

38. Norrington, F. E. and James, A. M., The cell-wall lipids of cells of tetracycline-sensitive and tetracycline-resistant strains of *Streptococcus pyogenes, Biochim. Biophys. Acta*, 218, 269, 1970.

39. Hill, M. J., James, A. M., and Maxted, W. R., Application of particle electrophoresis to the detection of antibody bound to cells of *Streptococcus pyogenes, Nature*, 202, 187, 1964.

40. Schott, H. and Young, C. Y., Electrokinetic studies of bacteria. I. Effect of nature, ionic strength, and pH of buffer solutions in electrophoretic mobility of *Streptococcus faecalis* and *Escherichia coli, J. Pharm. Sci.*, 61, 182, 1972.

41. Few, A. V., Gilby, A. R., and Seaman, G. V. F., An electrophoretic study on structural components of *Micrococcus lysodeikticus, Biochim. Biophys. Acta*, 38, 130, 1960.

42. Einolf, C. W., Jr. and Carstensen, E. L., Bacterial conductivity in the determination of surface charge by microelectrophoresis, *Biochim. Biophys. Acta*, 148, 506, 1967.

43. James, A. M. and Brewer, J. E., Non-protein components of the cell surface of *Staphylococcus aureus, Biochem. J.*, 107, 817, 1968.

44. James, A. M. and Brewer, J. E., A protein component of the cell surface of *Staphylococcus aureus, Biochem. J.*, 108, 257, 1968.

45. Adams, D. M. and Rideal, E., The surface behaviour of *Mycobacterium phlei, Trans. Faraday Soc.*, 55, 185, 1959.

46. Dahlberg, A. E., Dingman, C. W., and Peacock, A. C., Electrophoretic characterization of bacterial polyribosomes in agarose-acrylamide composite gels, *J. Mol. Biol.*, 41, 139, 1969.

47. Marshall, K. C., Electrophoretic properties of fast- and slow-growing species of *Rhizobium, Aust. J. Biol. Sci.*, 20, 429, 1967.

48. Marshall, K. C., Interaction between colloidal montmorillonite and cells of *Rhizobium* species with different ionogenic surfaces, *Biochim. Biophys. Acta*, 156, 179, 1968.

49. Marshall, K. C., Orientation of clay particles sorbed on bacteria possessing different ionogenic surfaces, *Biochim. Biophys. Acta*, 193, 472, 1969.

50. Marshall, K. C., Studies by microelectrophoretic and microscopic techniques of the sorption of illite and montmorillonite to Rhizobia, *J. Gen. Microbiol.*, 56, 301, 1969.

51. Lester, K. S. and Boyde, A., The question of von Korff fibres in mammalian dentine, *Calcif. Tissue Res.*, 1, 273, 1968.

52. Friedlander, S., A new method for detecting changes in the surface of Ruman exfoliated cervical cells with the scanning electron microscope, *Acta Cytol.*, 13, 288, 1969.

53. Whittaker, D. K. and Drucker, D. B., Scanning electron microscopy of intact colonies of microorganisms, *J. Bacteriol.*, 104, 902, 1970.

54. Roth, I. L., Scanning electron microscopy of bacterial colonies, *Bacteriol. Proc.*, Abstr. A121, 21, 1972.

55. Drucker, D. B. and Whittaker, D. K., Examination of certain bacterial colonies by scanning electron microscopy, *Microbios*, 4, 109, 1971.

56. Drucker, D. B., Bacterial colonial microstructure, *Microbios*, 6, 29, 1972.

57. Higgins, M. L. and Shockman, G. D., Model for cell wall growth of *Streptococcus faecalis, J. Bacteriol.*, 101, 643, 1970.

58. Torten, M. and Schneider, A. S., Characterization of bacteria by ultraviolet spectroscopy, *J. Infect. Dis.*, 127, 319, 1973.

59. Sharpe, A. N. and Kilsby, D. C., A rapid, inexpensive bacterial count technique using agar droplets, *J. Appl. Bacteriol.*, 34, 435, 1971.

60. Sharpe, A. N., Dyett, E. J., Jackson, A. K., and Kilsby, D. C., Technique and apparatus for rapid and inexpensive enumeration of bacteria, *Appl. Microbiol.*, 24, 4, 1972.

61. Hartman, P. A., *Miniaturized Microbiological Methods*, Academic Press, New York, 1968.

62. Fung, D. Y. C. and Hartman, P. A., Rapid characterization of bacteria, with emphasis on *Staphylococcus aureus*, *Can. J. Microbiol.*, 18, 1623, 1972.

63. Lee, J. S. and Wolfe, G. C., Rapid identification of bacteria in foods: replica plating and computer method, *Food Technol.*, 21, 35, 1967.

64. Bowie, I. S., Loutit, M. W., and Loutit, J. S., Identification of aerobic heterotrophic soil bacteria to generic level by using multipoint inoculation techniques, *Can. J. Microbiol.*, 15, 297, 1969.

65. Nabbut, N. H. and Hakim, A., Quantitative determination of the number of bacteria in urine by dip-slide method, *Leban. Med. J.*, 26, 41, 1973.

66. MacKey, J. P. and Sandys, G. H., Laboratory diagnosis of infections of the urinary tract in general practice by means of a dip-inoculum transport medium, *Br. Med. J.*, 88, 838, 1962.

67. Leigh, D. A. and Williams, J. D., Method for the detection of significant bacteriuria in large groups of patients, *J. Clin. Pathol.*, 17, 498, 1964.

68. Arneil, G. C., McAllister, T. A., and Kay, P., Detection of bacteriuria at room-temperature, *Lancet*, 1, 119, 1970.

69. Mara, D. D., A single medium for the rapid detection of *Escherichia coli* at 44°C, *J. Hyg.*, 71, 783, 1973.

70. Moussa, R. S., Keller, N., Curiat, G., and DeMan, J. C., Comparison of five media for the isolation of coliform organisms from dehydrated and deep frozen foods, *J. Appl. Bacteriol.*, 36, 619, 1973.

71. Donnelly, C. B., Black, L. A., and Lewis, K. H., An evaluation of simplified methods for determining viable counts of raw milk, *J. Milk Food Technol.*, 23, 275, 1960.

72. Tatim, S. R., Dalbah, R., and Olson, J. C., Jr., Comparison of plate loop and agar plate methods for bacteriological examination of manufacturing grade raw milk, *J. Milk Food Technol.*, 30, 112, 1967.

73. Boling, E. A., Blanchard, G. C., and Russell, W. J., Bacterial identification by microcalorimetry, *Nature*, 241, 472, 1973.

74. Forrest, W. W., Microcalorimetry, in *Methods in Microbiology*, Vol. 6B, Norris, J. R. and Ribbons, D. W., Eds., Academic Press, New York, 1972, 285.

75. Fung, D. Y. C. and Miller, R. D., Effect of dyes on bacterial growth, *Appl. Microbiol.*, 25, 793, 1973.

76. Engbaek, H. C., Bennedsen, J., and Larsen, S. O., Comparison of various staining methods for demonstration of tubercle bacilli in sputum by direct microscopy, *Bull. Int. Union Tuberc.*, 42, 94, 1969.

77. Convit, J. and Pinardi, M. E., A simple method for the differentiation of *Mycobacterium leprae* from other mycobacteria through routine staining technics, *Int. J. Leprosy*, 40, 130, 1972.

78. Pital, A., Janowitz, S. L., Hudak, C. E., and Lewis, E. E., Direct fluorescent labeling of microorganisms as a possible life-detection technique, *Appl. Microbiol.*, 14, 119, 1966.

79. Winter, F. H., York, G. K., and El-Nakhal, H., Quick counting method for estimating the number of viable microbes on food and food processing equipment, *Appl. Microbiol.*, 22, 89, 1971.

80. Frost, W. D., A rapid method of counting living bacteria in milk, *J. Ames Med. Assoc.*, 66, 889, 1916.

81. Breed, R. S., The determination of the number of bacteria in milk by direct microscopic examination, *Zentralbl. Bakteriol. Parasitenkd. Infektionskr. Abt. 2*, 30, 337, 1911.

82. Brew, J. D., A comparison of the microscopical method and the plate method of counting bacteria in milk, *Bull. N.Y. Agric. Exp. Sta.*, 373, 1, 1914.

83. Brew, J. D., The comparative accuracy of the direct microscopic and agar plate methods in determining numbers of bacteria in milk, *J. Dairy Sci.*, 12, 304, 1929.

84. Frazier, W. C. and Gneiser, D. F., Short-term membrane filter method for estimation of numbers of bacteria, *J. Milk Food Technol.*, 31, 177, 1968.

85. The Public Health Laboratory Service Standing Committee on the Bacteriological Examination of Water Supplies (1972), Comparison of membrane filtration and multiple tube methods for the enumeration of coliform organisms in water, *J. Hyg.*, 70, 691, 1972.

86. Public Health Laboratory Service Standing Committee on the Bacteriological Examination of Water Supplies (1969), A mineral modified glutamate medium for the enumeration of coliform organisms in water, *J. Hyg.*, 67, 367, 1969.

87. The Public Health Laboratory Service Standing Committee on the Bacteriological Examination of Water Supplies (1968), Comparison of MacConkey broth, Teepol broth, and glutamic acid media for enumeration of coliform organisms in water, *J. Hyg.*, 66, 67, 1968.

88. Walsh, C. and Allwood, M. C., The isolation of microorganisms from creams and ointments by membrane filtration, *Lab. Pract.*, 22, 522, 1973.

89. Tsuji, K., Stapert, E. M., Robertson, J. H., and Waiyaki, P. M., Sterility test method for petrolatum-based ophthalmic ointments, *App. Microbiol.*, 20, 798, 1970.

90. White, M., Bowman, F. W., and Kirshbaum, A., Bacterial contamination in some nonsterile antibiotic drugs, *J. Pharm. Sci.*, 57, 1061, 1968.

91. Buhlmann, X., Method for microbiological testing of nonsterile pharmaceuticals, *Appl. Microbiol.*, 16, 1919, 1968.

92. Pederson, E. A. and Szabo, L., Microbial content in non-sterile pharmaceuticals. II. Methods, *Dan. Tidsskr. Farm.*, 42, 50, 1968.

93. Public Health Service Laboratory Working Party, *Pharm. J.*, 207, 96, 1971.

94. Moore, W. E. C., Cato, E. P., and Holdeman, L. V., Fermentation patterns of some *Clostridium* species, *Int. J. Syst. Bacteriol.*, 16, 383, 1966.

95. Holdeman, L. V. and Moore, W. E. C., Eds., *Anaerobic Laboratory Manual*, The Virginia Polytechnic Institute and State University Laboratory, Blacksburg, 1972.

96. Slifkin, M. and Hercher, H. J., Paper chromatography as an adjunct in the identification of anaerobic bacteria, *Appl. Microbiol.*, 27, 500, 1974.

97. Charles, A. B. and Barrett, F. C., Detection of volatile fatty acids produced by obligate Gram-negative anaerobes, *J. Med. Lab. Technol.*, 20, 266, 1963.

98. Guillaume, J., Beerens, H., and Osteux, R., La chromatographie sur papier des acides aliphatiques volatils de C₁ a C₆. Son application a la détermination des bactéries anáerobies, *Ann. Inst. Pasteur*(Lille),8, 13, 1956.

99. Guillaume, J., Beerens, H., and Osteux, R., Étude des acides volatils aliphatiques de C₁ a C₆ produits par 215 souches de bactéries anáerobies, *Ann Inst. Pasteur* (Lille), 90, 229, 1956.

100. Fink, K. and Fink, R. M., Application of filter paper partition chromatography to qualitative analysis of volatile and non-volatile organic acids, *Proc. Soc. Exp. Biol. Med.*, 70, 654, 1949.

101. Thompson, A. R., Separation of saturated monohydroxamic acids by partition chromatography on paper, *Aust. J. Sci. Res.*, 4, 180, 1950.

102. Seligman, I. M. and Doy, F. A., Thin-layer chromatography of *N,N*-dimethyl-*p*-amino-benzolazophenacyl esters of volatile fatty acids and hydroxy acids, *Anal. Biochem.*, 46, 62, 1972.

103. Hiscox, E. R. and Berridge, N. J., Use of paper partition chromatography in the identification of the volatile fatty acids, *Nature*, 166, 522, 1950.

104. Lindqvist, B. and Storgards, T., Paper chromatographic separation of volatile fatty acids, *Acta Chem. Scand.*, 7, 87, 1953.

105. Roberts, H. R. and Bucek, W., Rapid procedure for separating C₂-C₆ volatile fatty acids by horizontal paper chromatography at elevated temperatures, *Anal. Chem.*, 29, 1447, 1957.

106. Slifkin, M. and Hercher, H. J., Paper chromatography as an adjunct in the identification of anaerobic bacteria, *Appl. Microbiol.*, 27, 500, 1974.

107. Taylor, J. J. and Whitby, J. L., *Pseudomonas pyocyanea* and the arginine dihydrolase system, *J. Clin. Pathol.*, 17, 122, 1964.

108. Williams, G. A., Blazevic, D. J., and Ederer, G. M., Detection of arginine dihydrolase in nonfermentative Gram-negative bacteria by use of thin-layer chromatography, *Appl. Microbiol.*, 22, 135, 1971.

109. Soru, E., Application of circular paper chromatography to the differentiation of bacteria by enzymic tests, *J. Chromatogr.*, 1, 380, 1958.

110. Moller, V., Simplified tests for some amino acid decarboxylases and for the arginine dihydrolase system, *Acta Pathol. Microbiol. Scand.*, 36, 158, 1954.

111. Pickett, M. J. and Pedersen, M. M., Characterization of Saccharoloytic nonfermentative bacteria associated with man, *Am. J. Microbiol.*, 16, 351, 1970.

112. Gilardi, G. L., Characterization of *Pseudomonas* species isolated from clinical specimens, *Appl. Microbiol.*, 21, 414, 1971.

113. Stanier, R. Y., Palleroni, N. J., and Doudoroff, M., The aerobic pseudomonads: a taxonomic study, *J. Gen. Microbiol.*, 43, 159, 1966.

114. Jenkins, P. A., Marks, J., and Schaefer, W. B., Thin-layer chromatography of mycobacterial lipids as an aid to classification: The Scotochromogenic Mycobacteria, including *Mycobacterium scrofulaceum, M. xenopi, M. aquae, M. gordonae, M. flavescens, Tubercle*, 53, 118, 1972.

115. Marks, J., Jenkins, P. A., and Schaefer, W. B., Thin-layer chromatography of mycobacterial lipids as an aid to classification: Technical improvements: *Mycobacterium avium, M. intracellulare* (*Battey bacilli*), *Tubercle*, 52, 219, 1971.

116. Szulga, T., Jenkins, P. A., and Marks, J., Thin-layer chromatography of mycobacterial lipids as an aid to classification: *Mycobacterium kansasii* and *Mycobacterium marinum* (*balnei*), *Tubercle*, 47, 130, 1966.

117. Tsukamura, M., Mizuno, S., and Tsukamura, S., Numerical classification of slowly growing mycobacteria, *Am. Rev. Respir. Dis.*, 99, 299, 1969.

118. Moore, W. E. C., in *Anaerobe Laboratory Manual*, Holdeman, L. V. and Moore, W. E. C., Eds., Virginia Polytechnic Institute and State University, Blacksburg, 1972.

119. Altenbern, R. A., Differentiation of *Escherichia, Enterobacter,* and *Klebsiella* by thin-layer chromatography, *Proc. Soc. Exp. Biol. Med.*, 144, 551, 1973.

120. Bleiweis, A. S., Reeves, H. C., and Ajl, S. J., Rapid separation of some common intermediates of microbial metabolism by thin-layer chromatography, *Anal. Biochem.*, 20, 335, 1967.

121. Oyama, V. I., Use of gas chromatography for the detection of life on Mars, *Nature*, 200, 1058, 1963.

122. Reiner, E., Identification of bacterial strains by pyrolysis-gas-liquid chromatography, *Nature*, 206, 1272, 1965.

123. Mitz, M. A., The detection of bacteria and viruses in liquids, *Ann. N.Y. Acad. Sci.*, 158, 651, 1969.

124. Neufeld, H. A., Conklin, C. J., and Towner, R. D., Chemiluminescence of luminol in the presence of hematin compounds, *Anal. Biochem.*, 12, 303, 1965.

125. **Levin, G. V., Clendenning, J. R., Chappelle, E. W., Heim, A. H., and Rocek, E.**, A rapid method for detection of microorganisms by ATP assay; its possible application in virus and cancer studies, *BioScience,* 14, 37, 1964.

126. **D'Eustachio, A. J., Johnson, D. R., and Levin, G. V.**, Rapid assay of bacterial populations, *Bacteriol. Proc.,* Abstr. A74, 13, 1968.

127. **Chappelle, E. W. and Levin, G. V.**, Use of the firefly bioluminescent reaction for rapid detection and counting of bacteria, *Biochem. Med.,* 2, 41, 1968.

128. **Technical Bulletin, E. I.** duPont de Nemours and Co., Instrument Products Division, Wilmington, Del. 19898.

129. **Neufeld, H. A.**, The Rapid Detection of Microorganisms in the Atmosphere, Meet. Inst. Food Technology, Chicago, May 1969.

130. **Levin, G. V., Harrison, V. R., Hess, W. C., and Gurney, H. C.**, A radioisotope technic for the rapid detection of coliform organisms, *Am. J. Public Health,* 46, 1405, 1956.

131. **Heim, A. H., Curtin, J. A., and Levin, G. V.**, Determination of antimicrobial activity by a radioisotope method, *Antimicrob. Agents Annu.,* p. 123, 1960.

132. **Levin, G. V.**, *Extraterrestrial Life Detection with Isotopes and Some Aerospace Applications. Radioisotopes for Aerospace, Part 2, Systems and Applications,* Plenum Press, New York, 1966.

133. **Levin, G. V. and Heim, A. H.**, Gulliver and Diogenes-exobiology antithesis. Life sciences and space research. III, *5th Int. Space Science Symp.,* North-Holland, Amsterdam, 1964.

134. **Levin, G. V., Heim, A. H., Clendenning, J. R., and Thompson, M.-F.**, "Gulliver" – a quest for life on Mars, *Science,* 138, 114, 1962.

135. **Levin, G. V., Heim, A. H., Thompson, M.-F., Beem, D. R., and Horowitz, M. H.**, An experiment for extraterrestrial life detection and analysis. Life sciences and space research. II, *4th Int. Space Science Symp.,* North-Holland, Amsterdam, 1963.

136a. **Levin, G. V. and Perez, G. R.**, Life detection by means of metabolic experiments. The search for extraterrestrial life, in *Advances in the Astronautical Sciences Series,* Vol. 22, American Astronautical Society, Tarzana, Cal., 1967, 223.

136b. **MacLeod, R. A., Light, M., White, L. A., and Currie, J. F.**, Sensitive rapid detection method for viable bacterial cells, *Appl. Microbiol.,* 14, 979, 1966.

136c. **MacLeod, R. A., White, L. A., and Currie, J. F.**, Detection of *Aerobacter aerogenes* by labelling with radioactive phosphorus, *Appl. Microbiol.,* 19, 701, 1970.

137. **Previte, J. J.**, Radiometric detection of some food-borne bacteria, *Appl. Microbiol.,* 24, 535, 1972.

138. **Adelman, S. L., Brewer, A. K., Hoerman, K. C., and Sanborn, W.**, Differential identification of microorganisms by analysis of phosphorescent decay, *Nature,* 213, 718, 1967.

139. **Schafer, W. J., Anderson, R. E., Morck, R. A., and Cassidy, W. E.**, Use of reagent tablets for rapid biochemical identification of salmonellae and other enteric bacteria, *Appl. Microbiol.,* 16, 1629, 1968.

140. **Smith, R. F., Jorgensen, J. H., Bettye, C. L., and Dayton, S. L.**, Evaluation of selective and differential media in the isolation and enumeration of airborne *Staphylococcus aureus, Health Lab. Sci.,* 9, 284, 1972.

141. **Bruun, J. N.**, Post operative wound infection. Predisposing factors and the effect of a reduction in the dissemination of staphylococci, *Acta Med. Scand. Suppl.,* 9, 514, 1970.

142. **Orth, D. S. and Anderson, A. W.**, Polymyxin-coagulase-deoxyribonuclease-agar: a selective isolation medium for *Staphylococcus aureus, Appl. Microbiol.,* 20, 508, 1970.

143. **Pavlova, M. T., Brezenski, F. T., and Litsky, W.**, Evaluation of various media for isolation, enumeration and identification of fecal streptococci from natural sources, *Health Lab. Sci.,* 9, 289, 1972.

144. **Isenberg, H. D., Goldberg, D., and Sampson, J.**, Laboratory studies with a selective *Enterococcus* medium, *Appl. Microbiol.,* 20, 433, 1970.

145. **Brooks, W. F., Jr. and Goodman, N. L.**, Rapid differential testing of the Enterobacteriaceae using modified standard media, *Am. J. Med. Technol.,* 38, 429, 1972.

146. **Hoben, D. A., Ashton, D. H., and Peterson, A. C.**, A rapid, presumptive procedure for the detection of *Salmonella* in foods and food ingredients, *Appl. Microbiol.,* 25, 123, 1973.

147. **Hargrove, R. E., McDonough, F. E., and Reamer, R. J.**, A selective medium and presumptive procedure for detection of *Salmonella* in dairy products, *J. Milk Food Technol.,* 34, 6, 1971.

148. **Naguib, K.**, Evaluation of Bacto-strip technique for the detection of coliforms in Pasteurized milk, *Zentralbl. Bakteriol. Parasitenkd. Infektionskr. Hyg. Abt. 2,* 128, 88, 1973.

149. **Goldin, M.**, A comparison of multiple-test systems for the presumptive identification of *Enterbacteriaceae, Am. J. Med. Technol.,* 38, 288, 1972.

150. **Gillies, R. R.**, An evaluation of two composite media for preliminary identification of *Shigella* and *Salmonella, J. Clin. Pathol.,* 9, 368, 1956.

151. **Edwards, P. R. and Ewing, W. H.**, *Identification of Enterobacteriaecae,* 2nd ed., Burgess, Minneapolis, 1962.

152. **Shaffer, J. and Goldin, M.**, Medical microbiology, in *Clinical Diagnosis by Laboratory Methods,* 14th ed., Davidsohn, I. and Henry, J. B., Eds., W. B. Saunders, Philadelphia, 1969, 794.

153. **Lapage, S. P.**, Thoughts on screening tests in bacteriology, *J. Clin. Pathol.,* 24, 404, 1971.

154. **Elston, H. R., Baudo, J. A., Stanek, J. P., and Schaab, M.**, Multi-biochemical test system for distinguishing enteric and other Gram-negative bacilli, *Appl. Microbiol.,* 22, 408, 1971.

155. **McIlroy, G. T., Yu, P. K. W., Martin, W. J., and Washington, J. A., II**, Evaluation of modified R-B system for identification of members of the family Enterobacteriaceae, *Appl. Microbiol.,* 24, 358, 1972.

156. Brown, W. J., Evaluation of the four tube R/B system for identification of *Serratia marcescens* and *Serratia liquefaciens, Am. J. Med. Technol.,* 39, 272, 1973.

157. Isenberg, H. D. and Painter, B. G., Comparison of conventional methods, the R/B system, and modified R/B system as guides to the major divisions of *Enterobacteriaceae, Appl. Microbiol.,* 22, 1126, 1971.

158. Smith, P. B., Tomfohrde, K. M., Rhoden, D. L., and Balows, A., Evaluation of the modified R/B system for identification of *Enterobacteriaceae, Appl. Microbiol.,* 22, 928, 1971.

159. Buissiere, J. and Nardon, P., Micromethode d'identification des bactéries. I. Interet de la quantification des characteres biochemiques, *Ann. Inst. Pasteur* (Paris), 115, 218, 1968.

160. Hartman, P. A., *Miniaturized Microbiological Methods,* Academic Press, New York, 1968.

161. Washington, J. A., II, Yu, P. K. W., and Martin, W. J., Evaluation of accuracy of multitest micromethod system for identification of *Enterobacteriaceae, Appl. Microbiol.,* 22, 267, 1971.

162. Smith, P. B., Tomfohrde, K. M., Rhoden, D. L., and Balows, A., API system: a multitube micromethod for identification of *Enterobacteriaceae, Appl. Microbiol.,* 24, 449, 1972.

163. Starr, S. E., Thompson, F. S., Dowell, V. R., Jr., and Balows, A., Micromethod system for identification of anaerobic bacteria, *Appl. Microbiol.,* 25, 713, 1973.

164. Dowell, V. R., Jr., Comparison of techniques for isolation and identification of anaerobic bacteria, *Am. J. Clin. Nutr.,* 25, 1335, 1972.

165. Jeans, B., The use of PathoTec strips in medical bacteriology, *Can. J. Med. Technol.,* 29, 114, 1967.

166. Blazevic, D. J., Schreckenberger, P. C., and Matsen, J. M., Evaluation of the PathoTec "Rapid I-D System," *Appl. Microbiol.,* 26, 886, 1973.

167. Washington, J. A., II, Yu, P. K. W., and Martin, W. J., Evaluation of the Auxotab Enteric 1 system for identification of *Enterobacteriaceae, Appl. Microbiol.,* 23, 298, 1972.

168. Rhoden, D. L., Tomfohrde, K. M., Smith, P. B., and Balows, A., Auxotab – a device for identifying enteric bacteria, *Appl. Microbiol.,* 25, 284, 1973.

169. Seal, S. C., A new simplified sugar medium for fermentation test with plague, pseudotuberculosis and diphtheria organisms, *Indian J. Med. Res.,* 60, 1414, 1972.

170. Baird-Parker, A. C., A classification of micrococci and staphylococci based on physiological and biochemical test, *J. Gen. Microbiol.,* 30, 409, 1963.

171. Chalmers, A., A modification of the oxidation/fermentation test for the classification of Micrococcaceae, *Med. Lab. Technol.,* 29, 279, 1972.

172. International Committee, Recommendations of the Subcommittee on taxonomy of staphylococci and micrococci, *Int. Bull. Bacteriol. Nomencl. Taxon.,* 15, 109, 1965.

173. Kellogg, D. S., Jr. and Turner, E. M., Rapid fermentation confirmation of *Neisseria gonorrhoeae, Appl. Microbiol.,* 25, 550, 1973.

174. Taylor, G. S., Jr. and Keys, R. J., Simplified sugar fermentation plate technique for identification of *Neisseria gonorrhoeae, Appl. Microbiol.,* 27, 416, 1974.

175. Wolf, P. L., Von der Muehll, E., and Ludwick, M., A new test to differentiate *Serratia* from *Enterobacter, Am. J. Clin. Pathol.,* 57, 241, 1972.

176. Schreier, J. B., Modification of deoxyribonuclease test medium for rapid identification of *Serratia marcescens, Am. J. Clin. Pathol.,* 51, 711, 1969.

177. Kedzia, W., The phosphatase activity of coagulase-positive *Staphylococcus aureus* strains isolated from patients and healthy persons, *J. Pathol. Bacteriol.,* 85, 528, 1963.

178. Kocka, F. E., Magoc, T., and Searcy, R. L., Evaluation of rapid tests for staphylococci characterization, *Am. J. Med. Technol.,* 39, 269, 1973.

179. Mitz, M. A., The detection of bacteria and viruses in liquids, *Ann. N.Y. Acad. Sci.,* 158, 651, 1969.

180. Closs, O. and Digranes, A., Rapid identification of *Proteus* species and *Providencia* by a simple two-step procedure, *Acta Pathol. Microbiol. Scand. Sect. B,* 81, 684, 1973.

181. Goldin, M. and Glenn, A., A simple phenylalanine paper strip method for identification of *Proteus* strains, *J. Bacteriol.,* 84, 870, 1962.

182. Lowe, G. H., The rapid detection of lactose fermentation in paracolon organisms by the demonstration of β-D-galactosidase, *J. Med. Lab. Technol.,* 19, 21, 1962.

183. Levin, J., Poore, T. E., Zauber, N. P., and Oser, R. S., Detection of endotoxin in the blood of patients with sepsis due to Gram-negative bacteria, *N. Engl. J. Med.,* 283, 1313, 1970.

184. Levin, J., Poore, T. E., Young, N. S., Margolis, S., Zauber, N. P., Townes, A. S., and Bell, W. R., Gram-negative sepsis: detection of endotoxemia with the linulus test. With studies of associated changes in blood coagulation, serum lipids and complement, *Ann. Intern. Med.,* 76, 1, 1972.

185. Nachum, R., Lipsey, A., and Siegel, S. E., Rapid detection of Gram-negative bacterial meningitis by the linulus lysate test, *N. Engl. J. Med.,* 189, 931, 1973.

186. Zolg, W. and Ottow, J. C. G., Thin layer chromatography methods for detecting hippurate hydrolase activity among various bacteria (*Pseudomonas, Bacillus, Enterobacteriaceae*), *Experientia,* 29, 1573, 1973.

187. Coons, A. H., Creech, H. J., and Jones, R. N., Immunological properties of an antibody containing a fluorescent group, *Proc. Soc. Exp. Biol. Med.,* 47, 202, 1941.

188. Goldman, M., *Fluorescent Antibody Methods,* Academic Press, New York, 1968.

189. Ellis, E. M. and Harrington, R., A direct fluorescent antibody test for *Salmonella, Arch. Environ. Health,* 19, 876, 1969.

190. Cherry, W. B. and Thomason, B. M., Fluorescent antibody techniques for *Salmonella* and other enteric pathogens, *Public Health Rep.,* 84, 887, 1969.

191. Garner, M. F. and Robson, J. H., A fluorescent technique for demonstrating treponemes in films made from suspected chancres, *J. Clin. Pathol.,* 21, 108, 1968.

192. Thomason, B. M., Rapid detection of *Salmonella* microcolonies by fluorescent antibody, *Appl. Microbiol.,* 22, 1064, 1971.

193. Smyster, C. F. and Snoeyenbos, G. H., Fluorescent antibody methods for detecting Salmonellae in animal by-products, *Avian Dis.,* 17, 99, 1973.

194. Georgala, D. L. and Boothroyd, M., A rapid immunofluorescence technique for detecting salmonellae in raw meat, *J. Hyg.,* 62, 319, 1964.

195. Georgala, D. L., Boothroyd, M., and Hayes, P. R., Further evaluation of a rapid immunofluorescence technique for detecting salmonellae in meat and poultry, *J. Appl. Bacteriol.,* 28, 421, 1965.

196. Haglund, J. R., Ayres, J. C., Paton, A. M., Kraft, A. A., and Quinn, L. Y., Detection of *Salmonella* in eggs and egg products with fluorescent antibody, *Appl. Microbiol.,* 12, 447, 1964.

197. Silliker, J. H., Schmall, A., and Chiu, J. Y., The fluorescent antibody technique as a means of detecting Salmonellae in foods, *J. Food Sci.,* 31, 240, 1966.

198. Insalata, N. F., Schulte, S. J., and Berman, J. H., Immunofluorescence technique for detection of salmonellae in various foods, *Appl. Microbiol.,* 15, 1145, 1967.

199. Schulte, S. J., Witzeman, J. S., and Hall, W. M., Immunofluorescent screening for *Salmonella* in foods: comparison with culture methods, *J. Assoc. Off. Anal. Chem.,* 51, 1334, 1968.

200. Reamer, R. H., Hargrove, R. E., and McDonough, F. E., Increased sensitivity of immunofluorescent assay for *Salmonella* in nonfat dry milk, *Appl. Microbiol.,* 18, 328, 1969.

201. Fantasia, L. D., Accelerated immunofluorescence procedure for the detection of *Salmonella* in foods and animal byproducts, *Appl. Microbiol.,* 18, 708, 1969.

202. Barkate, J. A., Screening of feed components for *Salmonella* with polyvalent H-agglutination, *Appl. Microbiol.,* 16, 1872, 1968.

203. Laramore, C. R. and Moritz, C. W., Fluorescent-antibody technique in detection of salmonellae in animal feed and feed ingredients, *Appl. Microbiol.,* 17, 352, 1969.

204. Gibbs, P. A. and Hamilton, W. J., Evaluation of the fluorescent antibody technique for the detection of *Salmonella* in animal feeding-stuffs, *Rec. Agric. Res. North. Irel.,* 19, 1, 1971.

205. Thomason, B. M., Cherry, W. B., and Edwards, P. R., Staining bacterial smears with fluorescent antibody. VI. Identification of salmonellae in fecal specimens, *J. Bacteriol.,* 77, 478, 1959.

206. Thomason, B. M., McWhorter, A. C., and Sanders, E., Rapid detection of typhoid carriers by means of fluorescent antibody, *Bacteriol. Proc.,* p. 56, 1964.

207. Bissett, M. L., Powers, C., and Wood, R. M., Immunofluorescent identification of *Salmonella typhi* during a typhoid outbreak, *Appl. Microbiol.,* 17, 507, 1969.

208. Cohen, F., Page, R. H., and Stulberg, C. S., Immunofluorescence in diagnostic bacteriology. III. The identification of enteropathogenic *E. coli* serotypes in fecal smears, *Am. J. Dis. Child.,* 102, 82, 1961.

209. Chadwick, P., The relative sensitivity of fluorescent antibody and cultural methods in detection of small numbers of pathogenic serotypes of *Escherichia coli, Am. J. Epidemiol.,* 84, 150, 1966.

210. Whitaker, J., Page, R. H., Stulberg, C. S., and Zuelzer, W. W., Rapid identification of enteropathogenic *Escherichia coli* 0127:B$_8$ by the fluorescent antibody technique, *Am. Med. Assoc. J. Dis. Child.,* 95, 1, 1958.

211. Abshire, R. L. and Guthrie, R. K., Fluorescent antibody as a method for the detection of fecal pollution: *Escherichia coli* as indicator organisms, *Can. J. Microbiol.,* 19, 201, 1973.

212. Gibbs, P. A., Patterson, J. T., and Murray, J. G., The fluorescent antibody technique for the detection of *Salmonella* in routine use. II, *J. Appl. Bacteriol.,* 35, 415, 1972.

213. Baird-Parker, A. C., Methods for classifying staphylococci and micrococci, in *Identification Methods for Microbiologists, Part A,* Gibbs, B. M. and Skinner, F. A., Eds., Academic Press, New York, 1966.

214. Haglund, J. R., Ayres, J. C., Paton, A. M., Kraft, A. A., and Quinn, L. Y., Detection of *Salmonella* in eggs and egg products using fluorescent antibody. *Appl. Microbiol.,* 12, 447, 1964.

215. Gibbs, P. A., Patterson, J. T., and Murray, J. G., The fluorescent antibody technique for the detection of *Salmonella* in routine use, *J. Appl. Bacteriol.,* 35, 405, 1972.

216. Jones, W. D., Beam, R. E., and Kubica, G. P., Fluorescent antibody techniques with mycobacteria. II. Detection of *M. tuberculosis* in sputum, *Am. Rev. Respir. Dis.,* 95, 516, 1967.

217. Jones, W. D., Jr., Saits, H., and Kubica, G., Fluorescent antibody techniques with mycobacteria, *Am. Rev. Respir. Dis.,* 92, 255, 1965.

218. Kawamura, A., *Fluorescent Antibody Techniques and Their Applications,* University of Tokyo Press, Tokyo, 1969.

219. Martins, R. R., Walker, W. E., Batayias, G. E., and Gales, P. W., Type specific antibody for the identification of mycobacteria, *Lab. Med.,* 4, 28, 1973.

220. Martins, R. R., Walker, W. E., Batayias, G. E., and Gales, P. W., The production of type specific fluorescent antibody for the identification of mycobacteria, *Am. Rev. Respir. Dis.,* 108, 979, 1973.

221. Gales, P. W. Martins, R. R., and Walker, W. E., Production of multivalent fluorescent antisera for identification of organisms in the *Mycobacterium avium-Mycobacterium intracellulare* complex, *Appl. Microbiol.,* 27, 753, 1974.

222. Forsum, U., Characterization of FITC-labelled F (ab¹)₂ fragments of IgG and a rapid technique for separation of optimally labelled fragments, *J. Immunol. Methods,* 2, 183, 1972.

223. Nisonoff, A., Enzymatic digestion of rabbit gamma globulin and antibody and chromatography of digestion products, *Methods Med. Res.,* 10, 134, 1964.

224. McCracken, A. W. and Mauney, C. U., Identification of *Corynebacterium diphtheriae* by immunofluorescence during a diphtheria epidemic, *J. Clin. Pathol.,* 24, 641, 1971.

225. Holmberg, K. and Forsum, U., Identification of *Actinomyces, Arachnia, Bacterionema, Rothia,* and *Propionibacterium* species by defined immunofluorescence, *Appl. Microbiol.,* 25, 834, 1973.

226. Reamer, R. H. and Hargrove, R. E., Twenty-four-hour immunofluorescence technique for the detection of salmonellae in nonfat dry milk, *Appl. Microbiol.,* 23, 78, 1972.

227a. Danielsson, D., A membrane filter method for the demonstration of bacteria by the fluorescent antibody technique. I. A methodological study, *Acta Pathol. Microbiol. Scand.,* 63, 597, 1965.

227b. Danielsson, D. and Laurell, G., Application of membrane filter method for detection of small number of bacteria in water, *Acta Pathol. Microbiol. Scand.,* 63, 604, 1965.

228. Closs, O., Rapid identification of small numbers of bacteria after short-time cultivation in the presence of FITC-conjugated antiserum, *Acta Pathol. Microbiol. Scand.,* 72, 412, 1968.

229. Guthrie, R. K. and Reeder, D. J., Membrane filter-fluorescent-antibody method for detection and enumeration of bacteria in water, *Appl. Microbiol.,* 17, 399, 1969.

230. Danielsson, D. and Zieminska, S. Zastosowanie immunofluorescnencji w mikrobiologii wody na przykladzie wybranych charebotwórczych szczepów — *Escherichia coli.* II. Badania z zakazana woda naturalina, *Rocz. Panstw. Zakl. Hig.,* 24, 365, 1973.

231. Jost, R. and Fey, H., Rapid detection of small numbers of airborne bacteria by a membrane filter fluorescent-antibody technique, *Appl. Microbiol.,* 20, 861, 1970.

232. Geck, P., India-ink immuno-reaction for the rapid detection of enteric pathogens, *Acta Microbiol. Acad. Sci. Hung.,* 18, 191, 1971.

233. Hales, C. N. and Randle, P. J., Immunoassay of insulin with insulin-antibody precipitate, *Biochem J.,* 88, 137, 1963.

234. Miles, L. E. M. and Hales, C. N., The preparation and properties of purified ¹²⁵I-labelled antibodies to insulin, *Biochem. J.,* 108, 611, 1968.

235. Strange, R. E., Powell, E. O., and Pearce, T. W., The rapid detection and determination of sparse bacterial populations with radioactively labelled homologous antibodies, *J. Gen. Microbiol.,* 67, 349, 1971.

236. Nairn, R. C., Standardization in immunofluorescence, *Clin. Exp. Immunol.,* 3, 465, 1968.

237. Holborow, E. J., *Standardization in Immunofluorescence,* Blackwell, Oxford, 1970.

238. Beutner, E. H., Defined immunofluorescent staining, *Ann. N.Y. Acad. Sci.,* 177, 1, 1971.

239. Ginell, R. and Feuchtbaum, R. J., Fluorescent spectrophotometry in the identification of bacteria, *J. Appl. Bacteriol.,* 35, 29, 1972.

240. Schacter, M. M. and Haenni, E. O., Automatic triparametric recording in fluorometry of polynuclear hydrocarbons, *Anal. Chem.,* 36, 2045, 1964.

241. Greenwood, B. M., Whittle, H. C., and Dominic-Rajkovic, O., Counter-current immunoelectrophoresis in the diagnosis of meningococcal infections, *Lancet,* 2, 519, 1971.

242. Dorff, G. J., Coonrod, J. D., and Rytel, M. W., Detection by immunoelectrophoresis of antigen in sera of patients with pneumococcal bacteraemia, *Lancet,* 1, 578, 1971.

243. Coonrod, J. D. and Rytel, M. W., Determination of aetiology of bacterial meningitis by counter-immunoelectrophoresis, *Lancet,* 1, 1154, 1972.

244. Gordon, M. A., Almy, R. E., Green, C. H., and Fenton, J. W., II, Diagnostic mycoserology by immunoelectro-osmophoresis: A general, rapid, and sensitive microtechnic, *Am. J. Clin. Pathol.,* 56, 471, 1971.

245. Alter, H. J., Holland, P. V., and Purcell, R. H., Counterelectrophoresis for detection of hepatitis-associated antigen: methodology and comparison with gel diffusion and complement fixation, *J. Lab. Clin. Med.,* 77, 1000, 1971.

246. Dajani, A. S., Rapid identification of beta hemolytic streptococci by counterimmunoelectrophoresis, *J. Immunol.,* 110, 1702, 1973.

247. Janicki, B. W., Aron, S. A., and Raychaudhuri, A., Separation and isolation of mycobacterial antigens by continuous-flow electrophoresis, *Am. Rev. Respir. Dis.,* 106, 779, 1972.

248. Fossieck, B., Jr., Craig, R., and Paterson, P. Y., Counterimmunoelectrophoresis for rapid diagnosis of meningitis due to *Diplococcus pneumoniae, J. Infect. Dis.,* 127, 106, 1973.

249. Ingram, D. L., Anderson, P., and Smith, D. H., Countercurrent immunoelectrophoresis in the diagnosis of systemic diseases caused by *Haemophilus influenzae* type b, *J. Pediatr.,* 81, 1156, 1972.

250. Bradshaw, M. W., Schneerson, R., Parke, J. C., Jr., and Robbins, J. B., Bacterial antigens cross-reactive with the capsular polysaccharide of *Haemophilus influenzae* type b, *Lancet,* 1, 1095, 1971.

251. Wadström, T., Nord, E. E., Lindberg, A. A., and Molby, R., Rapid grouping of streptococci by immunoelectrophoresis, *Med. Microbiol. Immunol.*, 159, 191, 1974.

252. Coonrod, J. D. and Rytel, M. W., Detection of type-specific pneumococcal antigens by counterimmunoelectrophoresis. II. Etiologic diagnosis of pneumococcal pneumonia, *J. Lab. Clin. Med.*, 81, 778, 1973.

253. Wright, G. L., Jr. and Roberts, D. B., Differentiation of mycobacterial species and strains and the detection of common and specific antigens by micro two-dimensional immunoelectrophoresis, *Immunol. Commun.*, 3, 35, 1974.

254. Roberts, D. B., Wright, G. L., Jr., Affronti, L. F., and Reich, M., Characterization and comparison of mycobacterial antigens by two-dimensional immunoelectrophoresis, *Infect. Immun.*, 6, 564, 1972.

255. Janicki, B. W., Chaparas, S. D., Daniel, T. M., Kubica, G. P., Wright, G. L., Jr., and Yee, G. S., A reference system for antigens of *Mycobacterium tuberculosis*, *Am. Rev. Respir. Dis.*, 104, 602, 1971.

256. Gibson, L. F., Cooper, G. N., Saragea, A., and Maximescu, P., A bacteriological study of strains of *Corynebacterium diphtheriae* isolated in Victoria and New South Wales, *Med. J. Aust.*, 1, 412, 1970.

257. Saragea, A. and Maximescu, P., Phage typing of *Corynebacterium diphtheriae*. Incidence of *C. diphtheriae* phage types in different countries, *Bull. WHO*, 35, 681, 1966.

258. Gundersen, W. B., Investigation on the serological relationships of *Corynebacterium diphtheriae* type mitis and *Corynebacterium belfanti*, *Acta Pathol. Microbiol. Scand.*, 47, 65, 1959.

259. Robinson, D. T. and Peeney, A. L. P., The serological types amongst *gravis* strains of *C. diphtheriae* and their distribution, *J. Pathol. Bacteriol.*, 43, 403, 1936.

260. Hewitt, L. F., Serological typing of *C. diphtheriae*, *Br. J. Exp. Pathol.*, 28, 338, 1947.

261. Tarnowski, C., The type classification of Park Williams strain no. 8, *Acta Pathol. Microbiol. Scand.*, 19, 300, 1942.

262. Razin, S. and Rottem, S., Identification of *Mycoplasma* and other microorganisms by polyacrylamide-gel electrophoresis of cell proteins, *J. Bacteriol.*, 94, 1807, 1967.

263. Sacks, T. G., Haas, H., and Razin, S., Polyacrylamide-gel electrophoresis of cell proteins of enterobacteriaceae, *Isr. J. Med. Sci.*, 5, 49, 1969.

264. Larsen, S. A., Bickham, S. T., Buchanan, T. M., and Jones, W. L., Polyacrylamide gel electrophoresis of *Corynebacterium diphtheriae*: a possible epidemiological aid, *Appl. Microbiol.*, 22, 885, 1971.

265. Rottem, S. and Razin, S., Electrophoretic patterns of membrane proteins of *Mycoplasma*, *J. Bacteriol.*, 94, 359, 1967.

266. Larsen, S. A., Webb, C. D., and Moody, M. D., Acrylamide gel electrophoresis of group A streptococcal cell walls, *Appl. Microbiol.*, 17, 31, 1969.

267. Nickolaeva, S. A. and Safonov, V. D., A study on the complex of soluble proteins in the cells of *Clostridium perfringens* by electrophoresis in polyacrylamide gel, *Mikrobiologiya*, 39, 87, 1970.

268. Kronvall, G., A rapid slide-agglutination method for typing pneumococci by means of specific antibody adsorbed to protein A-containing staphylococci, *J. Med. Microbiol.*, 6, 187, 1973.

269. Kronvall, G. and Williams, R. C., Jr., Differences in anti-protein A activity among IgG subgroups, *J. Immunol.*, 103, 828, 1969.

270. Kronvall, G. and Frommel, D., Definition of staphylococcal protein A reactivity for human immunoglobulin G fragments, *Immunochemistry*, 7, 124, 1970.

271. Christensen, P., Kahlmeter, G., Jonsson, S., and Kronvall, G., New method for the serological grouping of streptococci with specific antibodies adsorbed to protein A-containing staphylococci, *Infect. Immun.*, 7, 881, 1973.

272. Aho, K. and Sievers, K., Screening for gonococcal arthritis, *Scand. J. Rheumatol.*, 1, 84, 1972.

273. Swift, H. F., Wilson, A. T., and Lancefield, R. C., Typing group A hemolytic streptococci by M precipitin reactions in capillary pipettes, *J. Exp. Med.*, 78, 127, 1943.

274. Lancaster, L. J. and Sherris, J. C., An agar-diffusion grouping technic for beta hemolytic streptococci, *Am. J. Clin. Pathol.*, 34, 131, 1960.

275. Rabinowitz, S. B., A cellulose acetate membrane immunodiffusion typing technique for group A hemolytic streptococci, *J. Lab. Clin. Med.*, 64, 488, 1964.

276. Goldin, M. and Glenn, A., Grouping of beta-hemolytic streptococci on cellulose acetate membranes, *J. Bacteriol.*, 87, 227, 1964.

277. Michael, J. G. and Massell, B. F., Use of unabsorbed antisera in gel diffusion for grouping and typing of hemolytic streptococci, *J. Lab. Clin. Med.*, 65, 322, 1965.

278. Prakash, K., Chawda, S., and Sharma, K. B., An evaluation of micro-gel diffusion technique for grouping of beta hemolytic streptococci, *Indian J. Med. Sci.*, 26, 578, 1972.

279. Sperber, W. H. and Deibel, R. H., Accelerated procedure for *Salmonella* detection in dried foods and feeds involving only broth cultures and serological reactions, *Appl. Microbiol.*, 17, 533, 1969.

280. Boothroyd, M. and Baird-Parker, A. C., The use of enrichment serology for *Salmonella* detection in human foods and animal feeds, *J. Appl. Bacteriol.*, 36, 165, 1973.

Chapter 4

DETECTION AND IDENTIFICATION OF BACTERIA BY GAS CHROMATOGRAPHY

INTRODUCTION

Gas chromatography (GC) is a method of separating the components of a volatile mixture by partition (distribution) between two phases, a stationary phase, and a moving gaseous phase. A vaporous sample is introduced into the stream of the moving phase (carrier gas) at the inlet end of the column. The process of separation takes place inside the column (a length of tubing containing a packing material which consists of a stationary phase coated on an inert support). The carrier gas (such as nitrogen, helium, or argon) flows through the column continuously, and the components (solutes) are distributed between the mobile phase and the finely distributed liquid stationary phase in the column. The rate at which the solute molecules pass down the column depends on their affinity for the stationary phase; those with the stronger affinity are retained longer than those whose affinity for the stationary phase is weak. If a solute has no affinity for the stationary phase (e.g., air) and the column is being operated at a temperature in excess of the solute's boiling point, then that solute will pass through the column at the same rate at which the carrier gas is flowing. When a solute is distributed between two phases under a given set of temperature and pressure conditions, the ratio is called the distribution or partition coefficient (K). The separated components of a sample emerge as more or less separate bands at the column outlet; they are detected by specific detection systems. Subsequently, these bands are recorded as peaks on the chromatogram. The solute that interacts with the stationary phase emerges later than the unsorbed solute.

The position of such a peak is defined either by the retention time (Rt) or by the adjusted retention time. The Rt is simply the time elapsed between the injection of the sample and the emergence of the peak maximum. In an ideal situation, the peaks of the chromatogram should be completely separated. The resolution (separation of two components in a mixture) and the Rt values of the components are essentially determined by the column efficiency. The efficiency of column is usually given by the number of theoretical plates (N) and may be calculated by

$$N = 16 \left(\frac{Rt}{Wb} \right)^2$$

where

N = number of theoretical plates;
Wb = peak width at the base.

Since the retention time to peak width ratios can be obtained from the chromatogram as the ratios of the two distances, the measurement of the plate numbers is very convenient. Therefore, N is the most commonly used quantity to describe column performance. As the plate number is proportional to the column length, in practice the column efficiency is often expressed by the number of plates per unit column length, for example, by plate per foot. For purposes of comparison, efficiency is commonly expressed in terms of height equivalent to a theoretical plate (HETP). HETP is the column length (L) divided by the number of theoretical plates.[1,2]

In summary, the efficacy of the GC technique is determined mainly by the selectivity of the stationary phase for the sample components to be separated and by the efficiency of the column. The objective of this chapter is to review the applications of the GC technique to the detection and identification of bacteria in in vitro cultures and biological samples. A brief description of component parts of GC equipment and methods applicable in bacteriological determinations is also given. The detailed information on these and related topics may be found in recent publications.[3-5]

GAS-CHROMATOGRAPHIC EQUIPMENT AND METHODS

Basic Components of a Gas Chromatograph
Carrier Gas

An inert gas such as helium, argon, nitrogen, or hydrogen is used to move the sample through the column. This is the mobile phase in the GC system. The carrier gas is maintained at a constant pressure by a pressure and/or flow control apparatus on the gas bottles and on many commercial gas chromatographs. The most common method for introducing liquid samples into the chromatograph is by means of a microsyringe. The desired sample sizes are usually between 1 to 10 μl for analytical work; thus, 5- to 10- μl syringes are generally preferred. The sample is introduced into the vaporizer with the syringe through a silicone-rubber system.

Injection Port

In most gas chromatographs a sample inlet system is built for introduction of solid, liquid, or gas samples. The solid or liquid sample, introduced as a plug, is vaporized and mixed with the carrier gas near the injection port of the gas chromatograph. The injection port is maintained at a constant temperature above the column temperature in order to vaporize the sample and to prevent backlash due to condensation. In bacteriological analysis, a liquid sample may be introduced into a gas chromatograph with microliter syringes through a rubber septum in the apparatus.

Automated injection devices adaptable to gas chromatographs are available from commercial sources.* These injection ports are quite convenient to use and reduce errors due to problems such as microair bubbles in the syringe, syringe plugging, and fast or slow injections encountered in manual operations. However, the automated injection systems are expensive (the cost ranges from $700 to $3,000) for small-scale routine bacteriological testing by GC methods.

Column and Column Oven

A glass or metal tubing containing the stationary (liquid) phase (which is coated either on an inert solid support or on column walls) is used. The solid support is usually made of diatomacous earth material (or glass beads). For bacteriological analysis, the commonly used solid supports include Chromosorb® (G, W, or P), Celite® 545, Anakrom® (U or P), Gas Chrom® (P, R, or GL), or Poropak® (P, Q, R, S, or T). Most of the details dealing with important aspects of solid supports are described by Ottenstein[6] and Palframan and Walker.[7] The physical properties of solid supports and the coating procedures, mesh size, and selection of solid support material for biological separations have been described by Littlewood,[5] Street,[8] and Grant.[4]

The selection of the liquid phase can be the most important step in achieving the desired chromatographic results. In general, a liquid phase should have the following.

1. Chemical stability
2. Low vapor pressure at operating column temperature
3. Sufficient partition coefficient
4. Sufficient selectivity for the sample components to be separated
5. Adequate wetting properties at column temperature
6. Solubility in some volatile solvent.

Although several hundred types of liquid phases are available from commercial sources, not a single liquid phase meets all of the qualities desired for a particular application. Various liquid phases differ in the polarity of liquids; therefore, in the selection of a liquid phase one should keep in mind its activity toward the polar and nonpolar compounds to be separated. In general, a polar liquid phase (e.g., Carbowax® M) is used for the separation of polar compounds and a nonpolar liquid phase (e.g., Apiezon L) is used for the separation of nonpolar compounds. The most widely used liquid phases for bacteriological analysis are

1. Carbowax 20M or 4000

$$(OH - [CH_2 - CH_2 - O]_n H)$$

2. Diethyene glycol succinate (DEGS)

$$([CH_2 - CH_2 - O - CH_2 - CH_2 -$$
$$O \qquad O$$
$$O - C - CH_2 - CH_2 - C - O]_n)$$

3. Neopentyl glycol succinate (NEPGS)

$$([O - CH_2 - \overset{CH_3}{\underset{CH_3}{C}} - CH_2 - O - \overset{O}{C} - CH_2 -$$
$$O$$
$$CH_2 - C -_n O])$$

*Perkin Elmer Corporation, Norwalk, Conn.; Hewlett-Packard, Avondale, Pa.

4. SE-30 (silicone gum rubber), OV-1 (Methyl silicone), OV-17 (Methyl pentyl silicone), SE-52, or other silicone derivatives with the following structures

A. Methyl silicone

$$H_3C-\begin{bmatrix} CH_3 \\ | \\ Si-O \\ | \\ CH_3 \end{bmatrix}\begin{matrix} CH_3 \\ | \\ Si \\ | \\ CH_3 \end{matrix}\begin{matrix} CH_3 \\ | \\ O-Si-CH_3 \\ | \\ CH_3 \end{matrix}$$

B. SE-52

$$\begin{bmatrix} CH_3 & CH_3 & CH_3 \\ | & | & | \\ O-Si-O-Si-O-Si \\ | & | & | \\ CH_3 & \bigcirc & CH_3 \end{bmatrix}_n$$

C. XF-1150 or XE-60

$$Si(CH_3)_3-O\begin{bmatrix} CH_3 \\ | \\ Si-O \\ | \\ CH_2 \\ | \\ CH_2 \\ | \\ C\ N \end{bmatrix}Si(CH_3)_3$$

D. QF-1

$$Si(CH_3)_3\begin{bmatrix} CF_3 \\ CH_2 \\ | \\ CH_2 \\ | \\ O-Si \\ | \\ CH_3 \end{bmatrix}_y\begin{bmatrix} CH_3 \\ | \\ O-Si \\ | \\ CH_3 \end{bmatrix}_y O-Si(CH_3)_3$$

E. Squalane

$$\begin{matrix} CH_3 & CH_3 & CH_3 \\ | & | & | \\ CH_3-C-(CH_2)_3-CH-(CH_2)_3-CH-(CH_2)_4-CH- \\ | & & | \\ CH_3 & & CH_3 \end{matrix}$$

$$\begin{matrix} & & CH_3 \\ & & | \\ (CH_2)_3-CH-CH-(CH_2)_3-CH \\ | & | & | \\ CH_3 & CH_3 & CH_3 \end{matrix}$$

Column dimensions are also important to the sample size and speed of analysis. Various sizes of columns (from capillary to 1/16 in. [0.150 cm] to 3/8 in. [0.900 cm] outer diameters) have been used. The internal diameters of the columns also vary from capillary (0.02 to 0.05 cm) to 0.8 cm.

The capillary columns (0.1158 cm inner diameter and 0.317 cm outer diameter) are used for greatest efficiency. These columns have small inner diameters, use small particle-size solid supports or thin liquid phase films, and require very small sample sizes (0.1 μg). The 0.635-cm outer diameter columns sacrifice some efficiency for capacity in semipreparative GC work. The 0.952-cm outer diameter columns are used when capacity, not a difficult separation, is required. In addition to diameter, shape, and liquid phase (solid support), other factors affecting column efficiency are column length, carrier gas, column temperature, gas velocity, sample size, and column conditioning. A number of these parameters are described in the manuals published by commercial companies manufacturing gas chromatographs. High-efficiency prepacked and preconditioned columns of desired specifications are also available from these commercial suppliers.

The column oven in a commercial apparatus is generally of low thermal mass and is heated by circulating warm air via a turbolater in such a way that an even temperature is obtained throughout the oven. The column oven should not be affected by other heaters such as the inlet or detector oven, and it should be free from the influence of changing ambient temperatures. Most column ovens are constructed of low-mass stainless steel to permit rapid heat-up and cool-down and to allow a uniform temperature of $\pm 1°C$ of the isothermal desired value and $\pm 2°C$ of the desired temperature to be maintained during programmed operation. Other important factors to be considered in selecting column ovens are dimensions, degree of temperature control, ease of reproducibility of temperature setting, availability of temperature programmers, and program rates possible.

Detection Systems

The detection system in a gas chromatograph provides the means of detecting sample components as they elute from the column and also provides the basis for quantitative determinations. An electrometer amplifies the detector signal, which is then sent to the recorder, thus establishing a permanent record of the analysis. More than 20 different types of detectors have been developed for use in GC; however, relatively few of these detectors are in common use today.

There are two basic types of detection systems and methods.

1. The integral method of detection in which the detector responds to the total mass of material emerging from the chromatographic column. The integral detector gives a series of steps, and each step corresponds to the presence of at least one solute in the carrier gas stream as it passes through the detector. Step height is a function of the amount of that solute in the mixture. The common types of integral detection systems include the gas volume detector, the Brunel mass detector, and the limit of flammability detector.

2. The differential method of detection in which the detector responds to the instantaneous concentration of the solvents in the carrier gas. In the ideal case, the differential detector displays a series of Gaussian-shaped peaks wherein each peak corresponds to at least one component and the peak area is a function of the amount of a component in a mixture. Examples of differential detectors include the Katharometer, the gas density balance detector, the flame ionization detector (FID), the electron capture detector (ECD), the cross-section detector, and the argon detector. In bacteriological analysis, the FID, ECD, alkali flame ionization detector (phosphorus detector), and thermal conductivity (TCD) detector are commonly used. Most manufacturers of detectors provide details of the components of these detectors and the principles involved. A detector for the analysis of components of bacteriological samples should be selected on the basis of criteria such as sensitivity, signal-to-noise ratio, drift, linearity, incidence of extraneous variables, ease of calibration in terms of known properties, speed of response, chemical inertness, and range of application.

In addition to the common detection devices (such as the flame ionization detector and electron capture detector), many other chemical, mechanical, or electronic methods and devices have been used as part of the detection system with GC. Examples of such detection systems include the coulometric detector (for sulfur and halogen com-

pounds),[9] the gas density detector,[10] flame photometers,[11] radiation counters,[12-14] spectroscopic methods (including ultraviolet and infrared spectrophotometry),[15-17] mass spectrometry,[18-22] and chemical methods.[5,23] Detailed information on these and other ancillary techniques employed as detection systems for GC may be found in recent publications by Mitruka[3] and Walker et al.[24] Mass spectrometric techniques are described in some detail elsewhere in this chapter.

Recording and Integrating Systems

A recorder provides a permanent visual record of the gas chromatographic analysis. The recorder presents the signals from the detector in graphic form on chart paper. Many types, designs, and performance capabilities of recorders are available commercially. Dual-pen 1-mV recorders are very useful in the GC system. In most modern recorders, provisions are made for auxiliary connections (e.g., integrators).

Quantitative analysis by GC may be achieved to some extent by the components described above. The peaks recorded on the strip-chart recorder may be measured by triangulation, peak height, cutting out the peak and weighing, and planimeter. These measurements may give a good estimation of the concentration of the component separated from a mixture. However, there are many integrating devices commercially available that are very convenient to use and that provide more accurate measurement of GC peaks. Various types of mechanical integrators, digital integrators, and integrating amplifiers are available. The operation of the disk integrator has been fully described in the literature[25] and in manufacturers' brochures. The disk integrator is dependent on the recorder operation, and any trouble in the recorder will affect the accuracy of the disk integrator. Change of base line is not automatically compensated for and each peak must be attenuated to a peak height less than full scale, which may necessitate a trial run for each sample. However, a disk integrator is a rather inexpensive instrument capable of measuring peak areas with fair accuracy from samples with stable analytical conditions.

A commonly used mechanical integrator is comprised of the "ball and disk" system, which can be attached to a strip-chart recorder. A ball moves with the recorder pen at the same time as it is in contact with the disk revolving at a constant rate, the center of the disk corresponding to the recorder's zero. The disk causes the ball to rotate at a speed which is proportional to its distance from the disk center, which is also proportional to the pen reading. Thus, a count of the number of ball rotations gives the area of a peak. In many laboratories at the present time, digital integrators provide completely unattended integration of the chromatogram. A combination of detector-electrometer-integrator produces a digital read-out number which is truly indicative of the concentration.

Electronic integrators are automated instruments in which the peak area is printed in numerical form by an automatic adding machine. For a complete automatic operation, one required feature of an electronic integrator is a high count rate expressed in counts per second per millivolt of input. Coupled with the count rate is the machine's ability to print out a large number of counts. The linear dynamic range of the integrator is the ratio of the largest to the smallest peak which the instrument can handle within the system's linear region of response. A typical digital commercial integrator has a solid state V/f converter input system which has slope, amplitude, and frequency peak sensor logic. A slope as small as 0.01 μV/sec can be sensed as a peak. The majority have a linear dynamic range of 500,000, which is certainly adequate for most situations. Typical count rates of about 2,000 counts/sec/mV, with a maximum count rate of about 10^6 counts/sec with a capacity for eight digits are common. Also, a three-digit read-out is standard. Most units can handle two peaks/second if necessary. An autoranging device allows inputs up to 500 mV, with autoranging occurring at 50 mV. Such integrators can easily handle all commonly encountered GC input signals. Many output options are available, including paper tape, magnetic tape, card punch, and computer.

Modern automated electronic integrators have certain standard features such as compatibility with other equipment in regard to input and output circuits of the integrator. They are also designed with a wide dynamic range for integrating small and very large peaks (e.g., peaks with a height ratio of 10,000:1). The flexibility of peak detection circuits provides controls for various settings, including rejection of significant base line drift, high-frequency noise, and all peaks below a given count. Also, the integrators have independent controls to permit separation of shoulders on leading and trailing edges of peaks. The instru-

ments are built with solid state electronics and plug-in modular designs with flexibility of output options including digital listing printers, card punches, and digital magnetic recorders. The peak detection sensitivity ranges from 0.1 to 0.2 μV/sec; the resolution count rate at maximum signal is about 2,000 to 10^6 counts/sec/mV. A three-digit retention time read-out is a standard feature of commercial integrators. Most units can handle two peaks per second if necessary. An autoranging device allows inputs of up to 500 mV, with autoranging occurring at 50 mv. Such integrators can easily handle all commonly encountered GC input signals.

Computer Systems for Handling GC Data

Computer link-up may either be on-line, in which case the computer is interfaced directly to the GC unit, or off-line, in which case a recorded digital output is submitted for computer processing at a later time. Very few analytical laboratories have the facility for on-line computer handling of GC analytical data due to the high cost of such systems and due to a load of routine GC analysis insufficient to justify even a minimal on-line system. However, relatively inexpensive on-line GC data handling systems (approximate cost, $17,000) are available with required hardware and software components (Figure 4.1).

This system includes a general purpose digital computer with 12,288 words of memory, a self-contained power supply, and a total of 14 input/output (I/O) channels. Memory can be expanded in plug-in steps to 32K words. Two of the I/O channels are used for the teleprinter and loop controller. The remaining 12 I/O channels are available for additional teleprinters and terminal printers. Also available as plug-in expansion options are a high-speed punched tape reader and a tape punch. The teleprinter allows communication with the computer through a single unit typewriter, tape reader, and tape punch. Information is input from the tape or keyboard and is output as a printed report or it is recorded on punched tape at a speed of ten characters per second. The software includes a binary tape for complete diagnostic check-out of the interface and the teleprinter. The data transmission loop services the analytical instruments in the system. All communication between the computer and the instruments is in bit-serial form through this loop. The word size for loop-controlled communications is 22 bits. Pri-

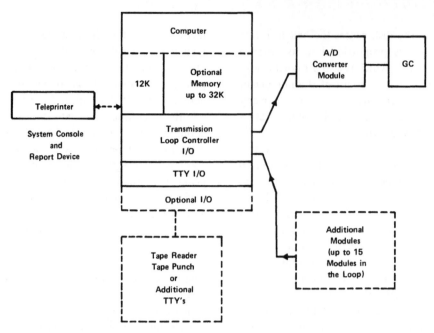

FIGURE 4.1. Block diagram of a GC data system. The broken lines indicate additional options to the system. (Courtesy of Hewlett-Packard Company.)

orities are determined by remote module placement in the loop, and additional modules (up to 15% loop) can be inserted in the loop at any time. The transmission loop controller printed-circuit board (PCB) normally plugs into the highest I/O priority slot (I/O address 10) in the computer. The I/O PCB sends poll messages out on the loop through program control. Poll messages circulate until one of the remote modules indicates, by setting a flag, that it has data ready for transfer. Data bytes are transferred back through the loop to the transmission loop controller PCB where they are collected as complete words; they are then processed by the computer.

The software is designed to allow turn-key operation of the system. The software attempts to detect and correct system hardware problems; if the problems are not correctable, the software permits the system to fail in such a manner that recovery is as simple as possible. The software system consists of the following modular blocks and approximate memory allocation.

9600 words	2688 words
Executive	Method storage
Dialogue	Peak data
Loop control	Storage
Data analysis routines	
Report generator	

Program control is carried out by the Executive block in the system software. Interrupt processing, the scheduling of tasks, and scheduling the use of peripherals are some of the jobs performed by the Executive. Peripheral operation is controlled by the I/O drivers in the Executive. Each I/O device is controlled by a driver related specifically to the individual device. Program control is transferred to the driver when an input or output operation is initiated. The primary means of communication with the system is through the Dialogue, which allows the operator to prepare, calibrate, and check methods; check method status; modify methods already stored; and clear out methods no longer needed. The operator can list (print-out) the status of the A and D modules data processed by data analysis routines and methods. The system software attempts to minimize typing requirements for the user. Many parameters have default conditions that are automatically set if the user elects to enter the parameter. A method does not have to be reassigned to a channel after each GC run. After a method is assigned to a channel, it will be used for every run until it is deleted or another method is assigned.

Data collection, detection of parity errors, and correction of data transmission are the functions of the Loop control block. It collects area counts from the A/D module outputs and assigns these data to channel buffers. Data from the A/D converter are obtained from the raw data buffers in the loop control block. The Data analysis routine tracks base line, detects peaks, calculates areas, and establishes retention times of peak maxima. Decisions and calculations are made in real time. Data analysis parameters which may be set in Dialogue include slope threshold, minimum area threshold, delay-start integrate, run time, and dead volume time. The output of Data analysis routine consists of component peak areas and retention times. The raw peak data from each A/D module during a run are stored in the Peak data buffer (four words per peak) and are available to be punched and listed at the teleprinter. Raw peak data are retained in the memory until another run is started at the A/D module assigned to the raw peak data buffer. The Report generator block analyzes raw peak data as defined by the assigned method block. The results of the calculations are presented to the user on the teleprinter as a printed, formated report. A short, medium, long, or extended form report can be selected by the user. A user-prepared method defines the calculation procedures in the Report generator block. Five types of methods are available, including zero area percent (no peaks identified), area percent (peaks identified), area normalization (peaks identified), internal standard (peaks identified), and external standard (peaks identified).

The Data storage block is dynamically assigned to a fixed (configured) number of methods. The use of memory is 47 words per method, 4 words per raw peak, and $(N/2 + 8)$ words per named peak. "N" is the number of characters used to identify a peak. The number of characters used to identify a peak must be $6 \leqslant N \leqslant 22$, and N must be even. When a method is prepared, storage is allocated for the named peaks defined by the method. As long as the method is in memory, 47 words plus the named peak storage requirement will be based in memory. The method can be cleared from memory by punching it on tape and then deleting it, after which it can later be loaded and used. The storage requirement raw peak is not allocated until the method is called (assigned) to a channel. Minimum hardware required to operate this chromatographic software includes a 12K memory computer, teleprinter and teleprinter I/O, transmission loop controller I/O, and A/D converter module.

Multifunction software systems may be used to interface with the basic chromatographic data analysis system. The multifunction systems provide capabilities of automatic liquid sampler control, simulated distillation, raw peak data files, timed events, and summing of designated peaks. In addition, the multifunction software allows the user to suppress unknown peaks from a report, change the sample name with the call command, automatically calibrate peak sort by retention time, and specify response factor for uncalibrated peaks.

The on-line large computer systems are extremely valuable because they achieve instant results at the end of an analysis. The computer can be programmed to control GC analysis parameters; more than one GC instrument can be linked up, and better quantitative results are obtained than those produced by digital integrators, since a much better job can be done in correcting peak areas of partially resolved peaks and in proportionally correcting for a drifting base line. However, a good knowledge of programming is a necessity to begin in this area (or immediate access to a system analyst); even with such precautions, the computer can only identify peaks positively with regard to peak patterns, elution times, and retention indicies.

The main characteristic of the off-line computer system is the indirect link between the digital output of the unit and the computer. A second feature is the lack of a direct control link from the computer to the GC system and the presence of a system control module. The chromatographer can choose from a variety of data acquisition systems, as well as from numerous computer facilities. The system usually consists of a dedicated coupler or specially built interfaces to measuring instruments and recording devices such as paper tape punch, magnetic tape, or IBM card punch. After the data are recorded, the tape or cards are taken to the computer for processing and print-out. One of the main advantages of this method of data acquisition is that the tape and/or cards provide a permanent record of the data. Also, the system is quite simple to operate. The recorded data may be taken to a large or dedicated computer for processing or (if paper tape is the data gathering medium) to a time-shared computer terminal. For routine analysis, the analog to the digital converter most widely applicable to off-line chromatographic systems is the digital integrator. If a time-sharing computer is available in the laboratory, a common teletype facility with keyboard print-out and a simultaneous paper tape punch can be used. The retention time and peak area data are printed out and punched on paper tape. The data on the paper tape are then fed into the time-shared computer. The only drawback to this system is that the teletype terminals yield slow data print-out (ten characters per second); for some rather involved programs for calculating solution properties from compositional data, it could take as long as 2 to 3 hr to print out the data. However, for most routine types of analysis, this is an ideal off-line system. Other frequently used systems for data acquisition include magnetic tape and the IBM card punch. However, a large computer is usually required to process a data card deck, and such a system is only useful when detailed calculations are required that necessitate the print-out of a large number of characters.

With the development of improved resolution capabilities of gas-chromatographic systems, analyses of complicated mixtures produce an ever-increasing number of separated species resulting in chromatograms with a higher information content. In addition, repetition of the separation using modified parameters increases the amount of data. A computer system is necessary to handle such data efficiently.

Evaluation of intensity and peak position is required for basic data processing in gas chromatography and requires a series of steps. Determination of peak area, detector response, base line drift, and peak overlapping must be taken into consideration. Standardization, normalization, and correction of peak positions and area values must be performed. Finally, results are displayed as tables, summaries, or in graphical representations.

In a large computer-GC system, contact-scanner and line-driver units permit the transmission of start or stop commands and the operation of relays for control functions at and in chromatographs. With the line-drive unit, it is possible to initiate commands at preset intervals after the start of a chromatogram, such as at the start of a temperature program. Teletypes permit the input of measurement parameters in a dialogue mode. A system of data flow has been described by Schomburg et al.[26] in which gas-chromatographic data from the interface are stored by a data acquisition program in core buffers. When the buffers are full, the data are transferred to disk storage. Raw data are processed by an analysis

program controlled by an organizing program, but remain available for processing by other programs such as the plotting program. Specific programs are described for data flow, specification of data acquisition parameters, production of basic reports, and refined analysis, as well as qualitative and quantitative analyses.

Mass Spectrometry Combined with Gas Chromatography (GC-MS)

The combined technique of GC-MS offers a powerful tool for the analysis of volatile organic mixtures. The effectiveness of this technique has been established for studies which involve structural elucidations, determinations of reaction mechanisms, and identification of trace components of a mixture of organic compounds. The combination of GC with MS permits direct identification of compounds. The advantage of this system is the combination of the great separation power of GC with the great identification power of MS. The identity of a compound is established by comparing its mass spectrum to that of a reference substance. The principles and applications of the combination of GC and MS in the biomedical sciences have been reviewed in detail.[3,24,27-37]

In mass spectrometric analysis, a sample containing organic compounds in vapor phase diffuses into the low pressure system; the sample is ionized with sufficient energy to cause fragmentation of the chemical bonds in the original molecule. The resulting positively charged ions are accelerated in a magnetic field which disperses them and permits measurement of the relative abundance of ions of a given mass-to-charge (m/e) ratio. Thus, the fragmentation patterns are a recording of ion abundance vs. mass, and this is known as mass spectrum. The GC-MS technique has been highly refined from the beginning, with a Benedix time-of-flight mass spectrometer (TOF-MS) monitoring the effluent of a packed column,[38] to present-day monitoring of capillary column effluent with fast-scanning, high-resolution mass spectrometers. In addition, the methods for interfacing GC-MS systems have been greatly improved and have been reviewed in some detail.[24,39,40] One of the methods for obtaining mass spectra is to trap the peak in a cooled (dry ice or liquid nitrogen) flow-through trap. The sample must be revaporized and transferred to the ionization region of the MS without contamination or loss. Such systems are necessary if only slow-scanning MS units are available. The Bendix TOF unit has been extremely useful with GC units as it produces 10,000 or 20,000 separate mass spectra per second in the 0- to 220-unit mass range, depending on the desired frequency of operation. Displayed on an oscilloscope, this appears as the complete spectrum for the eluting component and can be photographed with a Polaroid®-type camera for spectral interpretation. Later, units can be fitted with grafted-type scanning units that allow scanning a decade (mass 20 to 200) in about 3 sec, which is rapid enough to obtain a useable spectrum from rapidly eluting capillary peaks. Several modern techniques for interfacing GC with MS units now allow GC units to be connected directly to fast-scanning high-resolution mass spectrometers that have unit resolution values from m/e to 1 to 10,000.[3] However, for most bacteriological and medical analyses, an MS resolution value from m/e 1 to 1,000 is adequate.

The important features to consider in selecting a mass spectrometer for GC-MS are

1. Basic mass spectrometer sensitivity
2. Sample utilization or pumping efficiency
3. Mass resolution
4. Reproducible spectral pattern without mass discrimination
5. Computer compatibility

There are four common types of mass spectrometers used in GC-MS systems:

1. Single-focusing magnetic deflection
2. Double-focusing electrostatic sector/magnetic deflection
3. Time-of-flight
4. Quadrupole mass spectrometers

Single-focusing Magnetic Deflection Mass Spectrometer

This is the most common type of mass spectrometer used in organic analysis. The system is relatively simple in its operation, with a resolution of 300 to 1,000. In the single-focusing magnetic deflection system, ions formed in the source are accelerated through a source slit toward a homogeneous magnetic field. As the ions enter the magnetic field, they experience a force orthogonal to the field, which results in a curvature of the ion path. At a fixed radius, and for a singly charged ion, the mass focused at another slit and collected by the detector is proportional to the square of the magnetic field and inversely proportional to the accelerating voltage. By varying either of these two parameters, ions of a different mass-to-charge ratio can be deflected to the collector; the mass spectrum is scanned in this fashion. For most applications, it is preferable to vary the magnetic field and maintain a constant accelerating voltage. The resolution of a sector-field mass spectrometer is principally determined by the radius of curvature and by the width of the source and collector slits. The radius determines the mass dispersion, i.e., the distance by which two mass peaks are separated. The source slit defines the width of the ion beam. The collector slit determines the distance of the focused image or peak traversed in the course of scanning. For most GC-MS work, a minimum resolution of 300 to 400 should be available to assure maximum sensitivity and scan convenience.

Double-focusing Mass Spectrometer

A mass spectrometer containing both an electrostatic and a magnetic sector can give focus with respect to both velocity divergence and energy divergence for most electric and magnetic sector configurations. Ions of different mass achieve velocity focusing on one locus and direction focusing on another locus. The interaction of these two lines is the point of double focus for a given mass. Various configurations of the double-focusing mass spectrometer have been described.[31,41-44]

A double-focusing mass spectrometer is useful for achieving sufficient resolving power to separate doublet or multiplet peaks for accurate mass determination. A resolution of 10,000 to 18,000 is sufficient to give mass accuracy of a few parts per million. Occasionally, resolution as high as 40,000 to 70,000 is needed to separate certain mass doublets. In GC-MS studies, this can be achieved only by using photoplate detection. The double-beam system consists of two independent inlet and ion source assemblies, one electrostatic and magnetic sector, and two independent amplifiers and recorders. In this way, sample reference doublets do not occur, and accurate mass measurement can be attained at a resolution of 1,000 to 2,000, with increased sensitivity and scan rate. Mass measurement accuracy of 15 ppm has been reported for major peaks in the spectrum with 2×10^{-9} g of total sample eluted from the column.[4][5]

Time-of-flight Mass Spectrometer

Several mass spectrometers have been devised that select or disperse ions according to velocity. Ions are formed in the ionizing region either by a continuous or a pulsed electron beam. A repeller voltage pulse of a few nanoseconds duration is applied to the repeller electrode, and a bundle of ions is ejected out of the ion chamber through the grid system. This ion bundle is accelerated by a few seconds of voltage applied to the final accelerating grid. The ions enter into the field-free drift region where they continue with a constant velocity to the collector. The velocity of each individual mass is determined by the accelerating voltages. During the flight, ions of different masses are separated, and the arrival for each ion group at the detector is correlated with the time of the original voltage pulse. A typical flight time for a 1-m drift tube is in the range of 1 to 30 μm/sec. The repetition rate of the process may be 10,000 to 50,000 times per second, but 20,000 counts/sec is commonly used. Resolution in the range of 500 to 600 is attained, which is sufficient for most GC-MS studies. Occasionally, compounds such as highly substituted trimethyl silyl derivatives may be encountered at higher molecular weight. The time-of-flight mass spectrometer was the first system to receive widespread acceptance as a versatile instrument for GC-MS because the instrument was supplied with an electron multiplier. Thus, the fast response required for GC-MS scanning could be attained without sacrifice of sensitivity.

Quadrupole Mass Spectrometer

The quadrupole mass spectrometer consists of a quadrant of four parallel hyperbolic or circular rods which provide a specific radio frequency field. Opposite rods are electrically connected. Two variations of the quadrupole mass spectrometer used in GC-MS work are the monopole and the duodecapole mass spectrometers. The monopole mass spectrometer is designed as one quadrant of a quadrupole system. Because the equipotential lines in a quadrupole are symmetrical, a planar pole surface can be used to stimulate parts of the total quadrupole configuration. A monopole mass analyzer uses the same potential as the quadrupole, except that one quadrant is isolated by a right-angled bar which serves as the other electrode. The constant and oscillating voltages are applied between this bar and the monopole. An advantage of such a configuration is that only two parts must be carefully aligned. However, its disadvantages are that the angle and position of the ion entrance are more restrictive and the ion transmission is reduced for comparable geometric size.

The dodecapole mass analyzer is not really a 12-pole mass filter, but actually a quadrupole system with two trimmer electrodes in each quadrant. These trimmer poles permit easier adjustment of the equipotential field and thus overcome irregularities due to pole misalignment or nonconductive coatings. The performance appears to be about the same as that obtained from a conventional quadrupole.

A number of quadrupole MS units ideally suited for bacteriological analysis are available commercially. Figure 4.2 depicts the principle components of a quadrupole mass spectrometer. Molecules from the inlet system (i.e., the gas chromatograph batch inlet or probe inlet) enter the ion source, where they are bombarded with electrons which are emitted from a hot filament. The neutral molecules are ionized to form a variety of products, including positive ions. The positive ions are used in the analysis because they predominate over the other species present by several orders of magnitude. While ionization can occur at any bond in the molecule, it occurs at certain preferred locations, giving rise to a distribution of ions which is reproducible and which constitutes a fingerprint of the original molecule. The positive ions are electronically extracted from the ion source and injected into the quadrupole mass filter, where they are separated according to their mass. The ions passing through the quadrupole filter are quantitatively detected by an electron multiplier and amplified, with the resulting signal fed to an appropriate display. The oscilloscope provides a continuous display of the mass spectrum. The oscillographic recorder provides a permanent spectrum of the compound of interest. A potentiometric recorder can be used to provide a recording of total ion current vs. time when a GC inlet is used. This total ion current chromatograph is the same as a normal gas chromatogram; thus, MS is used as a GC detector. In many cases, this output signal is fed directly to

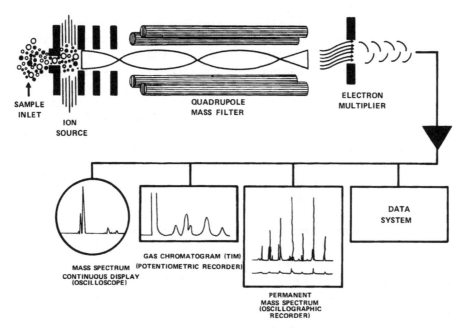

SAMPLE
INLET

ION
SOURCE

QUADRUPOLE
MASS FILTER

ELECTRON
MULTIPLIER

DATA
SYSTEM

MASS SPECTRUM
CONTINUOUS DISPLAY
(OSCILLOSCOPE)

GAS CHROMATOGRAM (TIM)
(POTENTIOMETRIC RECORDER)

PERMANENT
MASS SPECTRUM
(OSCILLOGRAPHIC
RECORDER)

FIGURE 4.2.   Functional diagram of quadrupole mass spectrometer. (Courtesy of Finnigan Corporation.)

an on-line data system for subsequent computer processing.

A quadrupole mass spectrometer usually has a linear mass scale, tolerates high operating pressures, gives very rapid mass scans, and easily adapts to computerization.[46] It has an ion source that can operate in either electron ionization (EI) or chemical ionization (CI) modes.[47] In some instances CI may identify a hydrocarbon that under EI conditions would show no useful molecule ion. A special method used to determine m/e values during a scan with the quadrupole MS involves performing a cross correlation between the known ion peak shape expected at a mass and that actually found during a scan. Noise peaks are rejected, making a threshold rejection of a spectral point unnecessary.

Distinction is made between a mass chromatogram and a mass fragmentogram. The mass chromatogram reconstructs a chromatogram by obtaining the ion current value for a given ion from each mass scan stored in the computer memory. For a 3-sec scan, this ion current occurs during a 0.01-sec interval. With a mass fragmentogram, a chromatogram is produced by specifically tuning the mass spectrometer to the given m/e value without scanning, so that the desired ion is detected continuously. It is possible to realize at least a tenfold greater sensitivity when the mass

fragmentogram technique is used to detect compounds through a specific ion.

Mass Fragmentography (MF)

Homlstedt and Palmér[34] reviewed the use of the mass spectrometer in biomedical analyses. They described it as an ion-specific detector wherein the intensity of ions produced when the effluent from the gas chromatograph is ionized in the ion source is recorded as a continuous curve call a mass fragmentogram, similar in appearance to a gas chromatogram. The mass fragmentogram yields retention time and peak area; the relative intensities of peaks are the same as the relative intensities of ionic fragments in the mass spectrum of the compound.

With mass fragmentography, the spectrometer is used exclusively for the mass specific recording of preselected ions; it is necessary that the mass spectrometer be focused beforehand on the masses at which ion abundance is to be measured. With mass chromatography, analysis is via recovery of a graph of a specific ion abundance against time. This may be accomplished from repetitively scanned mass spectra stored on a magnetic tape of disk. While any mass can be recalled, that of sensitivity is less than that of mass fragmentography which, for certain compounds, can only be matched by an electron capture system. Sensitivity of MF (10^2 to 10^3 times greater than FID and GC-MS) lies in selection of intense ions, wide optical slits, and a favorable signal-to-noise ratio which cannot be obtained in repetitive scanning.

Characterization of compounds by MF utilizes retention time data, presence of all investigated mass numbers and characteristic ratio between intensities of all mass numbers. MF can be applied to almost any compound that can be gas chromatographed and that has a suitable fragmentation pattern. It is both qualitative and quantitative and has been used in the analysis of drugs and endogenous compounds in biological samples. For quantitative metabolic studies, compounds labeled with stable isotopes can be given to humans; metabolites are recognized because of specific ion clusters.

Henneberg,[48,49] Gohlke,[50] and Henneberg and Schomburg[51] traced single ions appearing during gas-chromatographic runs even before the advent of molecule separators. Lindeman and Annis[18] combined the mass spectrometer and gas chromatograph for analysis of volatile organic mixtures to produce mass chromatograms. This led Sweeley et al.[22] to construct the accelerating voltage alternator (AVA), which enabled resolution of components that had been eluted simultaneously from the gas chromatograph. This suggested the idea of simultaneous mass fragmentography of drugs and metabolites, a technique reported by Hammar et al.[52] A multiple ion detector (MID) was described by Hammar and Hessling;[53] this had a mass range of 20% of the lowest mass within which three different masses could be focused. The MID described by Hammar[54] has four channels, and the mass range is increased approximately to 30% of the lowest mass. Wiesendanger and Tao,[55] Bonelli,[46] Cho et al.,[56] and Jenden and Silverman[57] described multiplexer devices for use with the quadrupole instruments which allow up to eight masses to be recorded over the entire mass range, with separate demodulated outputs.

Elkin et al.[58] described a flexible GC-MS laboratory computer system for on-line data collection and processing which has as part of the system a program to simultaneously sample four masses and record the resulting data. The system is not limited to hardware implemented channels as in previously used computer systems. Extra channel time in included in the software if additional channels are required.

Another use of mass spectrography is in field ionization mass spectrometry (FI-MS), which is characterized by the presence of prominent molecular ion ("parent ion") peaks and only a few minor fragment ion peaks.[59] FI-MS is a suitable technique for the analysis of multicomponent mixtures of volatile organic compounds.[60] FI-MS spectra of multicomponent mixtures are less complex than the corresponding electron impact ionization (EI) mass spectra.[61] Spectra produced by low-resolution FI-MS of multicomponent mixtures are difficult to interpret because molecular ions of different elemental composition may share the same nominal m/e value. High-resolution FI-MS permits the determination of the elemental compositions of the observed ions by accurate mass measurements.

Schulten et al.[35] explored the potentials of high-resolution FI-MS for the analysis of complex multicomponent mixtures and performed a general survey of the chemical nature of bacterial pyrolysis products using *Pseudomonas putida* bacteria. They compared their results with the data of Simmonds[62] who attempted a systematic chemical identification of bacterial pyrolyzates through direct coupling of a quadrupole mass spectrometer to a GLC system (GLC-MS). A double-focusing CEC 21-110B mass spectrometer equipped with an FI ion source and a specially designed emitter-adjusting manipulator were used.[63] Spectra were recorded on Ilford® Q2 plates and evaluated on a modified version of the apparatus described by Henneberg and Frexenius[64] for computer processing (by PDP-8E) of high-resolution data.

Isomeric compounds abundant in pyrolyzates were not differentiated by high-resolution FI-MS. Only pyrolysis products stable enough to survive storage in borosilicate glass reaction flasks and volatile enough to be transferred to the mass spectrometer at 150°C and 10^{-5} Torr were observed. Of the 66 bacterial pyrolysis products of the Gram-positive *Bacillus subtilis* and *Micrococcus luteus* strains studied by Simmonds,[62] 60 matched perfectly with the Gram-negative *Pseudomonas putida* strain. Hennenberg et al.[64] argued that both studies were qualitative and stressed the similarity of the chemical building blocks of bacteria. They felt that the results underline the importance of achieving a high level of quantitative reproducibility when using pyrolysis methods for differentiation and identification of such samples.

Shulten et al.[35] encountered bacterial pyrolysis products (such as free acids and amines) not previously observed in GLC-MS; these were apparently too polar to pass through most gas-chromatographic columns, but were favored in the FI-MS system by the reduced number of collisions of the pyrolysis products in the vacuum pyrolysis procedure employed.

It is important to remember that, apart from economic considerations, there are several fundamental limitations to the usefulness of mass spectrometry as an analytical tool, especially when it is used in a fast-scan mode on small samples heavily diluted with carrier gas. In general, therefore, it is necessary to supplement the mass spectrum with other information about the eluate before any confidence can be placed in an identification. The major difficulty to overcome when directly coupling GC and MS is the limited rate at which it is possible to admit gas to the source chamber while maintaining a sufficiently low pressure in the spectrometer. The actual value of the limiting flow rate for a particular spectrometer depends on the detailed design of the vacuum system, including such factors as the pumping rate and whether or not differential pumping of the source and analyzer is provided. Typical values are between 0.2 and 2 cm^3/min at atmospheric point pressure, compared with carrier gas flow rates of approximately 30 cm^3/min for conventional GC packed columns. It follows that at least 90% of the maximum possible MS response to a particular GC peak will be lost. This explains the great interest in molecular separators, which, by concentrating the eluate, allow a greater total percentage of the sample to enter the source. Two useful reviews concerning GC-MS interface have been published recently.[65,66]

With many separators, the enrichment factor is only about 100; as a result, carrier gas molecules entering the source chamber still commonly outnumber eluate molecules by at least 1000:1. However, in principle this modest enrichment can allow almost all of the eluate from packed columns to enter the mass spectrometer. In practice, however, the efficiency of the separator may be rather poor, so that in spite of a high enrichment factor, only a small percentage of the sample leaving the column actually reaches the spectrometer. Then, there may be little advantage in using the separator as compared to simply splitting the effluent from the GC column so that a suitable flow rate enters the source.

Many types of molecular separators have been designed. These include the following.

Fretted glass (or Biemann-Watson) separator[44] – This consists of a porous glass tube which acts as a stream splitter by moving the faster diffusion of the small carrier gas molecule through the fretted (porous) glass, resulting in an enrichment of the compound of interest. Such separators are most effective when operated at flow rates from 1 to 50 cm^3/min (which makes them useful with capillary columns) and at temperatures as high as 350°C in order to prevent heavy materials from condensing in the fretted glass or other porous surfaces.

Molecular jet (Ryhage) separator – This is constructed entirely of stainless steel.[20] The column effluent passes through a small orifice which is optically aligned with another orifice. The carrier gas (generally helium or hydrogen) diffuses away from the line of sight more rapidly than the heavier organic molecules, resulting in a good enrichment of the carrier gas entering the mass spectrometer with the organic material. The molecular jet separator is primarily used with packed columns directing higher carrier gas flows of 30 to 50 ml/min with 40 to 70% of methyl stearate passing into the ion source of the mass spectrometer. The two-stage separator is used in the LKB 9000 GC-MS system.

Silicone membrane (Llewellyen) separator[67] – This diffuses the eluted organic compounds through the diaphragm, which is relatively impermeable, to the carrier gas (hydrogen, helium, argon, or nitrogen). Both single-stage (one membrane) and dual-stage (two separate membranes) separators have been used. The desired flow rate range for the single stage is from 1 to 20 ml/min and about 60 ml/min maximum for the dual stage, with a maximum temperature of 250°C for either type. Although these separators are very efficient, polar components and high boilers tend to be partially absorbed on the silicone membrane, resulting in the tailing of these materials into the mass spectrometer and, in some cases, complete removal of very polar materials. One definite advantage of these separators is that the GC outlet is kept at atmospheric pressure, and no separate pump is required. A forepump is used in the dual-stage mode between Stage 1 and Stage 2 to enhance removal of the carrier. Routine maintenance of these silicone membrane separators is similar to that for other separators; that is, they must be kept very clean, although extreme high-temperature/vacuum backout is not possible.

Thin-walled Teflon® (Lipsky) separator[68] – This is a modification of the Biemann-Watson separator in that the porous glass tube is replaced by about 6 to 8 ft of thin-walled Teflon tubing. Both ends of the Teflon tubing are connected to stainless steel capillary tubes to reduce the pressure and to allow carrier gas flow rates of 15 ml/min or less. The Teflon tube is contained in a vacuum jacket connected to a rotary pump. The optimum operation temperature range in which the Teflon is most highly and selectively permeable to helium is 250 to 260°C above ambient temperature.

Porous silver frit (Blumer) separator[69] – This is a very small separator employing a porous silver frit to selectively remove the carrier gas. It is mainly designed for use with low-flow packed columns (10 to 15 ml/min). The separator consists of a ¼-in. stainless steel Swagelok tee with one arm (connected to a rotary pump) containing a porous silver membrane (6 mm diameter, 2 ml thickness, 3 μm maximum pore diameter). The other two arms of the tie are connected to the GC and the MS ion source through a needle-type metering valve. Satisfactory analysis may be obtained with saturated and unsaturated hydrocarbons (up to C-20), polar aldehydes, ketones, dicarbonyl compounds, and fatty acid esters (up to C-22). A disadvantage of this type of separator is the fact that silver-catalyzed surface reactions or organic molecules can be generated at high temperatures; however, the separator is very durable and has an inherent low dead volume which makes it ideal for smaller diameter columns.

Variable steel plate (Brunee) separator – The Brunee adjustable steel plate separator allows carrier gas/sample differentiation by a variable annular slit.[70] The outstanding characteristic of this separator is that it may be operated over a large range of flow rates (1 to 60 ml/min). The main disadvantage is the possible surface reactions of organic compounds on the high-temperature steel. However, the ruggedness of the separator, together with the simplicity of operation and the wide dynamic range of the unit, make it desirable to use in equipment where both packed and capillary columns will be employed.

Membrane frit (Grayson-Wolf) separator[71] – This is a two-stage membrane frit separator that is highly efficient between column flows of 5 to 60 ml/min. In addition, the GC outlet with this separator is maintained at atmospheric pressure.

Maximum operating temperature for this separator is 250°C, due to the silicone membrane portion.

APPLICATIONS OF GAS CHROMATOGRAPHY TECHNIQUES IN BACTERIOLOGY

The feasibility of utilizing gas chromatography for bacterial identification has been investigated by several workers in the field, and the results indicate that differentiation is possible among genera, species, and strains. Many methods have been used for identification of bacteria by GC.

1. Pyrolysis of isolated bacterial cells followed by GC analysis of pyrolysate.
2. Hydrolysis of cell component(s) of bacteria by acids or bases and extraction of hydrolyzed components with solvent, followed by GC analysis of extractants.
3. Direct GC analysis of bacterial culture for volatile products such as gases, acids, and alcohols.
4. Extraction of bacterial metabolites and products (acidic, neutral, or basic) from cultures followed, by GC analysis of products.
5. Isolation of specific metabolic products of bacteria produced by degradation of a substrate incubated with bacteria, and gas-chromatographic analysis of specific products associated with bacterial activity.
6. Gas-chromatographic characterization of bacterial activity in serum, other clinical specimens, or biological samples based on comparison with standard chromatograms (fingerprints) without necessary chemical identification of products or compounds of bacteria.

The application of GC methods in the analysis of bacterial cell components or products was first suggested by Oyama,[72] who proposed the pyrolysis of microorganisms for taxonomic differentiation. Since Oyama's proposal, GC techniques have been widely used for characterizing and identifying bacteria by analysis of volatile fatty acids, esters, nonvolatile acids, alcohol, ketones, volatile amines, gasses, and certain neutral compounds including acetoin, diacetyl sugars, amino sugars, amino acids, and glycoproteins. The analyses for volatile fatty acids and alcohols have been most useful for routine characterization of anaerobes.[73-77]

The main advantages of using GC in bacterial

analyses include its sensitivity, specificity, and rapidity and the multicomponent detection and identification capabilities of the technique. Gas chromatography is an ultrasensitive technique for determining bacterial components in nanogram or even picogram concentrations.[78,79] Specific components or products of bacteria can be detected qualitatively and quantitatively within a few hours, thereby facilitating the identification of organisms in a test sample.[80-83] Another advantage of GC is the relative uniformity of the technique. A simple extraction of a bacterial culture (such as ether extraction for 1 to 2 min) gives characteristic products in chromatograms which can readily identify the organism.[79,84-90] In many instances, the lengthy procedures of isolation of bacteria from clinical specimens or biological mixtures are not required with GC methods, since the presumptive identification of the organisms can be made by matching characteristic chromatographic patterns or by detecting one or more organism-specific peaks (compounds) in the chromatogram. A direct GC analysis of clinical specimens such as serum, urine, or tissue would indicate the presumptive identification of the infective agent.[3,85-92] The application of GC is also possible for identification of organisms in a mixed bacterial population in vitro and in vivo.[93] Finally, GC methods may be automated by qualitative and quantitative analyses and identification of specific compounds associated with bacteria.

The disadvantages inherent in the GC methods include the initial cost of the instruments, the many variations due to component parts of the equipment, the sample preparation and operational procedures, and the requirement of rather sophisticated instrumentation (e.g., mass spectrometer, infrared or nuclear magnetic resonance spectrometer, etc.) for identification of compounds and interpretation of large volumes of data. Substances which are not volatile at temperatures below those producing molecular fragmentation cannot be chromatographed. However, recent methods of derivatization permit analysis of nonvolatile compounds in bacteria by GC. Identification of unknown compounds appearing as peaks on the chromatogram and analysis of data and their interpretation become the limiting factors. Coupling GC with MS for recording the spectra of specific peaks eluting through the GC column often permits absolute identification of the key compound. However, when complex mixtures (such as sera, urines, or bacterial cultures) are analyzed by these techniques, they may cause difficulty in manual interpretation of the GC patterns due to large numbers of peaks. For this reason, various data processing techniques have been developed.[26,94-97] GC, in conjunction with MS and computer analysis, can be a very valuable tool for the rapid diagnosis and study of bacterial infections.

PYROLYSIS GAS-LIQUID CHROMATOGRAPHY (PGLC)

The pyrolysis gas-liquid chromatographic method is based on the thermal degradation of a small sample of dried material under carefully controlled conditions. The resulting breakdown products are separated by gas-liquid chromatography. Various types of pyrolyzers have been described that can be used in conjunction with GC (Table 4.1). The most satisfactory gas-chromatographic results in PGLC techniques are obtained by means of high-efficiency capillary columns.[62,98] These columns, which are used for the quantitative analysis of biological extracts, permit the complete separation of a very large number of components of a complex mixture and give better results. The technique of PGLC has been used to differentiate genera, species, and in many instances, strains of bacteria (Table 4.2). Reiner[99] first described the use of PGLC in 1965. Subsequent work by various investigators[62,96,98-106] has demonstrated the value of this technique for the typing of a wide range of bacteria and for the diagnosis of microorganisms. Each bacterial strain produces a characteristic fragment chromatogram (pyrochromatogram) which is used as a "fingerprint" to identify the microorganism. Automatic recognition of bacterial types by computer matching of pyrochromatograms has been described by Menger et al.[96]

The entire PGLC technique, including sample loading and analysis, is suited to automation. Although several workers in the field have described the successful application of PGLC to the classification and identification of bacteria, the quantitative and qualitative differences between the pyrochromatograms from different species are small. Since small differences can be attributed to changes in the conditions and the period of growth of the microorganisms and to pyrolysis conditions, little confidence can be placed in such results. The

TABLE 4.1

Comparison of Various Pyrolyzers Used with Gas Chromatography

Pyrolyzer type	Temperature		Sample size (μg)	Reproducibility	Advantages	Disadvantages	Reference
	Range	Rise time					
Conventional filament and ribbon	Up to 1200°C	2–10 sec	10–1000	Fair	Most widely used, wide temperature range, simple to apply sample	Slow rise time, aging of metal, solid material hard to apply	159
Curie point high power (1000 W)	Up to 985°C	40–130 msec	10–50	Good	Fast rise time, accurate temperature, good repeatability	Solid material hard to apply, pyrolysis chamber must be heated	160
Curie point low power (30–100 W)	Up to 985°C	0.5–2 sec	10–50	Fair	Accurate temperature, can be purchased	Solid material hard to apply, pyrolysis chamber must be heated, slow rise time	161
Tube reactor and boat	Up to 1500°C	30–50 sec	1–5000	Fair	Any material liquid or solid may be applied, wide temperature and sample size range, kinetic studies possible	Slow rise time, requires large samples, capillary columns cannot be used	162
Vapor phase	Up to 800°C	N/A	0.001–10	Good	Wide temperature range, good reproducibility, can be correlated with theory	Only volatile materials can be used	163
Electric discharge	Up to	10–50 sec	200	Poor	Any material (liquid or solid) may be used, wide temperature range, slow rise time	Slow rise time, large sample required, poor reproducibility	162

From Walker, J. Q., Jackson, M. T., Jr., and Maynard, J. B., *Chromatographic Systems*, Academic Press, New York, 1972. With permission.

TABLE 4.2

Identification of Bacteria by Pyrolysis Gas Chromatography (PGLC) and Mass Spectrometry (Py-MS)

Bacteria	Method	Reference
Soil microorganisms including species of *Azotobacter, Pseudomonas, Clostridium, Bacillus,* and *Cellulomonas*	PGLC (stainless steel loop 500°C for 2 min)	72, 164
Escherichia coli, Shigella species, and other enterobacteria (48 strains), Group A *Streptococcus pyogenes* (43 strains); mycobacteria (28 different nonpathogenic strains); *Staphylococcus* sp. (4 strains); and *Clostridium* sp. (53 strains)	PGLC (nickel filament 850°C for 10 sec)	99, 165
Gram-negative bacteria	PGLC (nickel filament 840°C for 10 sec)	100
Mycobacteria	PGLC (nickel filament 840°C for 10 sec)	102, 165
Vibrio cholerae	PGLC (800°C for 5 sec)	105
Salmonella (6 strains of Kaufmann-White serological groups)	PGLC (platinum ribbon 850°C for 10 sec)	103
Salmonella (6 strains)	PGLC (850°C for 10 sec), computer	96
Streptococcus mutans and other streptococcal strains; *Klebsiella* sp., and mycobacteria	Curie point-PGLC (pyrolysis temperature, 610°C; rise time, 100 m/sec)	108
Micrococcus luteus and *Bacillus subtilis* var. *niger*	PGLC-MS (electron impact quadrupole), pyrolysis temp, 500°C)	62, 137
Neisseria sicca, N. meningitidis, and *Leptospira* sp.	Direct pyrolysis-Mass Spectrometry in ion source (pyrolyzing temperature, 510°C)	107
Pseudomonas putida	Pyrolysis (temperature 500°C for 2 min) and high-resolution field-ionization mass spectrometry	35, 110
Streptococcal strains, *Klebsiella,* and *Vibrio* strains	Curie point-pyrolysis-mass spectrometry (Py-MS)	

most satisfactory GC results for this type of application are obtained by using high-efficiency capillary columns.[62,98] These columns permit the complete separation of a very large number of components in a complex mixture and give a better resolution of diagnostic components for one bacterial species. By using MS in conjunction with PGLC, analytical power may be further improved.[31] The mass spectrometer could be used as a gas-chromatographic detector to determine the chemical structure of compound eluted from the gas chromatograph. Thus, in comparison to the simple measurement of GC retention times (Rt), the addition of the mass spectrometer provides a more reliable and characteristic means of identification of eluted materials. If it is found that the amount of certain compounds in the pyrolysis products is the best indicator of bacterial species, the mass spectrometer can be set to act as a specific and quantitative detector of these compounds simply by monitoring ions of selected masses (mass fragmentography).

Samples can be subjected to mild pyrolysis in the mass spectrometer itself, without intervention of GC separation. The direct introduction of dried bacterium into the ion source of the mass spectrometer at 250 to 300°C produces some degree of thermal degradation. When this occurs, the mass spectrometer may be used to record ions characteristic of certain classes of compounds in the bacterium. The possible advantage of this technique lies in the in situ production and analysis of compounds that are not amenable to GC analysis because of their high molecular weight or polarity, but which, for the same reason, might be highly representative of cell wall polymer from different sources.[107]

The rapid and reproducible identification of bacteria may be performed by fingerprinting with Curie point PGLC,[108] Curie point hydrolysis MS,[109] or pyrolysis coupled with high-resolution field ionization and field desorption MS.[110] The high-resolution FI-MS can be used for the analyses of extremely complex multicomponent mixtures.

There are two major differences between the information available from the combined GLC-EI mass spectrometer and that available from high-resolution FI-MS. First, since compounds within a wide range of polarity are produced in the pyrolysis process, it is difficult to obtain ideal GC conditions. Thus, several of the more polar pyrolysis products (e.g., free acids and amines) do not appear to pass through the GC column; instead, they are identified by direct evaporation and the FI-MS of the mixture. Second, compounds such as branched-chain hydrocarbons, amines, and alcohols are difficult to identify in GLC-MS studies because they often fail to produce stable molecular ions on electron impact ionization. On the other hand, the softer field ionization process yields almost exclusively molecular ions or quasi-molecular ions from all pyrolysis products.

Curie Point Pyrolysis Gas Chromatography

In pyrolysis GLC, the samples to be studied are thermally fragmented in a stream of a carrier gas, and the reaction products are passed directly into a gas-chromatographic column where they are separated. There are two possible applications of this technique to bacterial identification.

1. The pyrolysis gas chromatogram is simply used as a "fingerprint" of the starting material.
2. Correlation between the structure of the fragments and the structure of the starting material is worked out and used for structural elucidation, identification, and quantitative analysis.

However, adequate long-term and interlaboratory reproducibility in thermal fragmentation can only be achieved by using a highly reproducible temperature-time profile for the fragmentation process. For nonvolatile or slightly volatile compounds, the application of a square-wave temperature-time profile to the highly diluted sample seems to be ideal. Bühler and Simon[111] reported that a fast and reproducible warm-up of samples in contact with cylindric ferromagnetic conductors can be obtained by high-frequency induction heating. Conductors of 0.5-mm diameter can reach the Curie temperature of the ferromagnetic material in less than 30.10^{-3} sec.

In Curie point pyrolysis GC, a ferromagnetic conductor in contact with the sample is heated inside a low-volume glass capillary which is inserted into an rf coil with the carrier gas flowing through the capillary. The intrinsic properties of the conductor permit manual control of the temperature to the Curie point. The optimum wire diameter for a rapid warm-up is a function of the rf frequency. In order to achieve a fast temperature drop after cutting the rf field, small wire diameters, preferable for the relatively high frequencies of such conductors, are inherently necessary for fast stabilization to the Curie point. Thus far, no differences in the self-controlled end temperatures have been detected for wires of different origin.

GAS-CHROMATOGRAPHIC ANALYSIS OF BACTERIAL CELL COMPONENTS

Identification of bacteria can be achieved by gas-chromatographic analyses of cell components. In this procedure, bacteria are grown in suitable media, cells are separated and washed, and the organisms are hydrolysed by acidic or alkaline solutions and compounds such as fatty acids, other organic acids, carbohydrates, hydroxy acids, amines, and amino acids. Other components are extracted and determined by GC. Abel et al.,[112] Yamakawa and Ueta,[113] and Steinhauer[114] reported extraction and transesterification of lipid material from selected species of various bacterial families and obtained characteristic GC elution patterns of the fatty acid methyl esters. Yamakawa and Ueta[113] introduced fragmentation of dried cells by HCl-catalyzed methanolysis. Two main groups of substances were released by this treatment, methyl esters of fatty acids and methyl glycosides of various monosaccharides. Extraction of the nonpolar fatty acid methyl esters by light petroleum ether and trimethyl silyl derivatization of the remaining methyl glycosides prior to GC analysis gave two sets of elution profiles from the selected species of *Neisseria* and the family *Micrococcaceae*.

Using a similar technique, Farsh and Moss[115,116] characterized a number of *Clostridium* species. Elution profiles were obtained from clostridia by comparing the trimethyl silyl derivative and whole-cell hydrolysates. Methanolysis fragmentation followed by volatilization and stabilization by a suitable derivatization reagent prior to GC is a common procedure in the identification of microorganisms. However, the usefulness of this type of fingerprint for taxonomic purposes is rather restricted if variations are large over a period. Bøvre et al,[117] Froholm et al.,[118] and Jantzen et al.[119] described trifluoroacetyl (TFA) and trimethyl silyl (TMS) derivatives of whole-cell methanolysates of several species of *Neisseria* and *Moraxella.* Bacterial cells (2 to 10 mg) suspended in normal HCl in methanol were heated at 90°C for 18 hr. The methanolysates were evaporated to dryness, and the residue was dissolved in anhydrous pyridine. Trimethyl silyl derivatives were prepared with the reagent (trimethyl silyl)-chlorosilane (TMCS) in *N,O*-bis-(trimethyl silyl)-acetamide (BSA). (TFA) deriva-

tives were prepared with trifluoracetic anhydride and acetonitrile reagents. Derivatized bacterial cell components were separated on 3% OV-1 Gaschrom Q and 10% UCCW982 on Chromosorb W (Applied Science Lab, Inc.) gas-chromatographic columns with temperatures programmed at 6°/min from 100 to 240°C. Characteristic GC profiles were obtained for each species of *Neisseria* and *Moraxella* examined. The profiles may be used for identification of these bacterial species. Although a general profile of microbial cell components gives useful information for taxonomic classification, analysis of specific cell constituents (such as fatty acids, carbohydrates, hydroxy acids, Krebs cycle intermediates, proteins, amino acids, and nucleic acids) provides a means of specific identification of bacterial species. Various investigations have been conducted on the GC analysis of bacterial cell components for characterization and identification of organisms (Table 4.3).

Drucker[120] analyzed fatty acids of various strains of streptococci by GC to differentiate the

TABLE 4.3

Gas-chromatographic Analysis of Bacterial Cell Components

Bacteria	Compound	Reference
Species of *Bacillus, Corynebacterium, Escherichia, Klebsiella, Neisseria, Pseudomonas, Serratia, Staphylococcus,* and *Streptococcus*	Fatty acids	114, 115, 121, 126, 127
Streptococcus (10 strains)	Fatty acid ($C_{14}-C_{18}$)	120, 124
Streptococcus mutans (D282) and (JC2), S. viridans (8 strains), S. pyogenes, S. faecalis, and Pneumococcus (12 strains)	Fatty acid ($C_{12}-C_{18}$); sugars (C_4-C_6 monosaccharides) and hexosamines	3, 81
Staphylococcus and *Micrococcus* (24 strains), *Staphylococcus aureus, Micrococcus* sp.	Fatty acids and neutral sugars; glycolipids, phospholipids, and neutral lipids	62, 167—169
Bacillus sp.	Fatty acids (long chain)	130, 131
Thiobacillus (18 strains of 10 species, including *T. neopolitanus* and *T. thioparus*)	Fatty acids (C_8-C_{20})	170
Propionibacterium (40 strains representing 7 species — *P. freudenreichii, P. shermanii, P. arabinosum, P. jensenii, P. pentosaceum, P. thoenii,* and *P. zeae*)	Fatty acids (long chain)	171, 172

TABLE 4.3 (continued)

Gas-chromatographic Analysis of Bacterial Cell Components

Bacteria	Compound	Reference
Gaffkya, Micrococcus cryophilus, and *Veillonella alcalescens*	Fatty acid ($C_9 - C_{20}$)	173
Vibrio cholerae	Fatty acids ($C_{14} - C_{19}$)	174
Arthrobacter globiformis	Fatty acid ($C_{14} - C_{17}$)	133
Neisseria sp. (*N. gonorrhoeae* and *N. meningitidis*)	Fatty acids	77, 113, 147, 168, 175–177
	Carbohydrates	113, 177
Escherichia coli and other Enterobacteriaceae	Phospholipids and neutral lipids	178
	Carboxylic acid ($C_{10} - C_{22}$)	112
	Citric acid cycle and related compounds	179
Proteus sp. (bacillary and L forms)	Fatty acids and purified lipopolysaccharides	114, 180
Mycobacterium (90 strains including 25 atypical strains)	Fatty acids ($C_{11} - C_{20}$)	181
Mycobacteria (58 strains of nonphotochromogenic [9 serotypes], 38 strains *Mycobacterium avium*, and 20 strains Battey strain	Fatty acids ($C_8 - C_{24}$)	182, 183
Mycobacterium marinum (10 strains) and *M. kansasii* (35 strains)	Fatty acids ($C_8 - C_{24}$)	184
Mycobacterium tuberculosis, M. bovis, M. avium, M. smegmalis, M. minetti, and *M. giae*	Fatty acids ($C_{12} - C_{22}$)	186
Mycobacterium phlei	3-*O*-Methyl-D-mannose	187
Mycobacterium tuberuculosis var. bovis (BCG strain)	Glutamic acid	188
Mycoplasma neurolyticum, M. pneumoniae, M. orale, M. fermentans, M. salivarium, and *M. hominis*	Glycolipids Fatty acids (long chain)	185 189
Listeria monocytogenes (33 strains, 10 serological types)	Fatty acids ($C_{12} - C_{23}$)	115, 132
Pseudomonas sp., *P. aeruginosa, P. cepacia,* and *P. maltophilia*	Organic acids Fatty acids	33, 190
Corynebacterium liquefaciens, C. granulosum, C. anaerobium, C. diphtheroides, C. acnes, and *C. pyogenes*	Fatty acids (long chain)	171
Corynebacterium acnes	Fatty acids (long chain)	172
Corynebacteria	Fatty acids	121
Corynebacterium xerosis	Fatty acids	126
Corynebacterium ulcerans	Fatty acids (long chain $C_{20} - C_{32}$) mycolic acid	191

TABLE 4.3 (continued)

Gas-chromatographic Analysis of Bacterial Cell Components

Bacteria	Compound	Reference
Clostridium (sp.) (41 strains and 13 species including *C. perfringens* and *C. bifermentans*)	Fatty acids	115, 172
36 strains and 10 species of *Clostridium*	Sugars (arabinose, glucose, and mannose)	116
5 strains of clostridia	Fatty acids ($C_{12}-C_{20}$)	192
Clostridium bifermentans	Carbohydrates (rhamnose)	142

organisms. Bacteria were grown under identical cultural conditions in brain-heart infusion medium supplemented with 1% glucose. The cells were harvested and washed. Methyl esters of cellular fatty acids were prepared by sealing 10 mg of dry cells and the methylating reagent with ampoules under vacuum and heating at 100°C for 3 hr. The methyl esters were extracted with ether at 40 to 60°C, and the extracts evaporated to a small volume. An aliquot was injected onto a 5-ft column packed with 10% polyethylene glycol adipate on 100 to 120 mesh celite in a chromatograph equipped with a flame ionization detector (FID). Peaks obtained were tentatively identified by comparison with pure reference mixtures. Methyl palmitate was used as an internal standard to calculate the relative retention times of the compounds of bacterial origin. The major peak in each chromatographic analysis could easily be identified as that of methyl palmitate. The next major peaks appeared to be myristate and oleate. Considerable strain differences were noted for stearate, palmitoleate, myristoleate, and an unidentified peak (relative retention time 0.68) which had similar retention characteristics to antisopentadecanote. Although fingerprints of the strains tested by the authors had similar fatty acid profiles, sufficient differences existed to enable GC identification. However, identification by GC with this method requires that standard strains be included for comparative purposes, since growth conditions and culture media have been shown to have considerable effect on fatty acid profiles of bacteria.[121-124]

In an attempt to obtain greater control over growth conditions of the organism and reproducibility of GC analyses, Drucker et al.[124] examined the fatty acid profiles of streptococci grown in a chemostat under a variety of defined conditions. It was found that changes in experimental conditions slightly altered the fatty acid composition of streptococcal strains, particularly oxygenation, which resulted in unsaturated fatty acid production. Changes in fingerprints in the JC2 strain of streptococci were less than the difference in fingerprints for JC2 and D282 strains grown under identical conditions. The major streptococcal fatty acids in the strains examined were palmetic, oleic, palmitoleic, and myristic acid.[125] Further differences in qualitative and quantitative fatty acid composition between various species of bacteria including streptococci were separated by Drucker and Owen.[126] In their studies, aerated cultures showed greater variation in fingerprints than nonaerated cultures, and distinction between the species tested was readily possible.

Amstein and Hartman[127] compared the relative fatty composition of 37 strains of enterococci. It was found that *Streptococcus faecalis, S. faecium,* and *S. faecium* var. *durans* yielded similar fatty acid patterns. However, strains of *S. faecium* var. *casseliflavus* and a motile yellow-pigmented streptococci contained very low levels of $C_{19:0}$ cyclopropane fatty acid and four unidentified components (these variations did not appear in the other strains of enterococci examined). No significant differences were found in the fatty acid patterns of enterococci grouped according to plant, animal, or human source.

Other groups of compounds of bacterial origin may be analyzed with comparative rapidity by means of GC, and the information thus obtained may be used for taxonomic differentiation. Bacterial components may be fractionated simultaneously for nucleic acids, carbohydrates, and lipids (Figure 4.3). After fractionation and separation, the desired component may be extracted, derivatized, and analyzed by GC both qualitatively and quantitatively.[3]

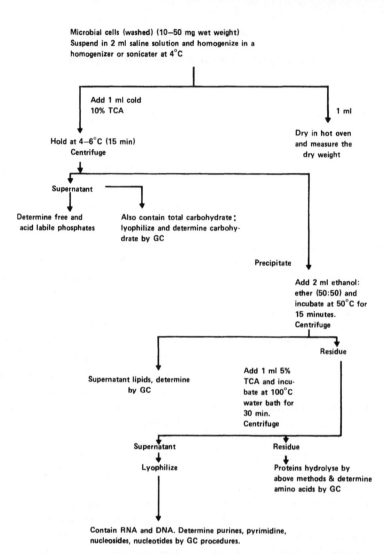

Microbial cells (washed) (10–50 mg wet weight)
Suspend in 2 ml saline solution and homogenize in a
homogenizer or sonicater at 4°C

Add 1 ml cold
10% TCA

1 ml

Hold at 4–6°C (15 min)
Centrifuge

Dry in hot oven
and measure the
dry weight

Supernatant

Determine free and
acid labile phosphates

Also contain total carbohydrate:
lyophilize and determine carbohydrate by GC

Precipitate

Add 2 ml ethanol:
ether (50:50) and
incubate at 50°C for
15 minutes.
Centrifuge

Residue

Supernatant lipids, determine
by GC

Add 1 ml 5%
TCA and incu-
bate at 100°C
water bath for
30 min.
Centrifuge

Supernatant

Residue

Lyophilize

Proteins hydrolyse by
above methods & determine
amino acids by GC

Contain RNA and DNA. Determine purines, pyrimidine,
nucleosides, nucleotides by GC procedures.

FIGURE 4.3.   Flow chart diagram for the GC analysis of microbial components.
(From Mitruka, B. M., *Gas Chromatographic Applications in Microbiology and Medicine*, John Wiley & Sons, New York, 1975. With permission.)

GAS-CHROMATOGRAPHIC ANALYSIS OF METABOLIC PRODUCTS OF BACTERIA IN IN VITRO CULTURES

Henis et al.[84] first reported the detection and identification of bacteria based on GC analysis of characteristic metabolic products. These investigators extracted cultures of a variety of bacteria with ether and chromatographed the extracts. Different metabolic product profiles of the organisms were obtained which were characteristic of the genera or the species of the organisms examined. Mitruka et al.[79,128,129] further explored the possibility of using GC methods for the detection and identification of bacteria by characterization of their specific metabolic

products in culture media. It was possible to differentiate organisms growing in cultures in vitro by means of GC. Taxonomic relationships among microorganisms were separated and, in some cases, could be subdivided into groups of species based on characteristic fatty acid profiles.[82,122,130-133] Fermentation patterns of anaerobic bacteria were studied by Moore et al.[134] and Brooks and Moore[74] using GC analysis of culture media. The found that *Clostridium* species can be identified by comparing GC patterns. Since these studies, GC methods have been used routinely as aids in the identification of bacteria through the analysis of fatty acids, alcohols, neutral compounds, and other volatile compounds produced by anaerobic bacteria in in vitro cultures.[135,136] Analyses that have proven useful for characterizing and identifying anaerobic organisms include those for volatile fatty acids, nonvolatile acids, alcohols, ketones, volatile amines, gases, and certain neutral compounds such as acetoin and diacetyl performed by means of a combination of thermal conductivity (TCD) and FID detectors. Analyses for volatile fatty acids and alcohols without derivitization of samples or with methylation (using boron trifluoride-methanol reagent, HCl-MeOH reagent, or diazomethane)[137,139] have by far been the most useful for routine characterization of anaerobes. These analyses have been particularly valuable in identifying nonsporulating Gram-positive bacilli, cocci and streptococci, but they are also very useful in identifying clostridia and Gram-negative bacilli.[75,82,83,128,140] Some taxa are defined on the basis of the above-mentioned types of metabolic products, for example, the family Lactobacillaceae, the genus *Propionibacterium*, and the species *Eubacterium alactoliticum*. In general, the types and relative amounts of products are similar from strain to strain and reproducible from culture to culture within a given strain.[75,77] Bricknell et al.[73] discussed various specific techniques useful for GC analysis of metabolic products of anaerobic bacteria.[141-145] Salanitro and Muirhead[146] described a quantitative method for the preparation and GC analysis of butyl esters of volatile (C_1 to C_7) and nonvolatile (lactic, succinic, and fumaric) acids in microbial fermentation media. Butyl esters were prepared from the dry salts of the acids. The esters were separated by temperature programming on a column of Chromosorb W coated with Dexsil® 300 GC liquid phase and analyzed with an FID. Chromatographic profiles of the butyl esters demonstrated that both volatile and nonvolatile acids can be detected and separated in 24 min on a single column. A number of anaerobic bacterial species (including *Bacteroides melaninogenicus*, *Lactobacillus vitulinus*, *Eubacterium ruminantium*, and *B. ruminicola*) were differentiated by this means.

Organic acids have been analyzed by GC from

various aerobic bacterial cultures for characterization and differentiation of the organisms. Mitruka[3] differentiated a number of Gram-negative and Gram-positive aerobic bacteria by the procedure outlined in Figure 4.4. A total of 36 ATCC stains were analyzed in replicate samples for organic acid products in the spent media. The chromatographic patterns of the identified organic acids were apparently distinct for each bacterial species examined (Figure 4.5). This procedure was found to be highly reproducible. Various classes of bacteria may be differentiated by the presence or absence of organic acids in their culture media. Detection of *Neisseria* by GC analysis of hydroxy acids in spent culture media have been described by Brooks et al.[147] Different GC profiles were obtained by these methods for several serogroups of species of *Neisseria*.

Sugars in bacterial cultures have been analyzed by GC for characterization of organisms.[3] The spent culture media are treated with a mixture of 5% $ZnSO_4$, 8% $Ba(OH)_2$,, and 30% H_2SO_4 to precipitate proteins. The supernatant containing sugars is neutralized, dried, and then dissolved in pyridine. Silyl derivatives are prepared by adding reagents such as hexamethyl disilazane (HMDS), trimethyl chlorosilane (TMCS), Tri-Sil®-Z, and others.* Alternatively, sugars from bacterial culture may be analyzed by a simple procedure described by Miller et al.[148] and Davis and McPherson.[149] In this procedure, a portion of bacterial culture is treated with an equal volume of 8 N methanolic HCl. The reaction tube is sealed and incubated at 80°C for 1 hr. The sample is then neutralized with concentrated NH_4OH and dried under a stream of N_2. The dried material is then taken up in pyridine, treated with HMDS and TMCS reagents, and allowed to stand for 15 min at room temperature to convert the O-methyl derivatives to TMSi ethers. After sedimentation of the NH_4Cl precipitate at 3,000 rpm for 15 min, the supernatant fluid is removed and taken to dryness under a nitrogen stream at room temperature. The dry derivatives are taken up in 0.3 ml of ethyl ether, and 3 μl of the solution is injected onto a 3% SE-30 column. Sugars and fatty acid profiles from serum, cerebrospinal fluid, and cultures have been analyzed by this procedure.[148] Metabolic products of microorganisms (including *Candida, Escherichia coli, Enterobacter*

*Described by the Pierce Chemical Company, Rockford, Ill.

aeruginosa, Shigella shigae, and *Diplococcus pneumoniae*) were differentiated based on their characteristic chromatographic patterns. Brooks[92] reported differentiation of *Enterobacter cloacae, Citrobacter freundii, Staphylococcus epidermidis,* and *Escherichia coli* by electron capture-gas

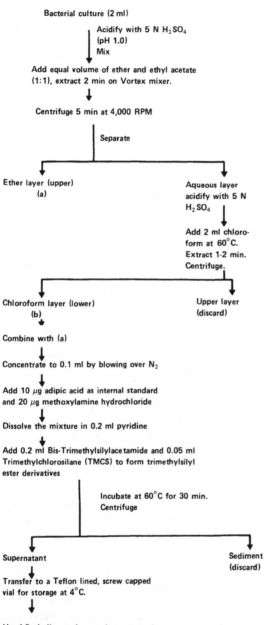

FIGURE 4.4. Flow chart diagram for the GC analysis of bacterial isolates. (From Mitruka, B. M., *Gas Chromatographic Applications in Microbiology and Medicine,* John Wiley & Sons, New York, 1975. With permission.)

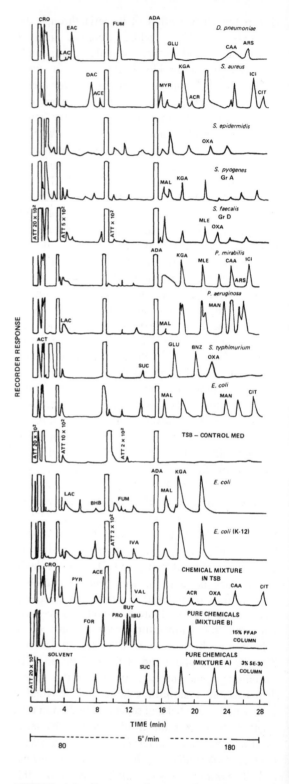

FIGURE 4.5. Representative gas chromatograms of standard organic acids and microbial cultures. (From Mitruka, B. M., *Gas Chromatographic Applications in Microbiology and Medicine,* John Wiley & Sons, New York, 1975. With permission.)

chromatographic (EC-GLC) analysis of derivatized spent culture media extracts. Electron capturing derivatives were prepared by carboxylic acids, hydroxy acids, alcohols, amines, and nitrosamines extracted from acidified or alkalinized cultures.

Using methods similar to those described by Miller et al.,[148] Mitruka and Jonas[150] and Mitruka[157] reported differentiation of several species of mycobacteria, streptococci, pneumococci, and other bacteria by GC analysis of carbohydrate components of bacteria in culture. A summary of work on the GC analysis of bacterial cultures reported by several investigators is given in Table 4.4.

DETECTION AND IDENTIFICATION OF BACTERIA IN CLINICAL SPECIMENS

Gas-chromatographic methods have been used for the detection and identification of bacteria by

1. Analysis of one or more specific metabolites of organisms present in serum, cerebrospinal

TABLE 4.4

Gas-chromatographic Analysis of Bacterial Products in Cultures

Bacteria	Compound	Reference
Species of *Bacillus*,	Fatty acids	84, 86
Escherichia, Aerobacter,	Volatile compounds including	79, 84, 88, 93
Pseudomonas, Staphylococcus,	ethyl alcohol, diacetyl,	
Streptococcus, Salmonella, and	acetoin, 2,3-butanediol, and	
coliform bacteria	ethanol	
Escherichia coli, E. freundii, Aerobacter	Volatile products (acetal-	169, 179, 194
aerogines, A. lipolyticum, Streptococcus	dehyde, ethanol, methyl	
diacetilactis, S. lactis,	sulfide, diacetyl, and organic	
S. faecalis, Pseudomonas fragi, Lactobacillus	acids)	
acidophilus, and *L. casei*		
Proteus mirabilis	Amines and nitrosoamines	85
Haemophilus vaginilis	Fatty acids $(C_{12}-C_{23})$	171
Salmonella typhimurium,	Halogenated acids	138
Escherichia coli, Staphylococcus		
aureus		
Neisseria sp.	Hydroxy acids	86
Clostridium species inclu-	Fatty acids: Short chain	77, 134, 140, 175
ding *C. botulinum* Type F	(C_2-C_6), alcohols (C_2-C_6),	
	and organic acids	
Clostridia (62 strains and	Neutral compounds:	74
and 13 species)	alcohols, phenols, aldehydes,	
	ketones, mercaptans, and	
	organic acids and amines	
Clostridia (10 species)	Organic acids, halogens,	145
	alcohols, and ketones	
Clostridia	Organic acids (C_2-C_8)	73, 134, 193
Clostridium bifermentans	Amines	152
and *C. sordellii* (31		
strains)		
Bifidobacterium	Short-chain fatty acids	73, 139, 141, 146, 155
(*Actinomyces eriksonii*),	(C_2-C_6), organic acids, and	
Bacteroides sp.,	alcohols	
Butyrivibrio, Eubacterium,		
Lactobacillus, and *Ruminococcus*		
Fusobacterium fusiforme,	Organic acids	73, 193
Peptostreptococcus anaero-		
bius, P. prevotti, Actino-		
myces viscosus, Arachnia		
propionica, Veillonella		
parvula, and *V. alcalescens*		

fluid, synovial fluid, or urine samples.[85-87,91, 92,145,148-151]

2. Comparison of metabolic profiles or fingerprints of clinical specimens with standard profiles prepared with known organisms.[3,80,85, 92,149,152,153]

3. Identification of specific components or metabolites of organisms after isolation (on selective media) from clinical specimens.[3,80,154,155]

In conventional methods of diagnosis of infectious disease, bacteria are isolated from clinical specimens and then identified by morphological, biochemical, and serological methods. A minimum of 16 hr is needed to carry out the isolation and identification procedures. In many instances, the number of organisms per milliliter of blood (or other body fluid) from patients is very low (frequently one organism or less per milliliter of blood).[156] Therefore, blood culture may not be positive by the current blood-culturing methods. Furthermore, bactericidal or bacteriostatic compounds present in blood may continue to exert their influence on bacterial growth in cultures. Occasionally, recognition of multiple bacteremia, which occur in 5 to 10% of cases, may be delayed if the microscopic morphology of the two (or more) infecting strains is similar. Certain fastidious organisms (such as mycobacteria, *Mycoplasma, Listeria, Leptospira,* anaerobes, and cell wall-deficient ["L" form] bacteria) may not grow in conventional media. From the foregoing, it is apparent that a rapid method of identification of etiologic agents would be very useful. In some cases, the potential speed (due to elimination or reduction of time needed for incubation) and specificity would represent a major advantage for gas-chromatographic techniques over cultural procedures. The real value of the GC method lies in the possibility of basing etiologic diagnoses on direct analysis of body fluids from infected patients. The technique may be applied by the analysis of one or more specific compounds that may be present in clinical specimens, correlating the specific activity of the organism.

Mitruka[3,80] and Mitruka et al.[89] reported a number of characteristic products in sera or tissues associated with bacterial activity in vivo. Cell wall and capsular polysaccharides of pneumococcal pneumonia were analyzed by GC in bacterial cultures, cell wall preparations, vaccine prepara-

tions (containing capsular polysaccharide), and sera of animals and patients with pneumococcal infections. Polysaccharides from cultures, bacterial cells, or clinical specimens were hydrolyzed with an equal volume of 8 N HCl in methanol for 1 hr at 100°C. After neutralization of the sample with ammonium hydroxide, the samples were evaporated to dryness under a stream of nitrogen. Silyl derivatives were prepared by adding a mixture of pyridine, trimethyl chlorosialane (TMCS), and hexamethyl disialazane (HMDS) (10:1:5). The sample was heated at 70°C for 30 min and centrifuged; a supernatant solution containing trimethyl silyl (MS) derivatives of sugars and amino sugars was concentrated under a stream of nitrogen. A 3-μl aliquot was introduced in a gas-chromatographic instrument equipped with a flame ionization detector (FID) and an electron capture detector (ECD). Components of polysaccharide were separated on a 6-ft stainless steel (or glass) column packed with 3% SE-30 (or OV-17) coated on 60/80 mesh Chromosorb W. The analyses were made with column temperature at 170°C, injector temperature at 230°, and detector temperature at 220°C. With these conditions, pneumococcal infections were characterized by the presence of components of polysaccharide antigens of pneumococci which were recognized by the cell wall analysis of rough (A66R2) or smooth (A66) varieties of the organism (Figure 4.6). Although the exact identity of the compounds was not determined, MS analysis indicated that the cell wall components (marked by arrows) were hexoses and *N*-acetyl aminosugars and the capsular polysaccharides were hexuronic acid. Thus, pneumococcal infection can be rapidly identified by the analysis of patient sera.

Gas-chromatographic profile (fingerprint) analysis of serum or culture media may be used for presumptive identification of infective agents without identification of individual peaks (compounds) in the chromatograph. This method has been utilized by Mitruka[3,157] for diagnosis of mycobacterial infections in patients (Figure 4.7). A standard chromatographic profile of tuberculosis was established by a GC analysis of *Mycobacterium tuberculosis* cultures and sera from 117 patients with pulmonary tuberculosis. Also, chromatographic patterns were established by analysis of replicate cultures of the so-called atypical mycobacteria (Figure 4.8). Because of the

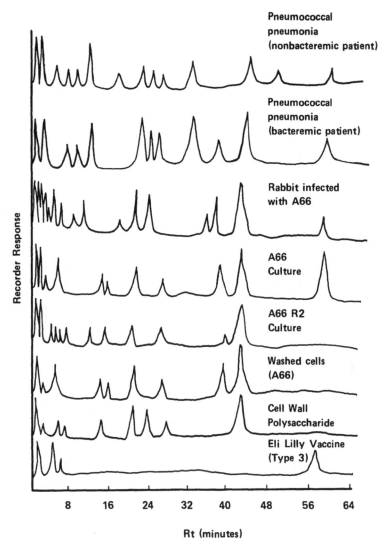

FIGURE 4.6. Gas-chromatographic patterns of pneumococcal components in culture and sera of infected animals and humans.

increasing incidence of mycobacterial infection in humans due to atypical mycobacterial strains, the chromatographic method may be very useful in the differential diagnosis of mycobacteriosis. GC profiles were also characterized by the analysis of cultures of various streptococcal strains and sera from rabbits and humans with streptococcal endocarditis (Figure 4.9). Distinctive patterns were obtained when sera from patients with endocarditis caused by *Streptococcus,* pneumococcus, *Staphylococcus,* or fungi were compared (Figure 4.10). Infectious endocarditis in humans has undergone a significant evolution during the past quarter century because of the availability of numerous antimicrobial agents, the use of new

techniques in cardiac surgery, and the longer survival of patients with ultimately fatal diseases. Although antibiotic therapy has changed an almost universally fatal disease to a serious disease that can be cured, infectious endocarditis still has an appreciable mortality and morbidity rate due to pneumococcal, staphylococcal, enterococcal, and sometimes fungal agents. Therefore, it is essential to establish the diagnosis and institute correct treatment early, since the sooner adequate therapy is begun, the less likely serious or irreparable damage to the heart or other organs is to occur. Gas-chromotographic procedures might be very useful in differentiating the etiologic agents involved in infectious endocarditis by comparing

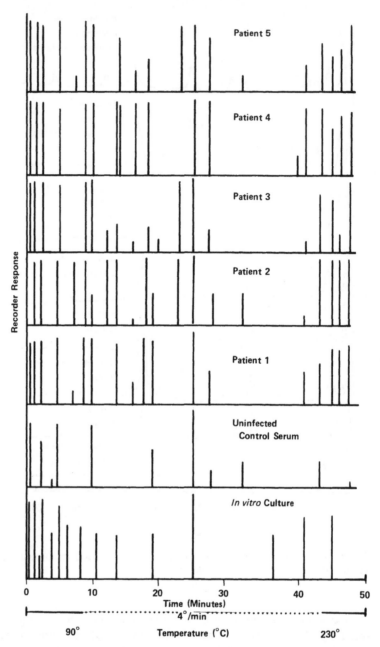

FIGURE 4.7. Gas-chromatographic patterns of sera from patients infected with *Mycobacterium tuberculosis*.

the standard chromatographic profiles (Figure 4.10). Nonbacteremic streptococcal endocarditis was established based on the clinical diagnosis of patients with unexplained fever, changing heart murmurs, and other classical signs of the disease.

One of the problems in the rapid identification of etiologic agents is that the organism may not be isolated from blood cultures. Diagnosis is then made by the clinician, based on the patient's history and clinical signs. Gas chromatography may be a valuable adjunct in such cases, whereby it can be used as an aid in the etiologic diagnosis of various infectious diseases. Mitruka[157] studied the application of GC in the diagnosis of nonbacteremic infections. Serum samples were collected over a 7-year period from nonbacteremic patients who had otherwise specific bacterial infections established by clinical, serological, and

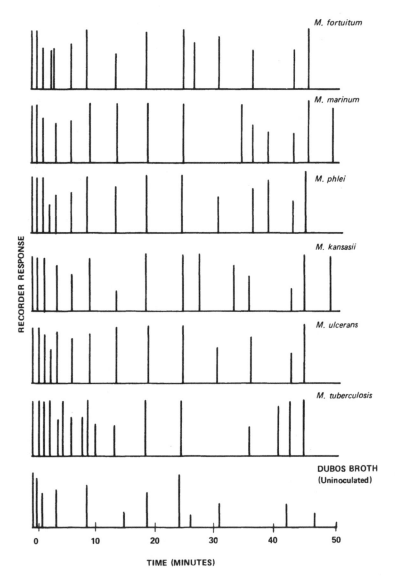

FIGURE 4.8. Gas-chromatographic patterns of mycobacterial cultures.

radiological evidence. The GC profiles of sera from such patients (presented in Figure 4.11) were comparable to those of bacteremic patients and specific bacterial cultures analyzed by this method. Rapid diagnosis of bacterial diseases by comparing distinct profiles has also been reported for urinary tract infections due to Gram-negative bacteria;[85] arthritis due to streptococcal, staphylococcal, or gonococcal bacteria;[91,92] and meningitis due to various Gram-negative and Gram-positive bacteria.[148,149]

Although the application of GC methods by the direct analysis of clinical specimens is potentially very useful, it is occasionally complicated by individual variations in patients, virulence of organisms in different hosts, multiple infections, and therapeutic measures. Another approach may be used to overcome these difficulties in the application of GC identification of bacteria. This approach involved characterization of metabolic products of organisms after primary isolation from clinical material. Colonies isolated on the surface of solid media may be incubated for brief periods (1 to 4 hr) to yield metabolic products that are detectable by highly sensitive GC means.[3,158]

The full potential or limitations for effective use of GC methods in the diagnosis of infectious

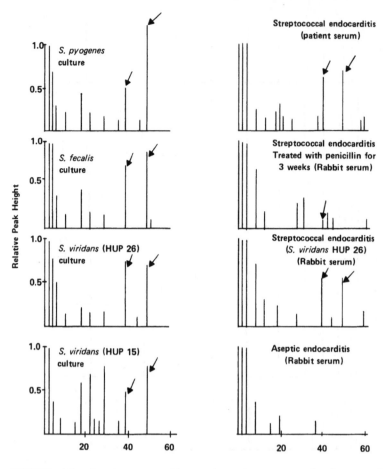

FIGURE 4.9. Gas-chromatographic patterns of streptococcal cultures and infected sera.

disease have not been established at present. However, many investigators in this field are currently engaged in refining the technique by standardization of GC conditions, instrumentation, and procedures so that the method may be included as part of the routine bacteriological procedure in clinical microbiology laboratories. If the GC technique can be used as a rapid test to differentiate between infectious and noninfectious diseases or to differentiate among bacterial, viral, fungal, or noninfectious diseases, the analysis of body fluids will be exceedingly valuable.

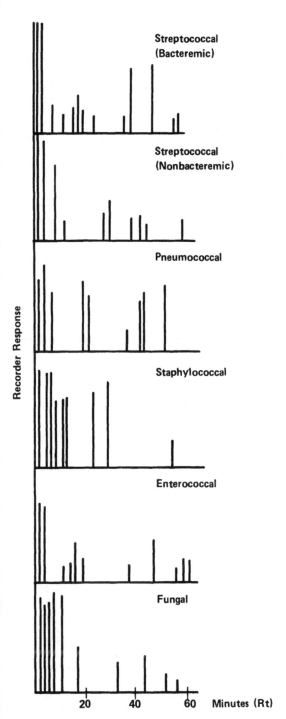

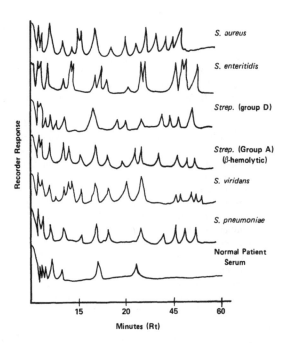

FIGURE 4.11. Gas-chromatographic patterns of sera from patients with nonbacteremic infections.

FIGURE 4.10. Gas-chromatographic patterns of sera from patients with infectious endocarditis.

REFERENCES

1. **Horvath, C.,** Theory and practice of gas chromatography, in *Gas Chromatographic Applications in Microbiology and Medicine,* Mitruka, B. M., Ed., John Wiley & Sons, New York, 1975.
2. **Ettre, L. S. and Zlatkies, A.,** Eds., *The Practice of Gas Chromatography,* John Wiley & Sons, New York, 1968.
3. **Mitruka, B. M.,** Ed., *Gas Chromatographic Applications in Microbiology and Medicine,* John Wiley & Sons, New York, 1975.
4. **Grant, D. W.,** *Gas Liquid Chromatography,* Van Nostrand Reinhold, London, 1971.
5. **Littlewood, A. B.,** *Gas Chromatography, Principles, Techniques and Applications,* Academic Press, New York, 1970.
6. **Ottenstein, D. M.,** Column support materials for use in gas chromatography, *J. Gas Chromatogr.,* 6, 129, 1968.
7. **Palframan, J. F. and Walker, E. A.,** Technique in gas chromatography. I. Choice of solid supports, a review, *Analyst,* 92, 71, 1967.
8. **Street, H. V.,** Application of gas chromatography in clinical chemistry, *Adv. Clin. Chem.,* 12, 217, 1969.
9. **Thielman, H., Behrens, V., and Leibuitz, E.,** Quantitation of gas chromatography by titration — evaluation by coulometric titration, *Chem. Technol.,* 14, 162, 1962.
10. **Phillips, C. S. G. and Timms, P. L.,** Molecular weight determination with the Martin Density Balance, *J. Chromatogr.,* 5, 131, 1961.
11. **Brody, S. S. and Chaney, J. E.,** Flame photometric detector: The application of a specific detection for phosphorous and for sulfur compounds sensitive to subnanogram quantities, *J. Gas. Chromatogr.,* 4, 42, 1966.
12. **Karmen, A. and Giuffrida, L.,** Enhancement of the response of the hydrogen flame ionization detector to compounds containing halogens and phosphorous, *Nature,* 201, 1204, 1964.
13. **Wolfgang, R. and Rowland, F. S.,** Radioassay by gas chromatography of thitium and carbon 14-labelled compounds, *Anal. Chem.,* 30, 903, 1958.
14. **Kroman, H. S. and Bender, S. R.,** *Theory and Application of Gas Chromatography in Industry and Medicine,* Grune and Stratton, New York, 1969.
15. **Kay, W. I.,** Far ultraviolet spectroscopic detection of GC effluent, *Anal. Chem.,* 34, 287, 1962.
16. **Haahti, E. A. and Fales, H. M.,** Infrared functional group detection of gas chromatographic eluates, *Chem. Ind.* (London), 16, 507, 1961.
17. **Behrendt, S. L.,** Infrared microspectrometer using molecular beam to monitor capillary gas chromatography effluents, *Nature,* 201, 70, 1964.
18. **Lindeman, L. P. and Annis, J. L.,** Use of a conventional mass spectrometer as a detector for gas chromatography, *Anal. Chem.,* 32, 1741, 1960.
19. **Watson, J. T.,** in *Ancillary Techniques of Gas Chromatography,* Ettre, L. S. and McFadden, W. H., Eds., Interscience, New York, 1969.
20. **Ryhage, R.,** Use of a mass spectrometer as a detector and analyzer for effluents emerging from high temperature gas liquid chromatographic columns, *Anal. Chem.,* 36, 759, 1964.
21. **Brooks, C. J. W. and Middleditch, B. S.,** The mass spectrometer as a gas chromatographic detector, *Clin. Chim.,* 34, 145, 1971.
22. **Sweeley, C. C., Elliott, W. H., Fries, I., and Ryhage, R.,** Mass spectrometric determination of unresolved components in gas chromatographic effluents, *Anal. Chem.,* 38, 1549, 1966.
23. **Walsh, J. R. and Merrit, C.,** Qualitative functional group analysis of gas chromatographic effluents, *Anal. Chem.,* 32, 1378, 1960.
24. **Walker, J. Q., Jackson, M. T., Jr., and Maynard, J. B.,** *Chromatographic Systems,* Academic Press, New York, 1972.
25. **Dalnagase, S. and Juvet, R. S.,** *Gas Liquid Chromatography, Theory and Practice,* Interscience, New York, 1962.
26. **Schomburg, G., Weeke, F., Weimann, B., and Ziegler, E.,** Data processing in gas chromatography, *Angew Chem.,* 2, 366, 1972.
27. **Hammar, C. G., Holmstedt, B., Lindgren, J. E., and Tham, R.,** The combination of gas chromatography and mass spectrometry in the identification of drugs and metabolites, *Adv. Pharm. Chemother.,* 7, 53, 1969.
28. **Hammar, C. G.,** Mass fragmentography and elemental analysis by means of a new and combined multiple ion detector peak matcher device, *Acta Pharm. Suec.,* 8, 129, 1971.
29. **Holmstedt, B. and Linnerson, A.,** Chemistry and means of determination of hallucinogens and marihuana, in *Drug Abuse — Proceedings of the International Conference,* Zarafonetis, C. J. D., Eds., Lea & Febiger, Philadelphia, 1972, 291.
30. **Jenden, D. J. and Cho, A. K.,** Applications of integrated gas chromatography/mass spectrometry in pharmacology and toxicology, *Annu. Rev. Pharmacol.,* 13, 371, 1972.
31. **McFadden, W. H.,** *Techniques of Combined Gas Chromatography Mass Spectrometry: Applications in Organic Analysis,* John Wiley & Sons, New York, 1973.
32. **Ryhage, R. and Stenhagen, E.,** in *Mass Spectrometry of Organic Ions,* McLafferty, F. W., Ed., Academic Press, New York, 1973, chap. 9.
33. **Moss, C. W. and Dees, S. B.,** Identification of microorganisms by gas chromatographic mass spectrometric analysis of cellular fatty acids, *J. Chromatogr.,* 112, 595, 1975.

34. Holmstedt, B. and Palmér, L., Mass fragmentography: principles, advantages, and future possibilities, *Adv. Biochem. Psychopharmacol.*, 7, 1, 1973.
35. Schulten, H. R., Beckey, H. D., Meuzelaar, H. L. C., and Boerboom, A. J. H., High resolution field ionization mass spectrometry of bacterial pyrolysis products, *Anal. Chem.*, 45, 191, 1973.
36. Nau, H. and Biemann, K., Computer assisted assignment of retention indices in gas chromatography mass-spectrometry and its application to mixtures of biological origin, *Anal. Chem.*, 46, 426, 1974.
37. Oswald, E. O., Albron, P. W., and McKinney, J. D., Utilization of gas liquid chromatography coupled with chemical ionization and electron impact mass spectrometry for the investigation of potentially hazardous environmental agents and their metabolites, *J. Chromatogr.*, 98, 363, 1974.
38. Gohlke, R. S., Time of flight mass spectrometry + gas liquid partition chromatography, *Anal. Chem.*, 31, 535, 1959.
39. Rees, D. I., Interfacil systems for the coupling of gas chromatography with mass spectrometry, *Talanta*, 16, 903, 1969.
40. Wallar, G. R., *Biochemical Applications of Mass Spectrometry*, Interscience, New York, 1972.
41. Mattauch, J., A double focusing mass spectrograph and masser of $N^{15} + O^{18}$, *Phys. Rev.*, 50, 617, and 1089, 1936.
42. Johnson, E. G. and Nier, A. O., Angular aberrations in sector shaped electromagnetic lenses for focusing beams of charged particles, *Phys. Rev.*, 91, 10, 1953.
43. Nier, A. O. and Roberts, T. R., The determination of atomic mass doublets by means of a mass spectrometer, *Phys. Rev.*, 81, 507, 1951.
44. Watson, J. T. and Bieman, K., High resolution mass spectra of compounds emerging from gas chromatography, *Anal. Chem.*, 36, 1135, 1964.
45. Barber, M., Chapman, J. R., Green, B. N., Merren, T. O., and Riddock, R. G., 18th Annu. Conf. Mass Spectrometry and Allied Topics, San Franscisco. June 1970. B299.
46. Bonelli, E. J., Gas chromatography/mass spectrometer techniques for determination of interferences in pesticide analysis, *Anal. Chem.*, 44, 603, 1972.
47. Strong, J. M. and Atkinson, A. J., Simultaneous measurement of plasma concentration of lidocaine and its disethylated metabolites by mass fragmentography, *Anal. Chem.*, 44, 2287, 1972.
48. Henneberg, D., Ein kontinuierliches Verfahren zur Massenspektrometrischen bestimmung Gaschromatographisch vorgetrennter Substanzgemische, *Z. Anal. Chem.*, 170, 365, 1959.
49. Henneberg, D., Eine Kombination von Gaschromatograph und Mesenspektrometer zur Analyse Organischer Stoffemische, *Z. Anal. Chem.*,, 183, 12, 1961.
50. Gohlke, R. S., Time-of-flight mass spectrometry: Application to capillary column gas chromatography, *Anal. Chem.*, 34, 1332, 1962.
51. Henneberg, D. and Schomburg, G., Mass spectrometric identification in capillary gas chromatography, *Gas Chromatogr. Int. Symp.*, (U.S.), p. 191, 1962.
52. Hammar, C. G., Holmstedt, B., and Ryhage, R., Mass fragmentography: Identification of chlorpromazine and its metabolites in human blood by a new method, *Anal. Biochem.*, 25, 532, 1968.
53. Hammar, C. G. and Hessling, R., Novel peak matching technique by means of a new and combined multiple ion detector peak matcher device, *Anal. Chem.*, 43, 298, 1971.
54. Hammar, C. G., Qualitative and quantitative analysis of drugs in body fluids by means of mass fragmentography and a novel peak matching device, in *Proc. Symp. Gas Chromatography Mass Spectrometry*, Frigerio, A., Ed., Raven Press, New York, 1974, 1.
55. Wiesendanger, H. U. D. and Tao, F. T., in *Recent Developments in Mass Spectroscopy*, Ogata, K. and Haikawa, T., Eds., University of Tokyo Press, Tokyo, 1970, 290.
56. Cho, A. K., Lindeke, B., Hodson, B. J., and Jenden, D. J., A gas chromatography/mass spectrometry assay for amphetamine in plasma, in Proc. 5th Int. Congr. Pharmacology, San Francisco, 1972, 41.
57. Jenden, D. J. and Silverman, R. W., A multiple specific ion detector and analog data processor for gas chromatograph/quadrupole mass spectrometer system, *J. Chromatogr. Sci.*, 11, 68, 1973.
58. Elkin, K. L., Peirrou, U. G., Ahlborg, B., Holdstedt, B., and Lindgren, J. E., Computer controlled mass fragmentography with digital signal processing, *J. Chromatogr. Sci.*, 11, 117, 1973.
59. Beckey, H. D., *Field Ionization Mass Spectrometry*, Oxford and Akademie Verlag, Berlin, 1971.
60. Beckey, H. D., Knoppel, H., Metzinger, G., and Schulze, P., in *Advances in Mass Spectrometry*, Vol. 3, Mead, W. H., Ed., Institute of Petroleum, London, 1966.
61. Schüddemage, H. D. R. and Hemmel, D. O., in *Advances in Mass Spectrometry*, Vol. 4, Kendrick, E., Ed., Institute of Petroleum, London, 1968, 857.
62. Simmonds, P. G., Whole microorganisms studied by pyrolysis-gas chromatography-mass spectrometry: Significance for extraterrestrial life detection experiments, *Appl. Microbiol.*, 20, 567, 1970.
63. Schulten, H. R. and Beckey, H. D., Field desorption mass spectrometry with high temperature activated emitters, *Org. Mass Spectrom.*, 6, 885, 1972.
64. Henneberg, D. and Frexenius, Z., High resolution mass spectra of compounds separated by capillary columns — use of plate scan technique, *Anal. Chem.*, 221, 321, 1966.
65. Freedman, A. N., The gas chromatograph-mass spectrometer interface, *Anal. Chem.*, 59, 19, 1972.
66. Simpson, C. F., Gas chromatography-mass spectroscopy interfacial systems, *CRC Crit. Rev. Anal. Chem.*, 3, 1, 1972.

67. Llewellyen, P. and Littlejohn, D., *Varian Tech. Q.*, Spring 1966.
68. Lipsky, S. R., Horvath, C. G., and McMurray, W. J., in *Gas Chromatography*, 2nd ed., Littlewood, A. B., Ed., Academic Press, New York, 1966.
69. Blumer, M., An integrated gas chromatography-mass spectroscopy system with carrier gas separator, *Anal. Chem.*, 40, 1590, 1968.
70. Brunee, C., Paper #46, 17th Annu. Conf. Mass Spectrometry and Allied Topics, Dallas, May 1969.
71. Grayson, M. A. and Wolf, C. J., Efficiency of molecular separators for interfacing a gas chromatograph with a mass spectrometer, *Anal. Chem.*, 39, 1438, 1967.
72. Oyama, V. I., Use of gas chromatography for the detection of life on Mars, *Nature*, 200, 1058, 1963.
73. Bricknell, K. S., Sutter, V. L., and Finegold, S. M., Detection and identification of anaerobic bacteria, in *Gas Chromatographic Applications in Microbiology and Medicine*, Mitruka, B. M., Ed., John Wiley & Sons, New York, 1975.
74. Brooks, J. B. and Moore, W. E. C., Gas chromatographic analysis of amines and other compounds produced by several species of *Clostridium*, *Can. J. Microbiol.*, 15, 1433, 1969.
75. Moore, W. E. C., Relationships of metabolic products to taxonomy of anaerobic bacteria, *Int. J. Syst. Bacteriol.*, 20, 535, 1970.
76. Moore, W. E. C., Cato, E. P., and Holdeman, L. V., *Ruminococcus bromii* n. sp. and emendation of the description of *Ruminococcus*, *Int. J. Syst. Bacteriol.*, 22, 78, 1972.
77. Moss, C. W., Howell, R. T., Farshy, D. C., Dowell, V. R., and Brooks, J. B., Volatile acid production of *Clostridium botulinum* type F, *Can. J. Microbiol.*, 16, 421, 1970.
78. Mitruka, B. M. and Alexander, M., Ultra sensitive detection of certain microbial metabolites by gas chromatography, *Anal. Biochem.*, 20, 548, 1967.
79. Mitruka, B. M. and Alexander, M., Rapid and sensitive detection of bacteria by gas chromatography, *Appl. Microbiol.*, 16, 636, 1968.
80. Mitruka, B. M., Biochemical aspects of *Diplococcus pneumoniae* infections in laboratory rats, *Yale J. Biol. Med.*, 44, 253, 1971.
81. Mitruka, B. M., Rapid and automated identification of microorganisms in clinical specimens by gas chromatography, *Symp. Applied Microbiology*, John Wiley & Sons, New York, 1973.
82. Moss, C. W., Dowell, V. R., Jr., Farshtchi, D., Raines, L. J., and Cherry, W. B., Cultural characteristics and fatty acid composition of *Propionibacteria*, *J. Bacteriol.*, 97, 561, 1969.
83. Moss, W. C., Kellogg, D. S., Jr., Farshy, D. C., Lambert, M. A., and Thayer, J. D., Cellular fatty acids of pathogenic *Neisseria*, *J. Bacteriol.*, 104, 63, 1970.
84. Henis, Y., Gould, J. R., and Alexander, M., Detection and identification of bacteria by gas chromatography, *Appl. Microbiol.*, 14, 513, 1966.
85. Brooks, J. B., Cherry, W. B., Thacker, L., and Alley, C. C., Analysis by gas chromatography of amines and nitrosamines produced *in vivo* and *in vitro* by *Proteus mirabilis*, *J. Infect. Dis.*, 126, 143, 1972.
86. Brooks, J. B., Kellogg, D. S., Thacker, L., and Turner, E. M., Analysis by gas chromatography of hydroxy acids produced by several species of *Neisseria*, *Can. J. Microbiol.*, 18, 157, 1972.
87. Brooks, J. B., Weaver, R. E., Tatum, H. W., and Billingsley, S. A., Differentiation between *Pseudomonas testosteroni* and *P. acidovorans* by gas chromatography, *Can. J. Microbiol.*, 18, 1477, 1972.
88. Newman, J. S. and O'Brien, R. T., Gas chromatographic presumptive test for coliform bacteria in water, *Appl. Microbiol.*, 30, 584, 1975.
89. Mitruka, B. M., Kundargi, R. S., and Jonas, A. M., Gas chromatography for rapid differentiation of bacterial infection in man, *Med. Res. Eng.*, 11, 7, 1972.
90. Mitruka, B. M., Rapid detection and identification of bacteria, Symp. 1st Int. Congr. Bacteriology, Jerusalem, September 1973.
91. Brooks, J. B., Kellogg, D. S., Alley, C. C., Short, H. B., and Handsfield, H. H., Gas chromatography as a potential means of diagnosing arthritis. I. Differentiation between staphylococcal, gonococcal, and traumatic arthritis, *J. Infect. Dis.*, 129, 660, 1974.
92. Brooks, J. B., Identification of disease and disease causing agents by analysis of spent culture media and body fluids with electron capture gas liquid chromatography, in *Microbiology — 1975*, Schlessinger, D., Ed., American Society for Microbiology, Washington, D.C., 1975.
93. Mitruka, B. M., Jonas, A. M., Alexander, M., and Kundargi, R. S., Rapid differentiation of certain bacteria in mixed populations by gas chromatography, *Yale J. Biol. Med.*, 46, 104, 1973.
94. Healy, M. J. R., Chalmers, R. A., and Watts, R. W. E., Reduction of data from the automated gas-liquid chromatographic analysis of complex extracts from human biological fluids using a digital electronic integrator and an off-line computer programme, *J. Chromatogr.*, 87, 365, 1973.
95. Hertz, H. S., Hites, R. A., and Biemann, K., Identification of mass spectra by computer-searching a file of known spectra, *Anal. Chem.*, 43, 681, 1971.
96. Menger, F. M., Epstein, G. A., Goldberg, D. A., and Reiner, E., Computer matching of pyrolysis chromatograms of pathogenic microorganisms, *Anal. Chem.*, 44, 423, 1972.

97. Finkle, B. S., Foltz, R. L., and Taylor, D. M., A comprehensive GC-MS reference data system for toxicological and biomedical purposes, *J. Chromatogr. Sci.*, 12, 304, 1974.

98. Meuzelaar, H. L. C. and In't Veld, P. A., A technique for Curie Point pyrolysis gas chromatography of complex biological samples, *J. Chromatogr. Sci.*, 10, 213, 1970.

99. Reiner, E., Identification of bacterial strains by pyrolysis-gas-liquid chromatography, *Nature*, 206, 1272, 1965.

100. Reiner, E. and Ewing, W. H., Chemotaxonomic studies of some gram negative bacteria by means of pyrolysis-gas-liquid chromatography, *Nature*, 217, 191, 1968.

101. Reiner, E. and Kubica, G. P., Predictive value of pyrolysis-gas-liquid chromatography in the differentiation of *Mycobacteria*, *Amer. Rev. Respir. Dis.*, 99, 42, 1969.

102. Reiner, E., Beam, R. E., and Kubica, G. P., Pyrolysis gas-liquid-chromatography studies for the classification of mycobacteria, *Amer. Rev. Respir. Dis.*, 99, 750, 1969.

103. Reiner, E., Hicks, J. J., Ball, M. M., and Martin, W. J., Rapid characterization of *Salmonella* organisms by means of pyrolysis gas-liquid-chromatography, *Anal. Chem.*, 44, 1058, 1972.

104. Vincent, P. G. and Kubits, M. M., Pyrolysis gas-liquid-chromatography of fungi: Differentiation of species and strains of several members of the *Aspergillus flavus* group, *Appl. Microbiol.*, 20, 957, 1970.

105. Haddadin, J. M., Stirland, R. M., Preston, N. W., and Collard, P., Identification of *Vibrio cholerae* by pyrolysis gas-liquid chromatography, *Appl. Microbiol.*, 25, 40, 1973.

106. Myers, A. and Watson, L., Rapid diagnosis of viral and fungal diseases in plants by pyrolysis and gas-liquid chromatography, *Nature*, 223, 964, 1969.

107. Meuzelaar, H. L. C. and Kistemaker, P. G., A technique for fast and reproducible fingerprinting of bacteria by pyrolysis mass spectrometry, *Anal. Chem.*, 45, 587, 1973.

108. Meuzelaar, H. L. C., Ficke, H. G., and den Harink, H. C., Fully automated Curie-Point pyrolysis gas-liquid chromatography, *J. Chromatogr. Sci.*, 13, 12, 1975.

109. Kistemaker, P. G., Meuzelaar, H. L. C., and Posthumus, M. A., Rapid and automated identification of microorganisms by Curie Point pyrolysis techniques, in *New Approaches to the Identification of Microorganisms*, Héden, C.G. and Illéni, T., Eds., John Wiley & Sons, New York, 1975.

110. Schulten, H. R., High resolution field ionization and field desorption mass spectrometry of pyrolysis products of complex organic material, in *New Approaches to the Identification of Bacteria*, Héden, C.G. and Illéni, T., Eds., John Wiley & Sons, 1975.

111. Bühler, C. and Simon, W., Curie point pyrolysis gas chromatography, *J. Chromatogr. Sci.*, 8, 323, 1970.

112. Abel, K., DeSchmertzing, H., and Peterson, J. I., Classification of microorganisms by analysis of chemical composition. I. feasibility of utilizing gas chromatography, *J. Bacteriol.*, 85, 1040, 1963.

113. Yamakawa, T. and Ueta, N., Gas chromatographic studies of microbial components. I. Carbohydrate and fatty acid constitution of *Neisseria*, *Jpn. J. Exp. Med.*, 34, 361, 1964.

114. Steinhauer, J. E., Flentge, R. L., and Lechowich, R. V., Lipid patterns of selected microorganisms as determined by gas-liquid chromatography, *Appl. Microbiol.*, 15, 826, 1967.

115. Farshy, D. and Moss, C. W., Characterization of bacteria by gas chromatography: Comparison of trimethylsilyl derivatives of whole-cell hydrolysates, *Appl. Microbiol.*, 17, 262, 1969.

116. Farshy, D. C. and Moss, C. W., Characterization of clostridia by gas chromatography: Differentiation of species by trimethylsilyl derivatives of whole-cell hydrolysates, *Appl. Microbiol.*, 20, 78, 1970.

117. Bøvre, K., Hytta, R., Jantzen, E., and Froholm, L. O., Gas chromatography of bacterial whole-cell methanolysate, *Acta Pathol. Microbiol. Scand. Sect. B*, 80, 683, 1972.

118. Froholm, L. O., Jantzen, E., Hytta, R., and Bøvre, K., Gas chromatography of bacterial whole-cell methanolysates, *Acta Pathol. Microbiol. Scand. Sect. B*, 80, 672, 1972.

119. Jantzen, E., Froholm, L. O., Hytta, R., and Bøvre, K., Gas chromatography of bacterial whole-cell methanolysates, *Acta Pathol. Microbiol. Scand. Sect. B*, 80, 660, 1972.

120. Drucker, D. B., The identification of streptococci by gas-liquid chromatography, *Microbios*, 5, 109, 1972.

121. Drucker, D. B. and Owen, I., Bacterial identification by gas liquid chromatography, *Proc. Soc. Gen. Microbiol.*, 69, 9, 1971.

122. Kaneda, T., Fatty acids in the genus *Bacillus*. I. Iso- and anteiso-fatty acids as characteristic constituents of lipids in 10 species, *J. Bacteriol.*, 93, 894, 1967.

123. Farshtelli, D. and McClung, N. M., Effect of substrate on fatty acid production in *Nocardia asteroides*, *Can. J. Microbiol.*, 16, 213, 1970.

124. Drucker, D. B., Griffith, C. J., and Melville, T. H., Fatty acid fingerprints of *Streptococcus mutans* grown in a chemostate, *Microbios*, 7, 17, 1973.

125. Consenza, B. and Girard, A. E., Fine structure and fatty acid composition of *Streptococcus faecalis* and a motile pigmented streptococcus, *Bacteriol. Proc.*, 45, 1970.

126. Drucker, D. B. and Owen, I., Chemotaxonomic fatty acid fingerprints of bacteria grown with, and without, aeration, *Can. J. Microbiol.*, 19, 247, 1973.

127. Amstein, C. F. and Hartman, P. A., Differentiation of some enterococci by gas chromatography, *J. Bacteriol.*, 113, 38, 1973.

128. **Mitruka, B. M. and Alexander, M.,** Cometabolism and gas chromatography for the sensitive detection of bacteria, *Appl. Microbiol.,* 17, 551, 1969.

129. **Mitruka, B. M., Jonas, A. M., and Alexander, M.,** Rapid detection of bacteremia in mice by gas chromatography, *Infect. Immun.,* 2, 474, 1970.

130. **Kaneda, T.,** Biosynthesis of branched-chain fatty acids. I. Factors affecting relative abundance of fatty acids produced by *Bacillus subtilis, Can. J. Microbiol.,* 12, 501, 1966.

131. **Kaneda, T.,** Fatty acids in the genus *Bacillus.* II. Similarity in the fatty acid compositions of *Bacillus thuringinesis, B. anthracis,* and *B. cereus, J. Bacteriol.,* 95, 2210, 1968.

132. **Raines, L. J., Moss, C. W., Farshy, D., and Pittman, B.,** Fatty acids of *Listeria monocytogenes, J. Bacteriol.,* 96, 2175, 1968.

133. **Walker, R. W. and Fagerson, I. S.,** Studies of the lipids of *Arthrobacter globiformis* 616. I. The fatty acid composition, *Can. J. Microbiol.,* 11, 229, 1965.

134. **Moore, W. E. C., Cato, E. P., and Holdeman, L. W.,** Fermentation patterns of some *Clostridium* species, *Int. J. Syst. Bacteriol.,* 16, 383, 1966.

135. **Holdeman, L. V. and Moore, W. E. C.,** *Anaerobe Laboratory Manual,* Virginia Polytechnic Institute and State University, Blacksburg, Va., 1972.

136. **Sutter, V. L., Attebery, H. R., Rosenblatt, J. E., Bricknell, K., and Finegold, S. M.,** *Anaerobic Bacteriology Manual,* Extension Division, University of California, Los Angeles, 1972.

137. **Simmonds, P. G., Pettit, B. C., and Zlatkis, A.,** Esterification, identification and gas chromatographic analysis of Krebs cycle keto acids, *Anal. Chem.,* 39, 163, 1967.

138. **Zaura, D. and Metcaff, J.,** Quantification of seven tricarboxylic acid cycle and related acids in human urine by gas-liquid chromatography, *Anal. Chem.,* 41, 1781, 1969. 139.

139. **Rogosa, M. and Love, L. L.,** Direct quantitative gas chromatographic separation of C_2-C_6 fatty acids, methanol and ethyl alcohol in aqueous microbial fermentation media, *Appl. Microbiol.,* 16, 285, 1968.

140. **Mayhew, J. W. and Gorbach, S. L.,** Rapid gas chromatographic technique for presumptive detection of *Clostridium botulinum* in contaminated food, *Appl. Microbiol.,* 29, 297, 1975.

141. **Lambert, M. A. and Moss, C. W.,** Gas liquid chromatography of short chain fatty acids on Dexsil 300 GC, *J. Chromatogr.,* 74, 335, 1972.

142. **Brooks, J. B., Dowell, V. R., Farshy, D. C., and Armfield, A. Y.,** Further studies on the differentiation of *Clostridium surdellii* from *Clostridium bifermentans* by gas chromatography, *Can. J. Microbiol.,* 16, 1071, 1970.

143. **Brooks, J. B., Moss, C. W., and Dowell, V. R.,** Differentiation between *Clostridium sordelli* and *C. bifermentans* by gas chromatography, *J. Bacteriol.,* 100, 528, 1969.

144. **Mitruka, B. M. and Alexander, M.,** Differentiation of *Brucella canis* from other *Brucellae* by gas chromatography, *Appl. Microbiol.,* 20, 649, 1970.

145. **Mitruka, B. M. and Alexander, M.,** Halogenated compounds for the sensitive detection of clostridia by gas chromatography, *Can. J. Microbiol.,* 18, 1519, 1972.

146. **Salanitro, J. P. and Muirhead, P. A.,** Quantitative method for the gas chromatographic analysis of short-chain monocarboxylic and dicarboxylic acids in fermentation media, *Appl. Microbiol.,* 29, 374, 1975.

147. **Brooks, J. B., Kellogg, D. S., Thacker, L., and Turner, E. M.,** Analysis by gas chromatography of fatty acids found in whole cultural extracts of *Neisseria* species, *Can. J. Microbiol.,* 17, 531, 1971.

148. **Miller, G. C., Witwer, M. W., Brande, A. I., and Davis, C. E.,** Rapid identification of *Candida albicans* septicemia in man by gas liquid chromatography, *J. Clin. Invest.,* 54, 1235, 1974.

149. **Davis, C. E. and McPherson, R. A.,** Rapid diagnosis of septicemia and meningitis by gas liquid chromatography, in *Microbiology — 1975,* Schlessinger, D., Ed., American Society for Microbiology, Washington, D.C., 1975.

150. **Mitruka, B. M. and Jonas, A. M.,** Rapid identification of bacteria in cultures, blood, and other body fluids of animals by gas chromatography, *Bacteriol. Proc.,* 72, 87, 1972.

151. **Amundson, S., Brande, A. I., and Davis, C. E.,** Rapid diagnosis of infection by gas-liquid chromatography: Analysis of sugars in normal and infected cerebrospinal fluid, *Appl. Microbiol.,* 28, 298, 1974.

152. **Kundargi, R. S. and Mitruka, B. M.,** Rapid detection of bacterial infections by gas chromatography, Proc. Int. Congr. Bacteriology, Jerusalem, 1973.

153. **MacGee, J.,** Characterization of mammalian tissues and microorganisms by gas-liquid chromatography, *J. Gas Chromatogr.,* 6, 48, 1968.

154. **Mitruka, B. M.,** Detection and identification of microorganisms by gas chromatography, Symp. Proc. Int. Congr. Food Microbiology and Hygiene, Kiel, Germany, September 1974.

155. **Carlsson, J.,** Simplified gas chromatographic procedure for identification of bacterial metabolic products, *Appl. Microbiol.,* 25, 287, 1973.

156. **Finegold, S. M.,** Early detection of bacteremia, in *Bacteremia,* Sonnenwirth, A. C., Ed., Charles C Thomas, Springfield, Ill., 1973.

157. **Mitruka, B. M.,** Diagnosis of infectious endocarditis, pneumonia, and mycobacterial infections by gas chromatography, unpublished data, 1976.

158. Mitruka, B. M., Rapid automated identification of microorganisms, in clinical specimens by gas chromatography, in *New Approaches to the Identification of Microorganisms*, Héden, C.-G. and Illéni, T., Eds., John Wiley & Sons, New York, 1975, 123.

159. Ettre, K. and Varidi, P. F., Pyrolysis gas chromatographic technique for direct analysis of thermal degradation products of polymers, *Anal. Chem.*, 34, 752, 1962.

160. Simon, W. and Giacoggo, H., Thermal fragmentation and structure determination of organic compounds, *Chem. Ing. Tech.*, 37, 709, 1965.

161. Jackson, M. T. and Walker, J. Q., Pyrolysis gas chromatography of phenyl polymers and phenyl ethers, *Anal. Chem.*, 43, 74, 1971.

162. Levy, R. L., Pyrolysis gas chromatography: A review of the technique, *Chromatogr. Rev.*, 8, 48, 1966.

163. Dencker, W. D. and Wolf, C. J., Vapor phase pyrolysis gas chromatography of methyl ethers, *J. Chromatogr. Sci*, 8, 534, 1970.

164. Oyama, V. I. and Carle, G. C., Pyrolysis gas chromatography application of life detection and chemotaxonomy, *J. Gas Chromatogr.*, 5, 151, 1967.

165. Reiner, E., Studies on differentiation of microorganisms by pyrolysis gas liquid chromatography, *J. Gas Chromatogr.*, 5, 65, 1967.

166. Meuzelaar, H. L. C., Posthumus, M. A., Kistemaker, P. G., and Kistemaker, J., Curie point pyrolysis in direct combination with low voltage electron impact ionization mass spectrometry, *Anal. Chem.* 45, 1546, 1973.

167. Ishizuka, I., Ueta, N., and Yamakawa, T., Gas chromatographic studies of microbial components. II. Carbohydrate and fatty acid constitution of the family *Micrococcaceae*, *Jpn. J. Exp. Med.*, 36, 73, 1966.

168. Girard, A. E., A comparative study of the fatty acids of some micrococci, *Can. J. Microbiol.*, 17, 1503, 1971.

169. Jantzen, E., Bergan, T., and Bøvre, K., Gas chromatography of bacterial whole cell methanolysates. VI. Fatty acid composition of strains within Micrococcaceae, *Acta Pathol. Microbiol. Scand. Sect. B*, 82, 785, 1974.

170. Agate, A. D. and Vishniac, W., Characterization of *Thiobacillus* species by gas-liquid chromatography of cellular fatty acids, *Arch. Mikrobiol.*, 89, 257, 1973.

171. Moss, C. W. and Dunkelberg, W. E., Jr., Volatile and cellular fatty acids of *Haemophilus vaginalis*, *J. Bacteriol.*, 100, 544, 1969.

172. Moss, C. W. and Lewis, V. J., Characterization of clostridia by gas chromatography. I. Differentiation of species by cellular fatty acids, *Appl. Microbiol.*, 15, 390, 1967.

173. Brown, J. P. and Cosenza, B. J., Fatty acid composition of lipids extracted from three spherical bacteria, *Nature*, 204, 802, 1964.

174. Brian, B. L. and Gardner, E. W., Cyclopropane fatty acids of rugose *Vibrio cholerae*, *J. Bacteriol.*, 96, 2181, 1968.

175. Lewis, V. J., Weaver, R. E., and Hollis, D. G., Fatty acid composition of *Neisseria* species as determined by gas chromatography, *J. Bacteriol.*, 96, 1, 1968.

176. Jantzen, E., Bryn, K., and Bøvre, K., Gas chromatography of bacterial whole cell methanolysates. IV. A procedure for fractionation and identification of fatty acids and monosaccharides of cellular structures, *Acta Pathol. Microbiol. Scand. Sect. B*, 82, 753, 1974.

177. Jantzen, E., Bryn, K., Bergan, T., and Bøvre, K., Gas chromatography of bacterial whole cell methanolysates. V. Fatty acid composition of Neisseriae and Moraxellae, *Acta Pathol. Microbiol. Scand. Sect. B*, 82, 767, 1974.

178. Rédai, I., Réthy, A., Sebessi-Gönczy, P., and Váczi, L., Lipids in *Staphylococcus aureus* and *Escherichia coli* cultured in the presence of human serum., *Acta Microbiol. Acad. Sci. Hung.*, 18, 297, 1971.

179. Rosenqvist, H., Kallio, H., and Nurmikko, V., Gas chromatographic analysis of citric acid cycle and related compounds of *Escherichia coli* as their trimethylsilyl derivatives, *Anal. Biochem.*, 46, 224, 1972.

180. Nesbitt, J. A., III and Lennarz, W. J., Comparison of lipids and lipopolysaccharide from the bacillary and L forms of *Proteus* P18, *J. Bacteriol.*, 89, 1020, 1965.

181. Lucchesi, M., Cattaneo, C., and de Ritis, G. C., The chromatographic separation of fatty acids (C11-C20) from mycobacteria, *Bull. Int. Union Tuberc.* (Technical Committee Meetings, Paris), p. 65, 1966.

182. Thoen, C. O., Karlson, A. G., and Ellefson, R. D., Comparison by gas-liquid chromatography of the fatty acids of mycobacterium avium and some other nonphotochromogenic Mycobacteria, *Appl. Microbiol.*, 22, 560, 1971.

183. Kando, E., Kanai, K., Nishimura, K., and Tsumita, T., Analysis of host originated lipids associating with *in vivo* grown tubercle bacilli, *Jpn. J. Med. Sci. Biol.*, 23, 315, 1970.

184. Thoen, C. O., Karlson, A. G., and Ellefson, R. D., Differentiation between *Mycobacterium kansasii* and *Mycobacterium marinum* by gas-liquid chromatographic analysis of cellular fatty acids, *Appl. Microbiol.* 24, 1009, 1972.

185. Smith, P. F., Lipid composition of *Mycoplasma neuralyticum*, *J. Bacteriol.*, 112, 554, 1972.

186. Tuboly, S., The lipid composition of pathogenic and saprophytic mycobacteria, *Acta Microbiol. Acad. Sci. Hung.*, 15, 207, 1968.

187. Gray, G. R. and Ballow, C. E., Isolation and characterization of a polysaccharide containing 3-0-methyl-D-mannose from *Mycobacterium phlei*, *J. Biol. Chem.*, 22, 6835, 1971.

188. Wietzerbin-Falszpan, J., Das, B. C., Gros, C., Petit, J.-F., and Lederer, E., The amino acids of the cell wall of *Mycobacterium tuberculosis* var. *bovis* strain BCG, *Eur. J. Biochem.*, 32, 525, 1973.

189. **Meyer, D. M. and Blazevic, D. J.,** Differentiation of human mycoplasma using gas chromatography, *Can. J. Microbiol.,* 17, 297, 1971.

190. **Wade, T. J. and Mandle, R. J.,** New gas chromatographic characterization procedure: Preliminary studies on some *Pseudomonas* species, *Appl. Microbiol.,* 27, 303, 1974.

191. **Yano, I. and Saito, K.,** Gas chromatographic and mass spectrometric analysis of molecular species of Corynomycolic acid from *Corynebacterium ulcerans, FEBS Lett.,* 23,, 352, 1972.

192. **Ellender, R. D., Hildalgo, R. J., and Grumbles, L. C.,** Characterization of five clostridial pathogens by gas-liquid chromatography, *Am. J. Vet. Res.,* 31, 1963, 1970.

193. **Werner, H.,** Anaerobierdifferenzierung durch Gaschromatographische Staffwechselanalysen, *Zentralbl. Bakteriol. Parasitenkd. Infektionskr. Hyg. Abt. 1 Orig. Reihe A,* 220, 446, 1972.

194. **Bassette, R., Bawdanand, R. E., and Claydon, T. J.,** Production of volatile materials in milk by some species of bacteria, *J. Dairy Sci.,* 50, 167, 1967.

INTRODUCTION

In the 1950s automation entered the area of the clinical laboratory with the first widely used automatic analytical instrument, the Autoanalyzer®.[1] Stevens[2] discussed two approaches to the problem of automation in the clinical laboratory, namely continuous flow and discrete analysis. The former system is the more successful, beginning as a single channel system in which a single result is produced from a single sample. Multichannel systems are now commonplace. An example is the Technicon instrument in which reactions take place in continuously flowing air-segmented streams. Means of detection, such as the colorimeter and the spectrophotometer, are available. The advantages of continuous flow include minimal cross contamination of samples (as regulated by air segmentation), built-in dialysis setups, and capability of continuously plotting concentrations of unknown against known standard concentrations. Discrete analysis is a system in which individual samples are processed in separate tubes, thereby reducing interaction between successive samples. Automatic diluters and dispensers of samples and reagents were forerunners of discrete analyzers.

When automation in the clinical chemistry laboratory is established with favorable results, it is just a matter of time until automated methodology becomes part of the bacteriology laboratory. The need for mechanization and automation in bacteriology is considerable, but bacteriology is still far from reaching the degree of development attained by clinical chemistry. The main problem in adopting the automated instruments and methodology in the bacteriology laboratory is the wide variety of methods utilized in bacteriological work. In particular, clinical bacteriology is highly individualistic and subjective. Concepts of quality control and standardization have not been widely applied, and training was not generally structured until about a decade ago. With recent advances in technology, the increasing work load, and the demonstration of application in clinical chemistry showing remarkable savings in time and labor, the new technical developments have been introduced into clinical bacteriological laboratories at an increasing rate since the mid-1960s. Although each new technique and automated procedure will require a thorough evaluation of its accuracy and precision, as well as an assessment of its benefit relative to cost, mechanization and automation of many procedures are now feasible. The automated methods in bacteriology can be expected to extend the capacity of the trained microbiologist and permit him to spend more time on the more difficult aspects of his work.

Rapid diagnosis of life-threatening infections due to the presence of bacteria or one or more of their products from clinical specimens can be of the utmost importance. In order to help improve reporting time to the clinician without sacrificing either economy or accuracy from the laboratory, a number of automated methods are in use or currently under investigation. In general, two approaches toward the ultimate technology of rapid diagnosis have been used, namely, the measurement and presentation of results of the test procedures and the manipulation of the specimens or other materials. These two aspects of automation in bacteriology will be discussed in this chapter. The data processing aspects of automated methods and computer identification of bacteria will be dealt with in the following chapter.

MECHANIZATION OF REPETITIVE PROCEDURES USED IN BACTERIOLOGY

Mechanization of routine bacteriological procedures (such as the preparation of dilutions, inoculations, culture transfers, and mixing reagents; streaking of plates; recording of growth; measurement of turbidity; preparation, dispensing, and sterilization of media; and cultivation and staining of bacteria) can result in improved analysis time, reduced materials consumption, or increased output. The main problem is to see the usefulness of innovations in the perspective of the complete test. This section briefly describes the mechanical devices used in the bacteriology laboratory to perform various types of repetitive procedures.

Preparation of Culture Media

In the media service sections of large bacteriology laboratories, much routine work is involved in making various types of culture media, which includes the sterilization, washing, and assembling of large numbers of various sizes of bottles, glassware, and equipment. Preparation of culture media involves heating, filtering, dispensing, coding, and sterilizing the completed medium. It also includes making poured medium plates and special media.

Elliott and Hurst[3] reported a complete media-making unit. Day et al.[4] further described the machine as consisting essentially of a steam-jacketed stainless steel vessel in which the medium is prepared. A pump is used to empty the vessel for cleaning purposes, to recirculate for mixing, or to transfer the contents to the filtration vessel. A vacuum pump is used to filter the medium through paper pulp from the filtration chamber to a holding vessel which is also steam jacketed. Thirteen liters of water can be boiled in the first vessel in 3 to 4 min. By using the circulating pump when adding the ingredients, even agar can be dissolved in about 3 min. Filtration of 13 l of agar medium can then be completed in a further 5 min.

Dispensing Culture Medium

Three methods of mechanizing this operation have been described by Day et al.[4]

1. Filament Vial Filler*

This instrument can be used to dispense 9-ml quantities in 1-oz bottles. It can be fitted with two syringes (glass or stainless steel), one on either side. A foot switch is available for operator control, and the speed of delivery can be regulated from 0 to 32 deliveries per minute, irrespective of the size of the syringe used or the volume dispensed. Different sized syringes ranging from 15 to 130 ml are available, and volumes delivered are accurate to within ±1%. The machine is not only reliable and simple, but also quite efficient due to the complete cutoff between deliveries, even of agar medium.

Media Filler **

This is a robust device which can be adjusted to dispense from 5 to 150 ml of medium with an accuracy of ±0.5%. It consists of two main parts, a reservoir and a filler unit. The reservoir is placed within a coil of pipe, and steam is passed through at 5 psi. After filling the reservoir, the lid is clamped down, and compressed air at 5 to 10 psi is used to lift the medium to the filler unit, which is operated by a foot switch so that the operator has both hands free to manipulate the glassware to be filled. The discharge rate can be regulated, although the speed is limited by splash back of the medium. Typical filling times per minute are 50 to 60 × 9-ml quantities of Ringer's solution and 30 to 40 × 90-ml quantities of yeast-dextrose-tryptone agar.

Brewer Automatic Pipetting Machine ***

This machine consists of a variable speed electric motor driving a plate via a sealed gear box at speeds between 10 and 60 rpm. The plate has five eccentric holes, into any of which a graduated brass cam is fitted. The setting of this cam gives fine adjustment of the volume delivered. The plunger of the syringe is attached to a syringe adaptor bearing on the brass cam, and the tip of the syringe barrel is screwed into a stainless steel needle valve assembly which bears on a valve post. The stainless steel valve assembly is fitted with rubber tubing on the inlet and outlet, which terminates in male and female Luer Lok adaptors. The machine may be used with a range of syringes from 0.025 to 60 ml with an average dispensed volume error of 1%. The Filamatic® filler or Brewer automatic pipetting machine can also be used for dispensing sterile agar medium into petri dishes.

Pour Plates

A fully automatic system for preparing poured plates for bacteriological analysis was constructed and tested by Sharpe et al.[5] The machine could prepare one filled, labeled dish every 15 sec, and the speed could be increased further if necessary. The machine also has the capabilities of making decimal dilutions of bacterial suspensions, dispensing measured amounts into petri dishes, adding molten agar, mixing the contents, and labeling the dishes with sample and dilution

---

*Model DAB 6, National Instrument Company Ltd., Baltimore, Md.
**Becker Equipment Ltd., Alperton, Middlesex, England.
***Becton, Dickinson Ltd., York Ho Empire Way, Wembley, London, England.

numbers. Using only the components of the media and sterile polystyrene petri dishes, the machine can be programmed to select different media for the analysis of different types of bacteria from the same sample.

Sykes and Evans[6] developed an instrument to overcome the problems associated with manual plating out, thereby improving assay precision. In this instrument, 12-in.² assay plates are used, with an 8 × 8 array of cups symmetrically punched. The 64 cups are automatically filled with antibiotic solutions. The instrument is programmed by punched tapes that dictate the order in which cups are to be filled. There are eight separate liquid handling channels; one end of each is dipped automatically into either a container of rinsing fluid or a container holding one of the eight antibiotic solutions. The other end of each channel terminates in a mobile filling head above the surface of the assay plate. The plate and the filling head can be moved in relation to each other so that any channel end may be brought above any of the 64 cups in the agar. The amount for each cup is delivered by means of a positive-displacement metering pump in each channel.

Plating Out Bacteria on Solid Medium

Williams and Bambury[7] described a device consisting of a rotating table on which the inoculated plate was placed. The inoculum was spread by the operator resting a bacteriological loop on the surface of the agar and carefully moving in a straight line along a radius of the plate as it was rotated. A similar device was developed by Trotman[8] in which an electrically sterilizable loop, resting gently on the agar, was mechanically drawn along a radius of a rotating plate. The inoculated plates are stacked in the machine. Each plate is automatically removed from the stack in turn, the lid is clamped, the bottom of the plate is lowered onto the rotary table of a culture spreading unit, and the culture is spread. The bottom of the agar plate is then removed from the spreading unit and raised into the lid (which is released), and the complete plate is transferred to another stack. The loop is electrically heated during this process and returned to its original place over the center of the rotary table, thus ready to receive the next culture. The cycle time is approximately 1 min.

Trotman[9] further modified this procedure by automatically transporting inoculated culture plates from a holder (into which they were placed manually) to a unit in which the culture was spread by means of the electrically sterilizable loop drawn along a radium and subsequently to another holder. The plates may be inoculated manually by streaking a swab along a radius or by placing a drop of liquid culture close to the center. This machine seems to be useful in handling a large work load in a diagnostic bacteriology laboratory.

Bradshaw et al.[10] described a machine for making bacterial pour plates to determine the viable bacteria counts of milk samples. The machine could be used to count 3000 to 300,000 bacteria per milliliter of sample at the rate of five samples per minute. Samples are automatically placed on a turntable conveyor which advances the sample under an arm that dips a 0.01-ml capacity cylinder and a 0.001-ml capacity loop into the sample three times. A divided petri dish is then positioned under the dipping arm, the milk in the cylinder and loop are flushed with 3 ml of distilled water, agar is pumped into the dish, and the sample and dish are automatically removed from the turntable. The agar is solidified by passing the dish over a cold plate. The machine may be used as a time- and labor-saving apparatus, as compared to the standard plate counting technique in dairy bacteriology.

An experimental prototype of automatic plating machines (Plating Automate) was used by Bürger and Quast.[11] Three loops worked simultaneously. One loop took up a drop of the liquid specimen from a test tube and placed it in a meandering path on the surface of the agar layer. The second loop passed through the lines made by the first and distributed bacteria fetched en route in a similar way. The third loop repeated the steps of the second loop. Thus, depending on the bacterial density of the sample, single colonies were obtained after incubation along the lines of the second loop or with very high bacterial densities along the lines drawn by the third loop. The machine was capable of performing 180 plates in 1 hr.

Multiple Inoculation Techniques

Various devices have been used to allow simultaneous inoculation of several cultures on to one substrate (or one or a few cultures simultaneously onto several devices). These devices may reduce time, labor, storage space, and cost in a large bacteriology laboratory. Hill[12] described a method of rapidly dispensing media into divided petri dishes and also described an automated multiple inoculator for simultaneously transferring up to 25 isolates into each dish.

The multiple inoculator consists of a beam carrying two sets of 25-wire inoculating loops (27-gauge platinum). The beam is operated by two motors, which give either rotation or vertical lowering and raising. The loops are free to slide in guides. Two sets of locating strips are on the base of the plates. Trays carrying the divided petri dishes (Dyos Replidish)* can be slotted into the plates. The loops are sterilized for about 10 sec by a timer motor. One of the Replidishes contains a series of "master cultures" of bacterial isolates in saline suspension. The loops are brought above the "master culture" by a revalve beam motor; the inoculum is deposited onto the medium automatically by another beam motor (lower/raise). If the medium is solid, the loops rest on the surface and their stems slide freely in the loop guides. By raising the beam, the loops are lifted and sterilized, and the procedure is repeated for other plates. The machine was tested for cross contamination and aerial contamination and compared with conventional inoculation techniques. It was found to be useful for more rapid identification of aerobic and anaerobic bacterial isolates.

Counting Colonies

Schoon et al.[13] studied various types of automatic apparatuses for counting microbial colonies. Yourassowsky and Schoutens[14] described a Feiss Micro-Videomat fitted with a Tessovar® lens for counting bacterial colonies, for measuring their diameters, and for automatically tracing the colonies. Pour plates more than 5-mm thick containing homogeneous colonies of *Staphylococcus aureus, Klebsiella pneumoniae, Escherichia coli, Proteus mirabilis* or *Pseudomonas aeruginosa* (⩽2,000 organisms per plate) were counted. One final count was based on the cumulative results of ten fields, making the number of bacteria counted between 500 to 2000. The precision of the automatic measurements was reported to be generally good.

The droplet technique for counting bacteria was developed by Sharpe and Kilsby.[15] The technique correlates well with counts made by using the standard pour plate method. It is essentially a miniature pour plate method. Each dilution in molten agar is plated as a series of 0.1-ml droplets in a petri dish. This step may be automated by using a diluter/dispenser. Colonies are counted under low magnification after incubation. The technique is suitable for samples containing over 1000 bacteria per gram and yields total viable count. In practice, three sets of dilutions are made in 45°C agar. A row of five droplets is made in a petri dish for each dilution. Usually, only about half of the time is required for incubation, as compared with conventional pour plates, and up to 200 colonies per droplet can be counted with ease. The diluter/dispenser is equipped with a foot pedal control. One pedal controls filling and mixing in the pipette and the other pedal delivers the material. A magnifier viewer or projector screen is available in which droplets are magnified ten times. Pour plate preparing machines have been designed and used experimentally for preparing up to eight dilutions in each of three media. All materials are handled within the instrument in bulk, and sterile pipettes are produced automatically within the instrument as needed. The steps performed by the machine include dispensing petri dishes, removing lids, performing dilutions, adding agar, mixing, replacing lids, labeling, and counting.

Recent reports suggest that success has been achieved in automation of various bacterial processes, including plate counting methods.[7,16-20] A fully automated system for preparing poured plates for bacteriological analyses was constructed and tested by Sharpe et al.[5] The machine could make decimal dilutions of bacterial suspensions, dispense measured amounts into petri dishes, add molten agar, mix the dish contents, and label the dishes with sample and dilution numbers at the rate of 2000 dishes per 8-hr day. In addition, the machine could be programmed to select different media so that plates for different types of bacteriological analysis may be made automatically from the same sample. It used only the components of the media and sterile polystyrene petri dishes; requirements for all other material, such as sterile pipettes and capped bottles of diluents and agar, were eliminated.

Alexander and Glick[21] described an apparatus for automatically counting bacterial colonies on solid agar plates. This was an electronic scanning

*Dyos Plastics Ltd., Surboton, Surrey, England.

device in which a cathode ray tube flying spot was focused onto a culture. The variations of light transmitted during a scan were measured on a photomultiplier, and its output was processed to give a count of the colonies present. Malligo[18] found that this machine gave a lower count than the visual (manual) method if more than 100 colonies of bacteria (*Bacillus subtilis* or *Serratia marcescens*) were present and a higher count than the visual method if less than 100 colonies were present. Ingels and Daughters[17] devised a machine based on the principles of the optical scanning method for counting colonies on agar dishes. However, there was considerable difficulty in distinguishing more than one type of colony and in the discrimination between two colonies overlapping or close together.

Gilchrist et al.[22] described a method for determining the number of bacteria in a solution by the use of a sample on a rotating agar plate in an ever-decreasing amount in the form of an Archimedes spiral. After the sample was incubated, different colony densities were apparent on the surface of the plate. A modified counting grid was described which relates area of the plate to volume of sample. By counting the appropriate area of the plate, the number of bacteria in the sample was estimated. This method was compared to the pour plate procedure with the use of pure and mixed cultures in water and milk. The results did not show a significant difference between duplicates at the L = 0.01 level when concentrations of $600 \times 12 \times 10^5$ bacteria per milliliter were used, but the spiral plate method gave counts that were higher than counts obtained by the pour plate method (Table 5.1). The time and materials required in the spiral plate method were substantially less than those required by the conventional aerobic pour plate procedure.

Measurement of Growth in Liquid Cultures

Spectrophotometers and nephelometers for measurement of the turbidity of growing cultures are commonly used in bacteriology. New techniques and instruments based on photometric principles have been devised in recent years.[23-25] Munson[26] described the design and operation of various types of turbidostats for continuous

TABLE 5.1

Statistical Summary of 17 Comparisons of the Spiral and Pour-plating Methods

| Organism | No. of observations | Variance | | % difference[a] |
		Spiral	Pour plate	
Staphylococcus aureus	18	0.00186	0.00170	4.8
	18	0.00084	0.00093	9.0
	16	0.00196	0.00050	14.7[b]
	16	0.00174	0.00355	−6.5
Lactobacillus casei	18	0.00332	0.00597	25.1[b]
	20	0.00306	0.00409	24.6[b]
Bacillus subtilis	18	0.00307	0.00562	34.7[b]
	18	0.00287	0.00060	17.7[b]
Escherichia coli	18	0.00413	0.00750	5.5
	18	0.00618	0.00424	36.6[b]
	18	0.00285	0.00180	11.5[b]
	18	0.00319	0.00275	7.5
	18	0.00184	0.00130	11.0[b]
Escherichia coli and Staphylococcus aureus	18	0.00280	0.00102	0.5
	18	0.00184	0.00174	21.5[b]
Pseudomonas aeruginosa	20	0.00207	0.00083	17.9[b]
	20	0.00079	0.00264	13.2[b]

[a] Percent difference = geometric mean (spiral-pour/pour) 100.

[b] Significant at $\alpha = 0.01$ level.

From Gilchrist, J. E., Campbell, J. E., Donnelly, C. B., Peeler, J. T., and Delaney, J. M., *Appl. Microbiol.*, 25, 244, 1973. With permission.

growth recording of bacterial cultures. Robrish et al.[27] designed a fiber optic probe attachment for a spectrophotometer to continuously measure the turbidity in growing bacterial cultures. Cobb et al.[28] described a nephelometer in which 12 independent cells containing lamp, photo cell, and culture are rocked in a water bath and sequentially monitored for turbidity. A growth curve for each culture is printed out on a single chart. The machine is suitable for at the most two cultures, and the two families of growth and curves could easily overlap on the record chart.

Norris et al.[29] developed an automatic growth recorder for measurement of growth in bacterial cultures. The machine measures the optical density of a culture, but could equally well be designed to perform a nephelometric measurement. Briefly, the instrument is described as follows. Samples are withdrawn from a culture through a glass capillary tube (1-mm internal diameter) via a catheter into an automatically operated gas-tight syringe (5-ml capacity). The syringe placed in a light-tight aluminum block fits between the light source-optical unit and the photomultiplier of a spectrophotometer. A light of 6-mm diameter passes through the part of the syringe barrel which is filled with liquid when the plunger is withdrawn to its maximum extent. Optical density is recorded by a modified recorder; the servomotor is activated only for a few seconds when the syringe plunger is withdrawn to its maximum. The apparatus was designed to fit the Vitatron spectrophotometer* and Vitatron recorder with logarithmic mode of operation.

The Coulter Counter® has been used as an automated method of particle counting.[30] A bacterial suspension is diluted in an electrically conducting fluid (usually saline), and the probe of the apparatus is immersed in the mixture. The probe consists of a flat-plate electrode and a glass tube enclosing a second flat-plate electrode. There is a small orifice in the wall of the glass tube, and an electrical potential is applied across the electrodes, thus producing an electrical field across the orifice. The diluted sample is drawn through the orifice; as each particle passes through, it displaces some of the conductive fluid, thus raising the impedance of the orifice contents and producing a voltage pulse. The amplitude of the pulse is proportional to the particle size. The output is processed to give the total count in digital form. Anderson and Whitehead[30] evaluated the Coulter Counter by comparing electronic particle counts and viable cell counts by a standard plate count technique. They found an exceptionally good correlation between the counts obtained by both methods. The applications of the Coulter Counter in the measurement of bacterial growth have been described previously.[31-34]

Staining Methods

Staining methods for differentiation of bacteria have been mechanized. A Gram-staining apparatus, the Shandon-Elliot slide staining machine, was developed by Cremer.[35] Drew et al.[36] described a machine which could perform the Gram stain. Comparison of slides stained by machine vs. hand revealed no difference in reproducibility or accuracy. The authors demonstrated that the machine staining not only provided clean, dry, and uniformly stained slides but also saved 24 sec per slide when compared with a hand-staining technique.

Nelson et al.[37] devised a machine for the rapid detection of bacteria by staining with dyes followed by microscopic examination. The instrument, Partichrome®, was designed to detect airborne microbes. With this method, samples of air are drawn in (17.5 l/min, and the particles are deposited by impaction onto a moving tape (Cronar®)** coated with a thin film of immersion oil. Impaction of particles larger than 5 μm is prevented by means of a cyclone-type separator placed ahead of the impactor. The deposited microbes are treated successively with HCl (1.9% aqueous) at 60°C for 1 min, water, ethyl violet (1% in 0.5% aqueous Triton® X-100 at 85°C for 45 sec), water, and (after drying) with nitrobenzene for 5 sec. The treated tape is then scanned with a white light through an oil immersion lens, and the transmitted light beam is split. The two beams are directed toward phototubes made sensitive by means of filters to blue and green light, respectively. A scanning pattern is obtained by rotation of a mirror system placed after a zirconium arc lamp is used as the light source. When white light passes through an unstained particle, both the blue and green components are

*Fisons Scientific Instruments, Ltd., Loughborough, Leics., England.

**E. I. DuPont de Nemours and Company, Inc., Wilmington, Del.

absorbed and the output of both phototubes is decreased. If the light passes through a blue-stained particle, the output of the green phototube is decreased more than that of the blue phototube; therefore, blue particles may be counted in the presence of other particles. Pulses from the photomultipliers are analyzed with a signal analyzer (base-line straightener, amplifier, discriminator, and anticoincidence circuit).

The Biosensor®, like the Partichrome instrument, is an automated detector based on staining microorganisms and observing them electronically with microscopic optics.[38]

MEASUREMENT OF BIOCHEMICAL ACTIVITY OF BACTERIA BY AUTOMATED METHODS

A great variety of automated instruments have been developed for the detection and identification of bacteria utilizing the biochemical properties of the organisms. Some of these techniques are applied to the determinations of products of bacteria (e.g., ammonia, amino acids, fatty acids, etc.); others measure the characteristic chemical components of bacteria (chromatography-mass spectrometry, ATP, etc.).

Ammonia, Amino Acids, and Enzymes in Bacteria

Bettelheim et al.[39] developed an automated technique for the determination of ammonia produced by bacteria. The technique was based on colorimetric determination of NH_3 produced by asparaginase or glutaminase activity in the organisms. Bacterial suspensions in phosphate buffer are incubated with asparagine or glutamine at 37°C for 20 min and dialysed against the buffer to collect the ammonia produced. The ammonia is then allowed to react with alkaline phenol reagent, and the intensity of the blue color is measured at 610 nm in a flow-through colorimeter while the optical density of the material remaining in the dialyzer is measured at 480 nm. Thus, a direct relationship can be obtained between optical density or bacterial concentration and ammonia production per sample. By using this technique with various concentrations of substrates, various kinetics can be determined and correlated for taxonomic classification of bacteria.

Bascomb and Grantham[40] described an automated method for the measurement of different ammonia-releasing enzymes, its application to cell suspension, and its potential use in bacterial identification. The automated assay system is based on the continuous flow principle which was first used in the field of microbiological assay of antibiotics.[41] It consists of three channels: a protein channel, an activity channel, and a control channel. These channels are operated simultaneously. The control channel estimates ammonia released from the substrate without organism; the activity channel estimates ammonia released from the substrate by the organism; the protein channel measures the protein content of the bacterial suspension. Ammonia estimation is based on colorimetric Nessler's reaction, and the sensitivity of the measurement ranges from 0.3 to 2.4 μmol ammonia per milliliter. The entire analytical system is run at the rate of 50 samples per hour. Technicon Autoanalyzer I modules were adopted in this automated method. Ultrasonicated suspensions of bacteria were used with short reaction time (15 min). Since the ingredients of culture medium determine the phenotypic expression of the bacterial genome and most of the enzymes detected are constitutive, the results obtained with this analytical system may differ from those obtained by traditional tests. Unique profiles displaying presence of individual enzymes and relative magnitudes were seen for different bacterial strains. This system may be linked with a computer for data processing with an identification program similar to that described by Wilcox et al.[42]

ATP Determination

Bacteria detection by means of determination of adenosine triphosphate (ATP) is a sensitive method because ATP is present in all living cells. The du Pont luminescence Biometer® is specifically designed for microbiological ATP analysis. This instrument is a sensitive photometer with a digital readout. The procedure involves placing cellular extracts containing ATP into cuvettes which have luciferin, luciferase, and magnesium ions at pH 7.4. The emitted light is detected by a photomultiplier tube and converted into an electric current which is measured and displayed. The reaction is specific for ATP. Each species or organism has a concentration of ATP of a narrow range, and the value obtained is the biomass related to cell numbers through a factor for the organism under study. However, the concept of biomass leads to difficulties with mixed cultures, i.e., the limited range expands. A basic assumption in this technique is that dead cells contain no ATP. The reagents used are supplied by the manufacturers. Certain specimens (e.g., food suspensions) must be prepared so that ATP is extracted either by a filtration method or by ultrasound techniques. The lower limit of sensitivity depends on an efficient extraction of ATP, which should be separated from the tissue or intrinsic ATP. The unit of ATP measurement of bacterial cells is the femtogram (1 fg = 10^{-15} g). The luminescence Biometer sensitivity is about 500 fg ATP, which is equivalent to about 1000 bacterial cells, and the analysis of a sample requires about 10 min.

Oleniacz et al.[43] described an automated Luminol® system for the detection of small numbers of bacteria and yeasts based on a Technicon Autoanalyzer. Samples of 11 representative species of bacteria activated Luminol to give a light emission peak that decayed in about 100 sec. In all cases, the light signal minus the assay blank value was directly proportional to microbial concentration. Of the Gram-positive bacteria tested, *Bacillus stearothermophilus* gave the most light emission per cell (about 500 cells could be detected). *Micrococcus lysodeikticus* and *Sarcina lutea* gave the smallest response; the minimum number of cells detectable was greater than 10^4. The response of Gram-negative bacteria also varied with species, but 5×10^4 or fewer of the species tested (Enterobacteriaceae, *Serratia marcescens*, *Pseudomonas fluorescens*, and *Neisseria catarrhalis*) were detected. Thus, in general, the automated Luminol system was sensitive to 10^3 to 10^4 cells of each bacterial species tested. Oleniacz et al.[43] suggest that if the system is attached to an efficient air collector, the detection of biological aerosols is feasible. The simplicity, high sensitivity, and short read-out of the procedure make it attractive for practical applications in the bacterial detection and identification of biological samples. However, the reaction is activated by some substances other than heme compounds, and interference can be a serious problem. Also, the method is not applicable to the detection of bacteria in clinical specimens (such, as blood or tissue) by direct analysis of the samples. The organisms have to be

isolated or the blank values have to be carefully standardized.

Measurement of Acids and Gas in Bacterial Cultures

Since bacterial growth in the presence of carbon energy sources such as glucose results in the production of organic acids that affect the pH value of the medium, measurement of changes in pH value in appropriately formulated growth medium is a potential method for detecting bacteria. An instrument called Wolf Trap® was developed based on the fermentation properties of bacteria. It consists of a culture tube containing nutrient medium that is automatically monitored for changes in acidity and/or turbidity with pH value probes and scattered light sensors, respectively. Samples are collected and delivered with water and nutrients into five culture tubes. In actual tests, an inoculum of 10 to 20 bacteria grew to 10^3 to 10^4 in a few hours to give a signal detectably greater than background. The growth is measured with the scattered light sensor. The details of the instrument technology and principles have been described by Mitz.[25]

Radioactive gases produced as a result of radioactive substrate (sugar) fermentation by bacteria have been automatically measured by a device called Gulliver®.[25] A medium was described in which detectable radioactive levels of $^{14}CO_2$ were evolved by the metabolic activity of represntative bacteria within a few minutes to several hours. Mixed microbial populations gave rise to several population curves that may be distinguished with time. However, Washington and Yu[44] evaluated the production of $^{14}CO_2$ as an index of bacterial growth and were unable to detect $^{14}CO_2$ within 6 hr from blood cultures with inoculum sizes ranging from 4 to 4250 colony-forming units.

An automated method for radiometric detection of bacteria was described by DeLand and Wagner.[45,46] The automated instrument, Bactec®,* is a bacterial growth detector. It is used to detect bacterial growth in blood cultures by monitoring $^{14}CO_2$ gas flushed from liquid culture bottles containing ^{14}C-labeled D-glucose. Bactec is designed to handle multiple samples simultaneously on a revolving core in the center of a constant temperature incubator. During the

*Johnson Laboratories, Cockeysville, Md.

sampling, the air of the culture bottle is flushed into an ionization chamber which produces an electrical current directly proportional to the amount of radioactivity detected. The electric potential is converted to a digital index between 0 and 99 and is simultaneously printed on paper. A sample cycle takes 30 min per tray of 25 vials. To be positive, a sample must have a digital index of 20 (2 × background). Organisms detected by this technique included *Pneumococcus, Staphylococcus aureus, Escherichia coli, Klebsiella,* beta streptococci, *Gonococcus, Pseudomonas, Haemophilus, Staphylococcus epidermidis, Proteus, Streptococcus faecalis,* and yeasts.

Washington and Yu,[44] using the method of DeLand and Wagner,[46] found that it was impossible to detect 4 to 4250 colony-forming units (CFU) within 6 hr using simulated blood cultures and a small sampling of patient blood cultures. They also found no advantage in earlier detection of bacteremia with the radiometric method compared to routine procedures. Washington and co-workers used room air to flush vials. The fact that 10% CO_2 in air is recommended for detecting all aerobic and microaerophilic organisms may account for the less than optimal findings.[47] DeBlanc et al.[48] studied 2967 blood cultures by both methods and found that bacteria were detected more rapidly by the radiometric method 70% of the time. In a comparative study of the radiometric method and conventional bacteriological techniques, 2 ml of blood suspected of being bacterimic was inoculated into a culture medium containing C^{14}-glucose. Release of $^{14}CO_2$ by bacterial metabolism was checked hourly for 18 to 24 hr, then daily 2 days, and with an automated instrument on the 12th day. A 10-ml sample of blood was studied by conventional bacteriological techniques.[48] Among 57 patients studied by one or both methods, routine techniques detected bacteria in 87%, and radiometric methods detected bacteria in 85%. Using the radiometric method, 70% of the cultures were detected first, 65% on the day of inoculation. Waters[49] and Previte[50] noted an inverse relationship between various bacteria detection times and cell numbers. Depending on the species, 6 to 14 hr were required with an inocula of 100 cells.

Waters[49] prepared decimal serial dilutions of eight common bacterial species: *Escherichia coli, Enterobacter cloacae, Staphylococcus aureus, Proteus vulgaris, Streptococcus pyogenes, Pseudomonas aeruginosa, Staphylococcus epidermidis,* and *Salmonella typhimurium.* Detection times were determined by measuring the $^{14}CO_2$ metabolized from the ^{14}C-labeled glucose substrate. Detection time was proportional to the logarithm of the initial inoculum; a high sensitivity, down to one colony-forming unit, was demonstrated (Figure 5.1). Waters[49] found that detection time is affected by unlabeled dextrose in the medium, carbohydrates in the peptones used, volume of medium, quality of stirring, and specific activity of the labeled glucose. For five of the eight species, 1000 bacterial CFU were detected in 5 to 8 hr; *Pseudomonas, Streptococcus pyogenes,* and *Staphylococcus epidermidis* took about 11 hr. This technique may be used for detecting life on other planets[51] and for clinical detection of bacteremia.[46,48] It also has applications for approximate quantification of bacteria in foods and cosmetics.

Schrot et al.[52] described a modification of their coliform procedures specifically for detection of low numbers of bacteria in pure cultures and in blood after lysis and filtration. Lysis of blood with a Rhozyme®-Triton X-100 solution and filtration prior to radiorespirometric detection of bacteria eliminates antibacterial agents present in blood; $^{14}CO_2$ is evolved by blood cells which otherwise would mask detection of low numbers of bacteria. Other advantages include low liquid volume and high specific activity of ^{14}C-labeled substrates to promote rapid response. Isolated colonies which can be used for identification and sensitivity determinations appear on the filter subsequent to positive detection. The method was developed for detection of bacteremia. It can also be used on fluids (CSF) and for sterility testing.

Cell suspensions are filtered with membrane filters. The filter is then moistened with 0.1 ml of labeled medium in a small, closed apparatus. $Ba(OH)_2$-moistened filter pads are used to collect evolved $^{14}CO_2$ (Figure 5.2). $^{14}CO_2$ is assayed with conventional radioactivity counting equipment. Fewer than 100 colony-forming units of most species tested were detected in 2 hr or less. Bacteria studied include *Escherichia coli, Staphylococcus aureus,* hemolytic *Streptococcus, Salmonella paratyphi B, Haemophilus aphrophilus, Cardiobacterium* sp., *Pseudomonas aeruginosa, Klebsiella* sp., *Salmonella typhi, Shigella dysenteriae, Enterobacter aerogenes, Serratia marcescens,* and *Proteus vulgaris.*

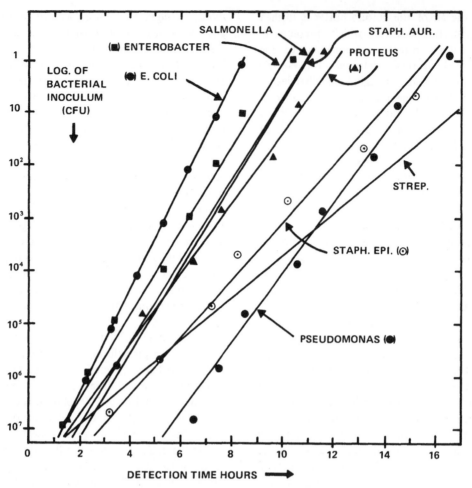

FIGURE 5.1. Radiometric detection times for eight different bacteria as a function of the initial inoculum. (From Waters, J. R., *Appl. Microbiol.*, 23, 198, 1972. With permission.)

Brooks and Sodeman[47] studied 1261 aerobic patient blood cultures using the Bactec system in parallel with a blood culture system containing Columbia broth (BBL) and sodium polyanethol-sulfonate routinely used in their laboratory. Any vial with a growth index of 20 or greater was Gram stained and subcultured onto the appropriate medium. Of the total number of vials tested, 43% gave a growth index of 20 or greater but were negative when subcultured. This proved to be the chief problem with the system. Brooks and Sodeman[47] found 311 blood cultures from 55 patients to be positive by one or both methods. Bactec detected 91% of all positive cultures; routine methods detected 88%. Both methods detected 88% of all positive patients. Of the positive cultures, 59% were detected first by Bactec; 11% were detected first by the routine method, and 30% required the same time for

detection by both methods. Brooks and Sodeman[47] found Bactec to be equal to their conventional technique in terms of positive cultures detected and in detecting positive cultures 1 to 2 days sooner (on an average) than the conventional system.

Renner et al.[53] are in disagreement with DeBlanc and other groups. Table 5.2 compares their results of the detection of bacteremia by the radiometric method and by conventional methods.

1. False-positive culture vials (primarily aerobic) with 30+ growth indices but with repeated negative smears and subcultures equivalent to 12% of cultures.

2. Time required to process 50 sets of blood cultures in the Bactec system is twice that required by the conventional system.

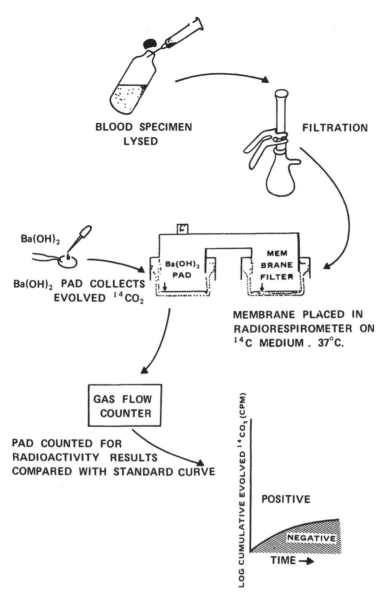

BLOOD SPECIMEN
LYSED

FILTRATION

Ba(OH)$_2$

Ba(OH)$_2$ PAD COLLECTS
EVOLVED $^{14}CO_2$

Ba(OH)$_2$
PAD

MEM
BRANE
FILTER

MEMBRANE PLACED IN
RADIORESPIROMETER ON
^{14}C MEDIUM . 37°C.

GAS FLOW
COUNTER

PAD COUNTED FOR
RADIOACTIVITY RESULTS
COMPARED WITH STANDARD CURVE

LOG CUMULATIVE EVOLVED $^{14}CO_2$ (CPM)

POSITIVE

NEGATIVE

TIME →

FIGURE 5.2. Procedure for rapid radiorespirometric detection of bacteria in blood. (From Schrot, J. R., Hess, W. C., and Levin, G. V., *Appl. Microbiol.*, 26, 867, 1973. With permission.)

3. Degree of automation not commensurate with price ($35,000).

4. Not reliable for anaerobes of *Haemophilus influenzae.*

Approximately 20% of positive cultures detected by the Bactec system were detected on the first day of incubation vs. 7% by conventional methods. DeBlanc detected 65% of positive cultures on the first day of incubation by the Bactec system. However, Renner et al.[53] argue

that a large population of *Diplococcus pneumoniae* distorted their results. They concluded that the radiometric method does not have much advantage over available conventional methods.

Improved identification of streptococci results when a 10% CO_2 wash is used instead of sterile room air. *Haemophilus influenzae* is not detectable by the standard radiometric technique. Iron protoporphyrine-(X factor) and reduced pyridine nucleotide-(V factor) impregnated strips (BBL Taxo® strips) added to aerobic culture vials

TABLE 5.2

Numbers of Cultures and Mean Detection Times (by Organism)

Organism	No. of cultures		Mean detection time (days)	
	BACTEC	Conventional	BACTEC	Conventional
Escherichia coli	23	24	2.2	2.5
Klebsiella sp.	2	4	1.5	3.0
Enterobacter aerogenes	2	6	2.5	2.5
Proteus mirabilis	1	1	3.0	3.0
Streptococcus, group D	8	8	1.9	1.9
Streptococcus, viridans group	8	10	5.1	3.6
Streptococcus pneumoniae	2	2	1	2
Staphylococcus aureus	14	16	3.4	4.5
Pseudomonas sp.	8	3	10.3	3.2
Pseudomonas aeruginosa	4	2	1.5	2
Pseudomonas maltophilia	1	1	1	2
Alcaligenes faecalis	1	0	3.0	
Haemophilus aphrophilus	1	1	4	4
Bacteroides fragilis	3	7	7	3.7
Bacteroides melaninogenicus	0	2		5.5
Fusobacterium nucleatum	1	0	10	
Torulopsis glabrata	7	2	5	13
Serratia marcescens	1	0	4	

From Renner, E. D., Gatheridge, L. A., and Washington, J. A., II, *Appl. Microbiol.*, 26, 368, 1973. With permission.

containing 30 ml of TSB and ^{14}C-labeled substrates enabled rapid radiometric detection of all strains tested in all concentrations[54] (i.e., 10^1, 10^2, and 10^4 per milliliter).

Metabolic Activity Measurement
Electrical Impedance

A change in the electrical impedance of a medium can be measured to indicate the metabolic activity of bacteria. As nutrients such as carbohydrates are metabolized into products such as lactates and carbonates; large, electrically inert molecules are replaced by a large number of electrically active molecules and ions, thus increasing the electrical conductivity of the medium. By monitoring changes in electrical impedance, the activity of a growing bacterial culture can be detected. The electrical impedance method has certain advantages over conventional methods.

1. High sensitivity of the system which allows very early detection of activity in a culture

*Stratton and Company Ltd., Hatfield, Herts., U.K.,

of microorganisms, usually within 2 hr of incubation with 10^5 cells/ml.

2. Ease of partial automation of the system.
3. Relative simplicity of the apparatus.

However, temperature fluctuation, evaporation of water, and spontaneous reactions in the media irrelevant to growth cause noise and drift in impedance measurements. A Strattometer[®]* is used to reduce such disturbances.[55] The impedance of the test sample is continuously compared with that of a control medium sample by using an impedance bridge circuit in which the test and control sample form opposing arms. The two samples are housed in a pair of conductivity measuring cells which are matched in their physical parameters. Well-sealed cells and a temperature-stable cell holder eliminate noise drift caused by temperature fluctuation and evaporation. The bridge arrangement cancels a large part of the remaining noise and drift. The balance of the bridge is immediately upset if the inoculation is greater than 10^2 bacteria per milliliter. Large inocula of 10^4 microorganisms per milliliter offset the balance of the bridge, enabling immediate estimation of the number of microorganisms in the inoculum.

Ur and Brown[56,57] studied several species of bacteria by the electrical impedance technique. With amplification of the signal, the activity of 300 *Escherichia coli* was detected within 2½ hr. With an inoculum of 3×10^4 organisms, activity was detected within 20 min. Other organisms detected by this method include species of *Staphylococcus, Klebsiella, Serratia, Streptococcus, Pseudomonas, Lactobacillus, Pediococcus, Acetobacter, Candida,* and *Mycoplasma* (Figure 5.3). Characteristic curves obtained with different organisms in different media may allow rapid identification of pathogenic species.

Buffered media are undesirable, as they reduce the effect of the electrically active products of metabolism. The advantages of the "Strattometer" include

1. Quick detection of bacterial contamination in industrial processes.
2. Clinical tests involving detection of bacterial activity as affected by antibiotics, sugars, etc. which can be detected much earlier than by conventional techniques.
3. Sensitivity for detection of activity of slow-growing microorganisms such as *Mycobacterium tuberculosis.*

Cady[58] applied the impedance measurement technique for rapid automated identification of small numbers of *Neisseria gonorrhoeae, Streptococcus pyogenes,* and common urinary tract pathogens. By using specific inhibitors, certain organisms were detected in the presence of large numbers of other organisms in a mixed culture without prior separation into pure cultures. The method was found to be highly sensitive for the detection of a number of different bacterial species (Figure 5.4). It indicates that automated simultaneous measurements of an organism in many different selective media may provide the characteristic biochemical profiles necessary for identification. The technique may also be applied to perform certain immunologic tests such as complement fixation tests and to assess the metabolism of many cellular systems in a great variety of situations.

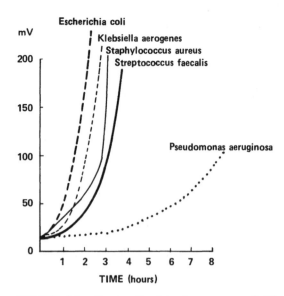

FIGURE 5.3. Comparative impedance curves of different organisms grown in PPLO broth. (From Ur, A. and Brown, D. F. J., in *New Approaches to the Identification of Microorganisms,* Héden, C. G. and Illéni, T., Eds., John Wiley & Sons, New York, 1975. With permission.)

Gas Chromatography and Mass Spectrometric Methods

Metabolic products of bacteria in growing cultures can be detected and identified by gas chromatography — alone or in conjunction with mass spectrometry.[59] The methods are described in detail in Chapter 4. Several automated instruments have been developed for the detection and identification of microorganisms.[60-63] Computer-assisted data analysis and integration have been employed by several workers and should prove helpful in rapid detection and identification of bacteria based on their metabolic product analysis.[59,64-67]

Measurements of Chemical Constituents of Bacteria

Curie Point Pyrolysis Techniques

Meuzelaar and Kistemaker[68] and Meuzelaar et al.[69-71] fully automated Curie point pyrolysis gas-liquid chromatography and mass spectrometry in combination with computer identification of the pyrograms for the analysis of streptococcal strains. The technique involves transferring bacterial colonies (equivalent to about 50 to 100 μg dry weight) to the ferromagnetic heating wires used in Curie point pyrolysis. The Curie point wires are then dried in a vacuum chamber at a pressure of 10 Torr for 3 to 5 min, mounted into a

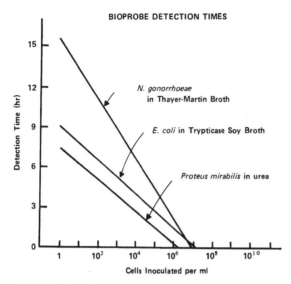

BIOPROBE DETECTION TIMES

N. gonorrhoeae
in Thayer-Martin Broth

E. coli in Trypticase Soy Broth

Proteus mirabilis in urea

Cells Inoculated per ml

FIGURE 5.4. Detection times measured in hours from inoculation plotted against the logarithm of cells inoculated for different bacteria in conventional culture media. Impedance measurements used in detection made at 100 mV and 2 kHz. (From Cady, P., *in New Approaches to the Identification of Microorganisms*, Hédén, C. G. and Illéni, T., Eds., John Wiley & Sons, New York, 1975. With permission.)

glass reaction tube, and subsequently transferred to an automatic sample exchanger. The sample exchanger transports the glass reaction tube with the sample into the Curie point pyrolysis reactor which is directly coupled to the inlet of the GLC column. Inside the reactor, the sample is centered in a high-frequency coil connected to a HF power supply (Fischer, 1.1 MHz, 1.5 kW). The sample is then pyrolyzed by energizing the HF coil, which causes rapid heating of the ferromagnetic wire. When the wire reaches the Curie temperature, the temperature of the wire stabilizes until the HF power supply is switched off. Meuzelaar et al.[70,71] pyrolyzed all samples on Fe/Ni wires of 0.5-mm diameter, with a Curie temperature of 610°C and a temperature rise time of 100 msec. Total heating time was 1 sec. The duration of a complete analysis cycle can be varied by means of the matrix board, but 45 to 60 min is normally needed for effective separation of pyrolysis products (including reconditioning, cooling, sample exchange, and restabilization) by GLC columns. The pneumatic sample exchanger has a capacity of 24 samples, and since the system is capable of continuous unattended operation, a maximum of 24 to 30 samples can be analyzed per 24 hr.

With analyses using Curie point pyrolysis gas chromatography (Py-GLC) techniques, streptococcal strains and other bacterial species gave characteristic fingerprints. Certain strains of streptococci such as sanguis showed relatively large differences, whereas other strains seemed to be more closely related (Figure 5.5). A computer was used to compare each peak height value of the unknown pyrogram with the corresponding value of the filed (known standards) strain, always dividing the smaller value by the larger one. It seems that automated Py-GLC in combination with computer evaluation of pyrograms is a powerful tool for rapid identification of streptococcal strains.

Kistemaker et al.[72] applied pyrolysis mass spectrometry (Py-MS) in conjunction with a Curie point pyrolysis system for the identification of microorganisms. The sample was introduced in the apparatus through a vacuum lock and pyrolyzed in a glass tube directly at the inlet of the expansion chamber. The pyrolysis product drifts through a 0.7-mm leak orifice into a quadrupole mass analyzer where the fragment molecules are ionized and selected to mass. The ionization occurs by impact of 15.eV electrons. Mass spectrometers are usually operated at electron energies of 50 to 70 eV to obtain maximum yields of ions. Most of the large molecules are ionized and fragmented at this electron energy, thereby complicating the analysis. In spite of the relatively low ion yield obtained with low-voltage electron impact ionization, the resulting mass spectrum is more characteristic because of the presence of large molecular ions and the absence of a bulk of small fragment ions. The mass spectra between m/e 30 and 140 were recorded, and the relative abundance of the ions was scanned by repeated fast mass scans (Figure 5.6). The signals from the electron multiplier detector (in the form of pulse) were amplified, discriminated against noise, shaped properly, counted, and stored in the appropriate channels of the signal averager. The final spectra accumulated in the memory of the signal averager were displayed on an oscilloscope and recorded on a strip chart for visual inspection. A paper punch tape read-out was used for off-line computer processing of the data.

Characteristic patterns of Py-MS analyses of various strains of streptococci were observed in the range of m/e values between 78 to 87 (Figure 5.7). The fingerprints, which were highly reproducible,

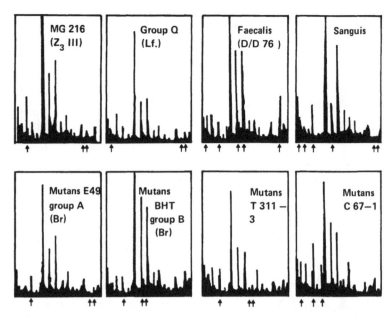

FIGURE 5.5.   Fingerprint regions from pyrograms of eight streptococcal strains.
The figures have not been corrected for sample weight. Characteristic combinations of relative peak amplitudes are indicated by arrows; (Br) refers to the subgroups of *Streptococcus mutans* proposed by Bratthåll. (From Meuzelaar, H. L. C., Kistemaker, P. G., and Tom, A., in *New Approaches to the Identification of Microorganisms,* Hédén, C. G. and Illéni, T., Eds., John Wiley & Sons, New York, 1975. With permission.)

were obtained with the analysis rate of one sample per minute using one bacterial colony in each analysis.

Electrophoretic Protein Profiles

Polyacrylamide gel electrophoresis has been used for characterization of various species of bacteria.[73-77] Many automatic instruments have been developed, including automatic scanning densitometers with data processing of multiple electrophoresis strips. Winkelman and Wyburga[78] described a system utilizing a disk integrator, automatic digital read-out, and a programmable disk top computer for calculating densitometer scans of electrophoresis strips. Larsen et al.[79] reported on polyacrylamide gel electrophoresis of various serotypes of *Corynebacterium diphtheriae.* This method may prove to be a useful epidemiological tool in establishing the distribution and occurrence of various *C. diphtheriae* types. Computers have also aided in the processing and interpretation of isoenzyme electrophoretic patterns.[80] Other automated electrophoretic instruments have been reported which may have

wide applications in the detection and identification of bacteria.[81]

Kersters and DeLey[82] described polyacrylamide gel electrophoresis and computer-assisted numerical analysis for the identification and classification of bacteria. Soluble protein extracts of *Zymomonas* and *Agrobacterium* species were analyzed by polyacrylamide gel electrophoresis. Densitometric tracings of these profiles were compared and grouped by computerized calculations of the relative mobility, sharpness, and relative protein concentration of bands and zones between bands. The method was compared and correlated with DNA-DNA hybridizations and phenotypic clustering of more than 200 genotypically well-known *Agrobacterium* strains. It seems that the technique may be quite useful for classification and rapid identification of strains within a species and species within a genus of bacteria.

AUTOMATED METHODS IN IMMUNOLOGY AND SEROLOGY

Recently, a number of automated procedures

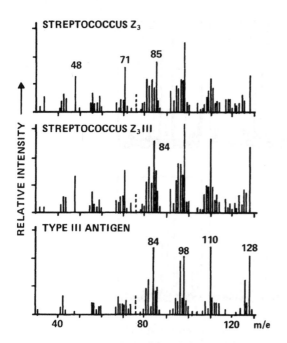

DIFFERENCE BETWEEN STREP. Z_3 AND Z_3 III
CURIE TEMP. 510°C: PYR. TIME 1 SEC.

FIGURE 5.6. Comparison of pyrolysis mass spectra of the *Streptococcus* strain Z_3 III and the mutant strain Z_3, lacking the type III polysaccharide. (From Kistemaker, P. G., Meuzelaar, H. L. C., and Posthumus, M. A., in *New Approaches to the Identification of Microorganisms,* Héden, C. G. and Illéni, T., Eds., John Wiley & Sons, New York, 1975. With permission.)

have been tested to perform serological and immunological procedures in clinical bacteriology, in public laboratories, and for the detection and identification of bacteria in food material, water, and soil samples. These techniques include the autoanalyzer, immunoelectrophoresis, immunofluorescence, fluorescent antibody staining, immunoabsorbence, and radioimmunoassay. The procedures are modifications of the conventional methods using the general principles of antigen-antibody reactions or measurement of their concentrations (precipitation, agglutination, complement fixation, immunodiffusion, etc.).

Based on the densitometry scanning of antigen-antibody precipitin reactions, a telemetric automated microbial identification system was described.[83-85] Radioimmune assays have been automated by Pollard and Waldron[86] and Haas et al.[87] The automation of a flocculation test for syphilis using the Technicon standard Autoanalyzer has been reported.[88,90] An automated

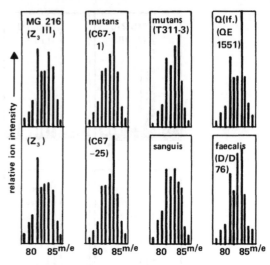

FIGURE 5.7. Characteristic Py-MS patterns observed for different streptococcal strains; asterisk indicates mutant strain. (From Kistemaker, P. G., Meuzelaar, H. L. C., and Posthumus, M. A., in *New Approaches to the Identification of Microorganisms,* Héden, C. G. and Illéni, T., Eds., John Wiley & Sons, New York, 1975. With permission.)

seroagglutination with somatic and flagellar suspension of *Brucella* and *Salmonella* species has also been described.[91,92] Automated hemagglutinin and complement-fixation reactions have been used to determine viral and bacterial concentrations during vaccine production.[93-98]

Autoanalyzer

Automated continuous-flow complement-fixation (CF) tests using simpler or more complicated flow designs have been described.[99-101] Nydegger et al.[102] used the continuous flow system of the Autoanalyzer* to measure quantitative complement fixation with hepatitis-associated antigen (HAA). The system consisted of sampler, manifold, water bath, colorimeter, and recorder (Figure 5.8). Sensitized red blood cells from sheep were pumped uninterruptedly and were incubated with identical volumes of serum or plasma dilutions of a standard healthy volunteer plasma pool and a single dilution of patient's plasma. Complement activity was measured turbidimetrically at 630 nm by loss in optical density through red cell lysis. The method consisted of incubation of 0.4 ml of antibody (Ab) dilution, 0.5 ml of guinea pig complement (GP-C), and 0.4 ml of antigen (Ag, patients serum),

*Technicon International Division, Technical Department, Tarrytown, N.Y.

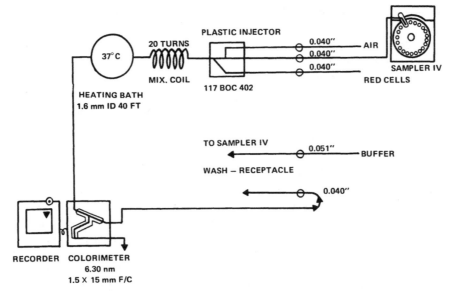

FIGURE 5.8. Flow design for automated complement dosage and quantitative CF. (From Nydegger, U. E., Lambert, P. H., Celada, A., and Miescher, P. A., in *Automation in Microbiology and Immunology*, Héden, C. G. and Illéni, T., Eds., John Wiley & Sons, New York, 1975. With permission.)

whereby the final GP-C dilution was the same as that which revealed 90% erythrocyte-amboceptor (EA) lysis when Ag and Ab were replaced by veronal buffered saline (1:300). The main advantage of this method over other automated techniques for complement estimation is the possibility of processing a large number of samples in a relatively short time (70 titrations per hour) because it avoids serial or exponential dilutions of each tested sample, requiring numerous manipulations. Assay results of identical plasma samples in the automated methods were compared with those of the classical CH_{50} method, and a significant correlation was found between the two methods.

Autoanalyzer has also been used for a performing Wasserman reaction and a flocculation test for syphilis.[88,98,103] The seroagglutination test was automated by Khoury et al.[91] to detect antibodies against the casual organisms of typhoid and paratyphoid fever and brucellosis. A Technicon Autoanalyzer was used for the sero-agglutination tests with somatic and flagellar suspensions. A series of 1000 human serum samples was tested with manual and automated methods in parallel with 6 suspensions of *Salmonella typhi* O and H; *S. paratyphi* A, H, and S; and *Brucella abortus* B, O, and H antigens. A 97.2 to 99.3% agreement was found between the results of the manual and automated tests, depending on

the antigen used. The automated test was relatively simple to perform and was rapid, accurate, and highly sensitive. As many as 60 samples per hour could be examined simultaneously for somatic and flagellar agglutination with a double-antigen probe system and a two-channel Auto-analyzer, with read-out based on percentage transmission. It seems that the Autoanalyzer could be adopted for rapid screening of a large number of samples for the presence of antibodies against Group A, B, and D *Salmonella* and *B. abortus* and possibly for other pathogenic bacteria.

Other Automated Instruments for Serological Testing

Miller et al.[92] described an automated screening test based on simultaneous agglutination and complement-fixation tests in the selective detection of *Brucella* antibodies. The automated system may provide a diagnostic result in many cases, without recourse to confirmatory conventional serological tests. It can detect a number of complement-fixation test reactors which are negative to a common test (Rose bengal plate test used for the diagnosis of Brucillosis). Miller[104,105] also described an automated acidified antigen test for specific agglutinin titration of *Brucella* antibodies. An automated application of the complement-fixation test has also been described previously by Joubert et al.[106] and Valette.[107]

Engelbrecht[108] described a Robot technician (Bioreactor) machine equipped with a multi-channel, high-precision serial dilutor and multi-channel, high-precision microdosage dispensers. It is a useful instrument for the measurement and transfer of samples in serological procedures. The machine is capable of delivering 24 fluids in any combination of 6; while working on 8 samples simultaneously, it can distribute each sample into three successive test tubes and can add different reagents to each tube, preparing rows of stepwise-increasing dilutions of the samples. It can also prepare the required control tests, mix the reacting fluids, add another series of reagents in a second cycle, and shake the flasks containing the reagents. The transfer systems and the serial dilutor are cleaned between treatments of samples. The machine is manipulated by a program consisting of a paper sheet on which the organs of the machine are represented by symbols. The order of the test tubes is indicated in squares following the symbols. The instructions to the machine are transferred to the programming sheet by punching holes in the corresponding squares. When a hole punched in a square allows contact to be established between a palp and the programming cylinder, the corresponding pilot circuit is connected to the low-voltage current. As soon as the working post gets down onto the test tubes, the information represented by the hole in the programming sheet is transformed into an order, which is then executed by the corresponding organ. By simply exchanging one programming sheet for another, programs can be changed in a few seconds. The Bioreactor can distribute 450 samples into three tubes each and perform 1350 single-dilution identification tests hourly, distribute up to 24 different solutions, and instill up to 2 ml into 1800 ampoules an hour or execute 300 semiquantitative reactions or 150 full titrations per hour. The machine may have wide applications in various serological tests such as complement fixation, hemagglutination inhibition, indirect hemagglutination, bacterial agglutination, globulin-sensitized erythrocyte agglutination, and antistreptolysin titration.

A Multiple dropper for a microtitration apparatus for serological tests was described by Sanderson.[109] The instrument consists of 12 syringes designed to dispense drops into the cavities of microtitration plates. The syringes were calibrated to deliver 0.25-ml drops. The instrument was devised to distribute drops of diluent into the cavities of the plates, titrate the test substances, and add the test reagents. The multiple dropper may be applicable to all tests which have been, or could be, adapted to the microtitration apparatus, for example, for virus hemagglutination inhibition (HI), complement fixation (CF), hemagglutination, and antiglobulin (Coombs') tests.

Automated Immunofluorescent Techniques

Immunofluorescence is widely used in diagnostic bacteriology for serological testing of toxoplasmosis, syphilis, lupus erythematosis, salmonellosis, and streptococcal infections. In recent years, methods have been developed in which immunofluorescence can be quantitatively determined with automatic microscopic scanning techniques[110] and microfluoremetric techniques.[111-113]

Holmberg and Forsum[114] reported that single cells of bacteria could be identified by immunofluorescence when admixed with morphologically similar bacteria and a large number of other contaminants. By using a semiautomated procedure for routine screening of antibody to syphilis in patients, O'Neill[115] applied the immunofluorescence test to many sera on a single slide with 2- to 5-mm diameter test areas. Microscopic examination of 12 slides (144 spots) requires about 2 hr by this method. O'Neill compared 1889 sera by the fluorescent-treponemal antibody absorption test (FTA-ABS), the cardiolipin Wasserman reaction, (CWR), and the Reiter protein complement-fixation test (RPCFT) and found that there was a significantly higher rate of antibody detection by the FTA system (Figure 5.9). The immunofluorescence method was found to be a rapid and economical screening procedure for syphilis. Stout and Lewis[90] evaluated an automated indirect fluorescent antibody test for syphilis using the seromatic system described by Binnings et al.[116] The FTA-ABS test developed by Hunter et al.[117] was subsequently evaluated by several workers.[118-120] A reliability range of 96% between the automated and manual tests using either human or rabbit sera was reported. It indicates that the FTA-ABS test can be used as an effective tool in the serodiagnosis of syphilis and that the automated test can be applied for a screening program.

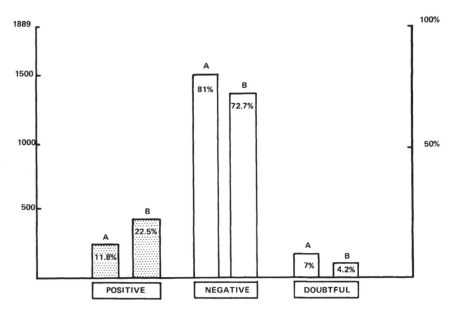

FIGURE 5.9. Screening of 1,889 sera for antibody to syphilis, a comparison of two procedures: (A) sera tested by CWR and RPCFT; (B) sera tested by modified FTA-ABS. (From O'Neill, P. and Johnson, G. D., *Ann. N. Y. Acad. Sci.*, 177, 446, 1971. With permission.)

Markovits and Burboeck[121] have described a series of kit systems. Fuoro-Kits®* contain interstandardized stabilized reagents, control serums, specially developed slides, and standardized dispensing units. Combined with the use of a new Zeiss halogen immunofluorescent microscope, these semiautomated kit systems permit a tenfold increase in operator productivity. The Fluoro-Kits utilize the direct staining procedure to test for *Salmonella* and group A Streptococci. The Fluoro-Kits utilizing the indirect technique contain slides having stabilized known antigen smears. The known antigens for use with the indirect method include virulent *Treponema pallidum* var. Nichols, *Toxoplasma gondii,* and cell nuclei.

Automated Fluorescent Antibody Technique

Mishuck and Roberts[122] defined a fully automated integrated antibody system as one that consists of (1) a specimen handling and processing subsystem, (2) a fluorescent antibody staining reaction subsystem, (3) a fluorescent read-out subsystem, and (4) a quantitation subsystem. At present, the only fully automated integrated system available is a prototype microbiological aerosol detector developed for the U.S. Army and based on the fluorescent antibody staining technique (FAST) to rapidly detect small concentrations of airborne microorganisms. Several other systems are partially automated. Potential areas of automation for fluorescent antibody tests are listed in Table 5.3. Kaufman[123] suggested the complete automation of the indirect fluorescent antibody (FA) test procedure for testing an individual blood sample for a large number of diseases. This may be accomplished by using an automated chemistry unit (SeroMatic®), which is capable of preparing slides from a number of such tests (e.g., fluorescent treponemal antibody-absorption test (FTA-ABS), malaria, toxoplasmosis, and antinuclear factor). However, no sample slide interpreting device is available because of the difficulty in distinguishing autofluorescence and debris fluorescence from positive test results.

The SeroMatic system consists of two components, a slide processor and a microscope attachment, which can be fitted onto any conventional fluorescent microscope. The total processing time required by the instrument is 85 min, 45 sec. The automated equipment has been used in bacteriology laboratories for syphilis testing and for antinuclear antibody tests.[116,124-126]

Mishuck and Roberts[122] reviewed the development of automated FA methods applied

*Clinical Sciences, Inc., Whippany, N.J.

Table 5.3

Potential Areas of Automation for Fluorescent Antibody (FA) Tests

FA test	Test for	Application
Direct	Group A streptococci	Screening for streptococcal infection
	Salmonella	Quality control of foodstuffs
Indirect	Syphilis and gonorrhea	Routine screening of all blood samples for venereal disease
	Neonatal diseases (IgM) (syphilis, toxoplasmosis, rubella, cytomegalovirus)	Screening all newborn babies
	Antibodies to cancer	Detection of cancer

From Kaufman, G. I., in *Automation in Microbiology and Immunology,* Héden, C. G. and Illéni, T., Eds., John Wiley & Sons, New York, 1975, 491. With permission.

for screening diseases of man and animals (e.g., syphilis, streptococcal infections, toxoplasmosis, mycobacteriosis, trichenellosis, etc.) and for detecting microbiological contaminants such as *Salmonella, Shigella, Escherichia coli* and *Staphylococcus aureus* in food, water, and air. A fully automated integrated system consists of at least four complete subsystems: (1) specimen handling and processing subsystem; (2) FA staining reaction subsystem (Figures 5.10 and 5.11); (3) fluroescent read-out subsystem — includes all optical equipment and electrical and electronic components, power supplies, amplifiers mechanical scanner, excitation and barrier filters, ocular viewers, phototubes, etc. (Figure 5.12); (4) quantitation subsystem — includes data-handling components, logic circuits and analyzers, plotters, printers, data storage, data displays, etc. A schematic diagram of such an integrated FA system (FAST system) is shown in Figure 5.13.

AUTOMATED ANTIBIOTIC SUSCEPTIBILITY TESTS

Various automated systems for the performance of antibiotic susceptibility tests have been described in recent years. Many of these systems use some form of particle-counting device as an indicator of bacterial proliferation and its inhibition by the antibiotic.

Bowman et al.[127] developed a technique for the automated detection and enumeration of bacterial colonies. It consists of culturing the

bacterial specimen as a suspension in nutrient agar contained in a glass capillary tube. If the tube is moved through an intense beam of light, the particles in the agar will scatter light as they pass through the beam. This light can be sensed by a photodetector positioned so as to receive none of the direct illumination from the source. The light scattered from discrete particles results in electrical pulses which increase in magnitude as the particles increase in size. This technique can be used for counting the particles as well as for sizing them on the basis of pulse amplitudes. With a fast-growing organism such as *Escherichia coli* or *Klebsiella,* growth can be detected in 2 hr or less. However, slow-growing organisms may require overnight incubation. Schoon et al.[13] developed an instrument based on these techniques and used it successfully for counting viable organisms in conjunction with their studies of the dynamics of microbial populations.

Blume et al.[128] described an automated system for the determination of bacterial antibiotic susceptibilities based on an automatic bacterial colony counter. The presence of viable bacteria in gelled medium contained in a narrow capillary tube was determined by passing them through an intense light beam and detecting the light scattered from discrete particles in the gel. Electrical pulses of certain magnitudes are counted to determine the presence of viable organisms. The procedure was partially automated but offers the possibility of fully automated operation. Application of this system may be extended to perform bioassays of

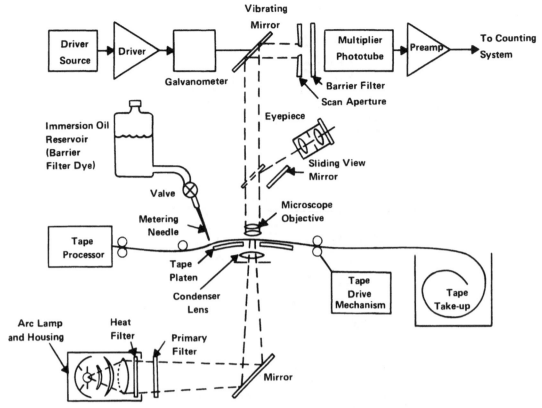

FIGURE 5.10. Fluorescent antibody stain technique, vibrating mirror image plane scanner. (From Mischuck, E. and Roberts, M., in *Automation in Microbiology and Immunology,* Héden, C. G. and Illéni, T., Eds., John Wiley & Sons, New York, 1975. With permission.)

serum levels of a particular antibiotic, screening of bacteria, and to some extent in the classification of bacteria based on antibiotic susceptibility data.

Wyatt[129] has also described the automation of antibiotic susceptibility testing using differential light scattering (DLS). The system (Differential III) is designed to simultaneously test up to nine antibiotics. A laser beam is used as the light source, and a portable microcomputer-controlled system yields hard-copy susceptibility reports within 2 min. At present, the procedure requires isolated colonies. It provides test results approximately 120 min after isolation, 30 min of this period being devoted to initiation of metabolic activity in broth. Results are based on a score range of up to 99 for the most susceptible organism. The Differential III is an open-ended system which can be further developed to permit determination of minimum inhibitory concentrations, rapid determination of antibiotic serum levels, and detection of bacteria in blood and urine specimens within a few hours, without the need for special reagents.

A fully automated device for determination of bacterial antibiotic susceptibility was described by Isenberg et al.[130] The system consisted of preparing an inoculum of bacteria isolated on enrichment or selective media.

One quarter strength Engonbroth, filtered through 0.45-μm membrane filters (Millipore) was used as the inoculum broth, and full-strength Engonbroth contained in a large plastic bottle was used in the automated machine as the test medium for the antibiotic susceptibility testing procedure. A 25% aqueous formalin solution stored within the machine was used to kill the bacteria at various test intervals. With this machine, data on 40 specimens tested against 13 antibiotics were obtained every hour after an initial 3-hr incubation period. The performance of the system was compared with the standardized Kirby-Bauer method. Substantial agreement was reported between the results obtained with the automated method and those obtained with the agar diffusion method (Table 5.4).

Praglin et al.[131] and McKie et al.[132] described

199

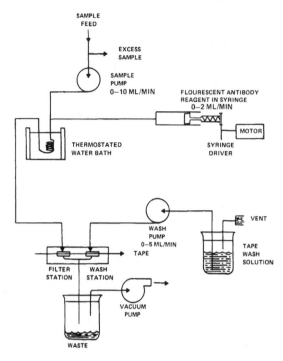

FIGURE 5.11. Basic flow diagram of FAST reactor sybsystem. (From Mischuck, E. and Roberts, M., in *Automation in Microbiology and Immunology,* Héden, C. G. and Illéni, T., Eds., John Wiley & Sons, New York, 1975. With permission.)

an automated antimicrobial susceptibility system (Autobac 1) which within 3 to 5 hr of colony selection could determine the susceptibility of a clinical isolate to a panel of up to 12 antibiotics. Correlation with ratings obtained by the Kirby-Bauer method was reported to be greater than 90%. The system, which embodies a light-scattering method, consists of four components.

1. A transparent plastic curette for automatic distribution of the prepared inoculum.
2. An incubation shaker.
3. An antibiotic-impregnated "elution-disk" dispenser.
4. An automated light-scattering photometer with print-out device.

The results are obtained in the form of a numerical ratio of the growth in each test to the growth in the control chamber; this numerical ratio is converted to a resistant, intermediate, or sensitive rating corresponding to the usual Kirby-Bauer interpretive result. The authors found that the Autobac 1 affords a high degree of reliability, simplicity of use, significant labor savings, and a

markedly higher day-to-day reproducibility and accuracy of results.

Wretlind et al.[133] described the possibility for automating the disk diffusion method of antibiotic sensitivity testing. Their technique, the Autoline system, is based on the use of 3 × 10 × 40 mm strips of nutrient agar deposited on glass slides and encased in tubes. A large number of diffusion centers can be easily handled in this machine. Various species of Gram-negative rods were tested against sulfonamide tetracycline and cephalosporin using a sensor for 65° scattered light. The Autoline system was reported to give reproducible and consistent results in good agreement with those of the manual disk diffusion method.

FUTURE CONSIDERATIONS

Bacteriology has been among the last of the clinical areas to be automated and computerized because of the nature and diversity of the laboratory procedures. One of the major problems in the development of automatic methods in bacteriology has been that all procedures (except

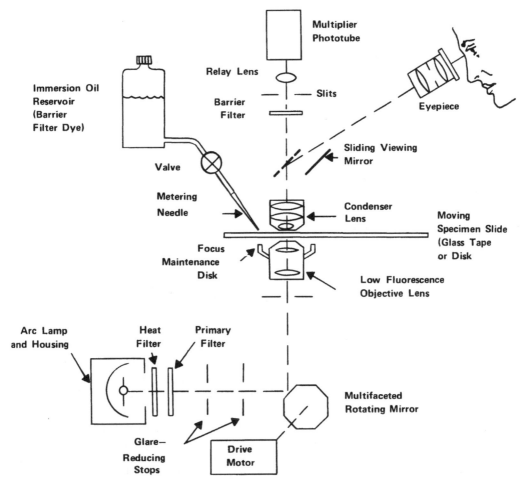

FIGURE 5.12. Flying spot scanner for fluorescent antibody scanning. (From Mischuck, E. and Roberts, M., in *Automation in Microbiology and Immunology*, Héden, C. G. and Illéni, T., Eds., John Wiley & Sons, New York, 1975. With permission.)

some serological procedures) involve handling infective material. Many technical methods have been devised for safe handling of this material.[9] It seems that further development of mechanical aids for performing the test procedures and for mechanical handling of specimens and cultures will be continued in the future, since there is undoubtedly a real need for such equipment in diagnostic, public health, and food bacteriology laboratories in order to cope with the increasing work load and cost involved in performing tests.

It seems that more emphasis will be placed on the direct methods of detection and identification of bacteria to expedite the diagnosis. For example, gas-chromatographic and mass-spectromatic methods, electric impedance measurements, and

radiometric methods have great potential in achieving this goal. In the future the instrumentation should be further refined and automated to be useful in bacteriology laboratories. At present, some of the newly automated instruments are very expensive; therefore, they might well remain research tools unless the demand makes mass production feasible. There is a need for a method that is organism specific and can rapidly detect growth by measuring some physical or chemical change (such as pH or the oxidation reduction potential, temperature, and metabolic products) in the medium in which the bacteria grow or a change in gas or other volatile emitted. If perfected, these methods would permit the bacteriologist to detect and identify the organism

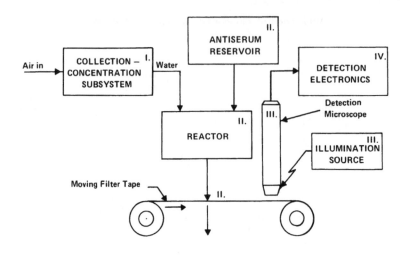

I. Sample Handling and Processing Subsystem   II. FA Staining Reaction Subsystem   III. Fluorescent Readout Subsystem

IV. Quantitation Subsystem

FIGURE 5.13. FAST system. (From Mischuck, E. and Roberts, M., in *Automation in Microbiology and Immunology,* Hédén, C. G. and Illéni, T., Eds., John Wiley & Sons, New York, 1975. With permission.)

on the same day as the specimen is sent to the laboratory, as well as eliminate the major obstacle to developing a fully automated identification system — the need to examine morphology of colonies or cultures and to recognize the distribution of colonies.

Another area of interest to the bacteriologist is the detection and identification of a small number of organisms in clinical material or biological samples. The current trend is toward development of some broad-spectrum methods including luminol chemiluminescence, determination of ATP, straining methods (e.g., Partichrome instrument, Biosensor, Fluorescein isothiocyanate staining, membrane filtration staining), physical methods (e.g., Royco® particle counter, Coulter Counter®), and methods relying on growth and metabolism (e.g., Wolf Trap, Gulliver, uptake of ^{32}P-labeled phosphate, etc.) to increase the sensitivity of methods for detection and identification of bacteria. A major problem in the application of these and other similar methods has been interference of extraneous matter present in complex biological mixtures. Therefore, these methods can only be applied in the bacteriology laboratory if samples are known to be free from or contain a constant amount of interfering matter; they can also be applied following fractionation of

samples to separate the bacteria. Efforts would be directed toward the improvement of these broad-spectrum methods to eliminate the problem of interference; for example, the inclusion of automatic sample-fractionation devices in fast read-out detectors based on broad-spectrum principles would at least partially resolve the interference problem. However, if the requirement is not for immediate detection of small numbers of organisms present in a sample, and results within several hours are acceptable, more precise and reliable methods depending on microbial growth and replication may be used. Considerable efforts are being directed toward automating such methods (e.g., immunofluorescence-membrane filtration techniques, gas chromatography-mass spectrometry-computer techniques, radioactive antibody techniques, radioassays, etc.).

Currently, data processing systems (data transmission, storage, retrieval, etc.) are being used to some extent in bacteriology laboratories, not only to the advantage of bacteriologists but also to the advantage of physicians and epidemiologists.[134-139] However, no one data processing system can be recommended at the present time that will meet the requirements of the individual bacteriology laboratory. Recently, the use of computers and other data processing equipment

TABLE 5.4

Comparison of Results Obtained by Automated and Agar Diffusion Methods of Antibiotic Susceptibility Testing

Drug	Escherichia coli			Staphylococcus auerus			Proteus mirabilis			Klebsiella pneumoniae		
	ADT zone[a]	AS ratio[b]	Interpretation[c]	ADT zone	AS ratio	Interpretation	ADT zone	AS ratio	Interpretation	ADT zone	AS ratio	Interpretation
Ampicillin	17.5	0.095	S	30	0.11	S	22.5	0.095	S	6.0	0.68	R
Cephalothin	19.0	0.05	S	32	0.095	S	17.0	0.095	E/S	19.0	0.09	S
Chloramphenicol	24.0	0.095	S	21.5	0.085	S	14.0	0.50	E/R	21.0	0.04	S
Colistin	14.0	0.03	S	6.0	0.78	R	6.0	1.00	R	12.0	0.02	S
Erythromycin	10.0	0.855	R	23.0	0.06	S	6.0	1.00	R	10.0	0.89	R
Kanamycin	20.0	0.085	S	21.0	0.06	S	21.0	0.07	S	17.0	0.06	E/S
Lincomycin	6.0	0.94	R	19.0	0.05	S	6.0	1.00	R	6.0	0.96	R
Methicillin	6.0	0.93	R	20.5	0.07	S	6.0	1.00	R	6.0	0.97	R
Oleandomycin	6.0	0.97	R	21.0	0.02	S	6.0	1.00	R	6.0	0.94	R
Penicillin	6.0	0.96	R	33.0	0.01	S	17.5	0.99	E/R	6.0	0.99	R
Tetracycline	20.0	0.85	S	22.0	0.07	S	9.0	0.99	R	19.0	0.05	S
Carbenicillin	21.5	0.02	E/S	32.0	0.01	NA	25.0	0.05	S	12.0	0.32	NA
Gentamicin	18.0	0.035	S	20.0	0.03	S	20.0	0.02	S	17.0	0.01	S

[a] Size (in millimeters) of the zone of inhibition obtained in the agar diffusion test (ADT).

[b] Ratio obtained with the automated system (AS).

[c] S, susceptible; R, resistant; E, equivocal; NA, official interpretation not available.

From Isenberg, H. D., Reichler, A., and Wiseman, D., *Appl. Microbiol.*, 22, 980, 1971. With permission.

203

(such as punch card sorting equipment for producing laboratory records and reports for clinicians) has been investigated.[140-143] It is apparent that in bacteriology few data are produced in a form suitable for direct transmission to computer. Most data have to be interpreted and coded by a trained bacteriologist before they can be transmitted to a computer. However, direct digital print-out and computer processing of bacteriological results are possible if the tests for the enumeration of bacteria or the determination of biochemical activity, serological reactions, and antibiotic susceptibility are specially designed to be used in fully automated systems. Prototypes of fully automated systems for antibiotic titrations, antibiotic sensitivities, quantitative urine cultures, complement-fixation procedures, etc. have been reported (see Chapter 6). Although applicable in bacteriology laboratories, the present systems are far from perfect; there is no doubt that the future development of computer techniques and data acquisition systems would provide tremendous advances in the operation of bacteriology laboratories.

REFERENCES

1. Skeggs, L. T., Jr., An automatic method for colorimetric analysis, *Am. J. Clin. Pathol.*, 28, 311, 1957.
2. Stevens, J. F., Current trends in automation: a review, *Med. Lab. Technol.*, 30, 139, 1973.
3. Elliott, E. C. and Hurst, A., A compact medium making unit, *J. Appl. Bacteriol.*, 27, 134, 1954.
4. Day, H. J., Elliott, E. C., Lovelock, D., Shapton, D. A., and Whitlex, D. C., Examples of mechanization in microbiology media service sections, in *Automation, Mechanization and Data Handling in Microbiology*, Society for Applied Bacteriology Tech. Ser. No. 4, Baillie, A. M. and Gilbert, R. J., Eds., Academic Press, London, 1970.
5. Sharpe, A. N., Biggs, D. R., and Oliver, R. J., Machine for automatic bacteriological pour plate preparation, *Appl. Microbiol.*, 24, 70, 1972.
6. Sykes, D. A. and Evans, C. J., An automatic plating-out machine for microbiological assay employing an 8 × 8 design, in *Some Methods for Microbiological Assay*, Society for Applied Bacteriology Tech. Ser. No 8, Board, R. G. and Lovelock, D. W., Eds., Academic Press, New York, 1975, 53.
7. Williams, R. F. and Bambury, J. M., Mechanical rotary device for plating out bacteria on solid medium, *J. Clin. Pathol.*, 21, 784, 1968.
8. Trotman, R. E., in *Automation, Mechanization and Data Handling in Microbiology*, Society for Applied Bacteriology Tech. Ser. No. 4, Baillie, A. M. and Gilbert, R. J., Eds., Academic Press, London, 1970.
9. Trotman, R. E., The automatic spreading of bacterial culture over a solid agar plate, *J. Appl. Bacteriol.*, 34, 615, 1971.
10. Bradshaw, J. G., Francis, D. W., Peeler, J. T., Leslie, J. E., Twedt, R. M., and Read, R. R., Jr., Mechanical preparation of pour plates for viable bacterial counts of milk samples, *J. Dairy Sci.*, 56, 1011, 1973.
11. Bürger, H. and Quest, R., Neue Verfahren der Bakteriologischen Diagnostik, 8th World Congr. Anatomic and Clinical Pathology, Munich, September 12—16, 1972.
12. Hill, I. R., Multiple inoculation technique for rapid identification of bacteria, in *Automation, Mechanization and Data Handling in Microbiology*, Society for Applied Bacteriology Tech. Ser. No. 4, Baillie, A. M. and Gilbert, R. J., Eds., Academic Press, New York, 1970, 175.
13. Schoon, D. J., Drake, J. F., Fredrickson, A. G., and Tsuchia, H. M., Automated counting of microbial colonies, *Appl. Microbiol.*, 20, 815, 1970.
14. Yourassowsky, E. and Schoutens, E., Automated counting of colonies by micro-videomat, in *Automation in Microbiology and Immunology*, Hedén, C. G. and Illéni, T., Eds., John Wiley & Sons, New York, 1975, 81.
15. Sharpe, A. N. and Kilsby, D. C., A rapid, inexpensive bacterial count technique using agar droplets, *J. Appl. Bacteriol.*, 34, 435, 1971.
16. Bradshaw, J. G., Francis, D. W., and Read, R. B., Automation of pour plate preparation, *J. Dairy Sci.*, 54, 755, 1971.
17. Ingels, N. B., Jr. and Daughters, G. T., II, New design for an automated bacterial colony counter, *Rev. Sci. Instrum.*, 39, 115, 1968.
18. Malligo, J. E., Evaluation of an automatic electronic device for counting bacterial colonies, *Appl. Microbiol.*, 13, 931, 1965.
19. Mansberg, H. P., Automatic particle and bacterial colony counter, *Science*, 126, 823, 1957.
20. Trotman, R. E., Automatic aerial diluting: an instrument for use in bacteriological laboratories, *J. Clin. Pathol.*, 20, 770, 1967.
21. Alexander, N. E. and Glick, D. P., Automated counting of bacterial cultures — a new machine, *IRE Trans. Med. Electron.*, 12, 89, 1958.
22. Gilchrist, J. E., Campbell, J. E., Donnelly, C. B., Peeler, J. T., and Delaney, J. M., Spiral plate method for bacterial determination, *Appl. Microbiol.*, 25, 244, 1973.
23. Forrest, W. W. and Stephan, V. A., A simple recording nephelometer for studies of bacterial growth, *J. Sci. Instrum.*, 42, 664, 1965.
24. Watson, B. W., A simple turbidity cell for continuously monitoring the growth of bacteria, *Phys. Med. Biol.*, 14, 555, 1969.
25. Mitz, M. A., The detection of bacteria and other viruses in liquids, *Ann. N.Y. Acad. Sci.*, 158, 651, 1969.
26. Munson, R. J., Construction, operation and application of surbidostats, in *Methods in Microbiology*, Vol. 2, Norris, J. R. and Ribbons, D. W., Eds., Academic Press, New York, 1970.
27. Robrish, S. A., LeRoy, A. F., Chassy, B. M., Wilson, J. J., and Krichevesky, M., Use of a fiber optic probe for spectral measurements and the continuous recording of the turbidity of growing microbial cultures, *Appl. Microbiol.*, 21, 278, 1971.
28. Cobb, R., Crawley, D. F. C., Croshaw, B., Hale, L. J., Healey, D. R., Pay, F. J., Spicer, A. B., and Spooner, D. F., The application of some automation and data handling techniques to the evaluation of antimicrobial agents, in *Automation, Mechanization and Data Handling in Microbiology*, Society for Applied Bacteriology Tech. Ser. No. 4, Baillie, A. M. and Gilbert, R. J., Eds., Academic Press, New York, 1970, 53.
29. Norris, J. R., Hewett, A. J. W., Kingham, W. H., and Perry, P. C. B., An automated growth recorder for microbial cultures, in *Automation, Mechanization and Data Handling in Microbiology*, Society for Applied Bacteriology Tech. Ser. No. 4, Baillie, A. M. and Gilbert, R. J., Academic Press, New York, 1970, 151.

30. **Anderson, G. E. and Whitehead, J. A.**, Viable cell and electronic particle count, *J. Appl. Bacteriol.*, 36, 353, 1973.
31. **Kubitschek, H. E.**, Electronic counting and sizing of bacteria, *Nature*, 182, 234, 1958.
32. **Curby, W. A., Swanto, E. M., and Lind, H. E.**, Electrical counting characteristics of several equivolume microorganisms, *J. Gen. Microbiol.*, 32, 33, 1963.
33. **Manor, H. and Haselkorn, R.**, Size fraction of exponentially growing *Escherichia coli*, *Nature*, 214, 983, 1967.
34. **Hobson, P. N. and Mann, S. O.**, Applications of the Coulter counter in microbiology, in *Automation, Mechanization and Data Handling in Microbiology*, Society for Applied Bacteriology Tech. Ser. No. 4, Baillie, A. M. and Gilbert, R. J., Eds., Academic Press, London, 1970, 91.
35. **Cremer, A. W. F.**, Automatic slide staining in clinical bacteriology, *J. Med. Lab. Technol.*, 25, 387, 1968.
36. **Drew, W. L., Pedersen, A. N., and Roy, J. J.**, Automated slide staining machine, *Appl. Microbiol.*, 23, 17, 1972.
37. **Nelson, S. S., Bolduan, O. E. A., and Shurcliff, W. A.**, The partichrome analyzer for the detection and enumeration of bacteria, *Ann. N. Y. Acad. Sci.*, 99, 290, 1962.
38. **Whittick, J. J., Movaca, R. F., and Cavanaugh, L. A.**, Analytical Chemistry Instrumentation, NASA Document No. CP 5083, U.S. Government Printing Office, Washington, D.C., 1967.
39. **Bettelheim, K. A., Kissin, E. A., and Thomas, A. J.**, An automated technique for the determination of ammonia produced by bacteria, in *Automation, Mechanization and Data Handling in Microbiology*, Society for Applied Bacteriology Tech. Ser. No. 4, Baillie, A. M. and Gilbert, R. J., Eds., Academic Press, New York, 1970, 133.
40. **Bascomb, S. and Grantham, C.**, Application of automated assay of asparaginase and other ammonia-releasing enzymes to the identification of bacteria, in *Some Methods for Microbiological Assay*, Board, R. G. and Lovelock, D. W., Eds., Academic Press, New York, 1975, 29.
41. **Gerke, J. R., Haney, T. A., Pagano, J. F., and Ferrari, A.**, Automation of the microbiological assay of antibiotics with an autoanalyzer instrument system, *Ann. N. Y. Acad. Sci.*, 87, 782, 1960.
42. **Wilcox, W. R., Lapage, S. P., Bascomb, S., and Curtis, M. A.**, Identification of bacteria by computer. III. Statistics and programming, *J. Gen. Microbiol.*, 77, 317, 1973.
43. **Oleniacz, W. S., Pisano, M. A., and Rosenfeld, M. H.**, Detection of microorganisms by an automated chemiluminescence technique, in *Technicon Symposium, Automation in Analytical Chemistry*, Vol. 1, Mediad, New York, 1966, 523.
44. **Washington, J. A., II and Yu, P. K. W.**, Radiometric method for detection of bacteremia, *Appl. Microbiol.*, 22, 100, 1971.
45. **DeLand, F. H. and Wagner, H. N., Jr.**, Early detection of bacterial growth with carbon-14-labeled glucose, *Radiology*, 92, 154, 1969.
46. **DeLand, F. and Wagner, H. N., Jr.**, Automated radiometric detection of bacterial growth in blood cultures, *J. Lab. Clin. Med.*, 75, 529, 1970.
47. **Brooks, K. and Sodeman, T.**, Rapid detection of bacteremia by a radiometric system: a clinical evaluation, *Am. J. Clin. Pathol.*, 61, 859, 1974.
48. **DeBlanc, H. J., Jr., DeLand, F., and Wagner, H. N., Jr.**, Automated radiometric detection of bacteria in 2,967 blood cultures, *Appl. Microbiol.*, 22, 846, 1971.
49. **Waters, J. R.**, Sensitivity of the $^{14}CO_2$ radiometric method for bacterial detection, *Appl. Microbiol.*, 23, 198, 1972.
50. **Previte, J. J.**, Radiometric detection of some food-borne bacteria, *Appl. Microbiol.*, 24, 535, 1972.
51. **Levin, G. V., Heim, A. H., Clendenning, J. R., and Thompson, M. F.**, "Gulliver," a quest for life on Mars, *Science*, 138, 114, 1962.
52. **Schrot, J. R., Hess, W. C., and Levin, G. V.**, Method for radiorespirometric detection of bacteria in pure culture and in blood, *Appl. Microbiol.*, 26, 867, 1973.
53. **Renner, E. D., Gatheridge, L. A., and Washington, J. A., II**, Evaluation of radiometric system for detecting bacteremia, *Appl. Microbiol.*, 26, 368, 1973.
54. **Larson, S. M., Charache, P., Chen, M., and Wagner, H. N., Jr.**, Automated detection of *Haemophilus influenzae*, *Appl. Microbiol.*, 25, 1011, 1973.
55. **Ur, A. and Brown, D. F. J.**, Detection of bacterial growth and antibiotic sensitivity by monitoring changes in electrical impedance, *IRCS Med. Sci.*, 1, 37, 1973.
56. **Ur, A. and Brown, D. F. J.**, Rapid detection of bacterial activity using impedance measurements, *Biomed. Eng.*, 9, 18, 1974.
57. **Ur, A. and Brown, D. F. J.**, Monitoring of bacterial activity by impedence measurements, in *New Approaches to the Identification of Microorganisms*, Hedén, C. G. and Illéni, T., Eds., John Wiley & Sons, New York, 1975, 61.
58. **Cady, P.**, Rapid automated bacterial activity by impedance measurements, in *New Approaches to the Identification of Microorganisms*, Hedén, C. G. and Illéni, T., Eds., John Wiley & Sons, New York, 1975, 73.
59. **Mitruka, B. M.**, *Gas Chromatographic Applications in Microbiology and Medicine*, John Wiley & Sons, New York, 1975,
60. **Mitz, M. A.**, Space-age automated analysis, in *Automation in Microbiology and Immunology*, Hedén, C. G. and Illéni, T., Eds., John Wiley & Sons, New York, 1975, 3.
61. **Oro, J. and Tornabene, D. W.**, Bacterial contamination of some carbonaceous meteorites, *Science*, 150, 1046, 1965; 167, 765, 1970.

62. Updegrove, W. S., Oro, J., and Zlatkis, A., 6-C quadrupole mass spectrometric analysis of organic compounds, *J. Gas. Chromatogr.*, 5, 359, 1967.

63. Updegrove, W. S. and Oro, J., Analysis of organic matter on the moon by gas chromatography-mass spectrometry: A feasibility study, in *Research in Physics and Chemistry*, Malina, F. J., Ed., Pergamon Press, New York, 1969, 53.

64. Mitruka, B. M., Detection and identification of microorganisms by gas chromatography, Symp. Proc. Int. Congr. Food Microbiology and Hygiene, Kiel, Germany, September 1974.

65. Gladney, H. M., Dowden, B. F., and Swalen, J. D., Computer-assisted gas-liquid chromatography, *Anal. Chem.*, 41, 883, 1969.

66. Grushka, E., Myers, M. N., Shettler, P. D., and Giddings, J. C., Computer characterization of chromatographic peaks by plate height and higher central moments, *Anal. Chem.*, 41, 889, 1969.

67. Horning, E. C. and Horning, M. G., Human metabolic profiles obtained by GC and GC/MS, *J. Gas Chromatogr.*, 9, 129, 1971.

68. Meuzelaar, H. L. C. and Kistemaker, P. G., A technique for fast and reproducible fingerprinting of bacteria by pyrolysis, *Anal. Chem.*, 45, 587, 1973.

69. Meuzelaar, H. L. C., Posthumus, M. A., Kistemaker, P. G., and Kistemaker, J., Curie point pyrolysis in direct combination with voltage electron impact ionization mass spectrometry, *Anal. Chem.*, 45, 1546, 1973.

70. Meuzelaar, H. L. C., Ficke, H. G., and den Harink, H. C., Fully automated Curie-point pyrolysis gas-liquid chromatography, *J. Chromatogr. Sci.*, 13, 12, 1975.

71. Meuzelaar, H. L. C., Kistemaker, P. G., and Tom, A., Rapid and automated identification of microorganisms by Curie Point pyrolysis techniques: I. Differentiation of bacterial strains by fully automated Curie Point pyrolysis gas liquid chromatography, in *New Approaches to the Identification of Microorganisms*, Héden, C. G. and Illéni, T., Eds., John Wiley & Sons, New York, 1975, 165.

72. Kistemaker, P. G., Meuzelaar, H. L. C., and Posthumus, M. A., Rapid and automated identification of microorganisms by Curie Point pyrolysis techniques: II. Fast identification of microbiological samples by Curie Point pyrolysis mass spectrometry, in *New Approaches to the Identification of Microorganisms*, Héden, C. G. and Illéni, T., Eds., John Wiley & Sons, New York, 1975, 179.

73. Razin, S. and Rottem, S., Identification of *Mycoplasma* and other microorganisms by polyacrylamide gel electrophoresis of cell protein, *J. Bacteriol.*, 94, 1807, 1967.

74. El-Sharkaway, T. A. and Huisingh, D., Differentiation among *Xanthomonas* species by polyacrylamide-gel electrophoresis of soluble proteins, *J. Gen. Microbiol.*, 68, 155, 1971.

75. Forshaw, K. A., Electrophoretic patterns of strains of *Mycoplasma pulmonis*, *J. Gen. Microbiol.*, 72, 493, 1972.

76. Morris, J. A., The use of polyacrylamide gel electrophoresis in taxonomy of *Brucella*, *J. Gen. Microbiol.*, 76, 231, 1973.

77. Theodore, T. S., Tully, J. G., and Cole, R. M., Polyacrylamide-gel identification of bacterial L-forms and *Mycoplasma* species of human origin, *Appl. Microbiol.*, 21, 272, 1971.

78. Winkelman, J. and Wyburga, D. R., Automatic calculation of sensitometer scans of electrophoresis strips, *Clin. Chem.*, 15, 708, 1969.

79. Larsen, S. A., Bickham, S. T., Buchanan, T. T., and Jones, W. L., Polyacrylamide gel electrophoresis of *Corynebacterium diphtheriae*: a possible epidemiological aid, *Appl. Microbiol.*, 22, 885, 1971.

80. Pribor, H. G., Kirkham, W. R., and Fellows, G. E., Programmed processing and interpretation of protein and lactic dehydrogenase isoenzyme electrophoretic patterns for computer or for manual use, *Am. J. Clin. Pathol.*, 50, 67, 1968.

81. DeMets, M., Lagasse, A., and Rabacy, M., A new apparatus for polyacrylamide gel electrophoresis, *J. Chromatogr.*, 43, 145, 1969.

82. Kersters, K. and DeLey, J., Identification and grouping of bacteria by numerical analysis of their electrophoretic protein profiles, in *New Approaches to the Identification of Microorganisms*, Héden, C. G. and Illéni, T., Eds., John Wiley & Sons, New York, 1975, 193.

83. Glenn, W. G., Ralston, J. R., and Russell, W. J., Quantitative analyses of certain enteric bacteria and bacterial extracts. I. Standardization and sonic disruption of eight bacterial species, each at five population levels, *Appl. Microbiol.*, 15, 1399, 1967.

84. Glenn, W. G., Ralston, J. R., and Russell, W. J., Quantitative analyses of certain enteric bacteria and bacterial extracts. II. Discrimination of sonic extracts by interfacial densitometry of precipitin systems, *Appl. Microbiol.*, 15, 1402, 1967.

85. Glenn, W. G., The telemetric automated microbiol identification system (TAMIS) and subsystems, *Aeromed. Rev.*, 1-68, 1-47, 1968.

86. Pollard, A. and Waldron, C. B., Automatic radioimmunoassay, in *Technicon Symposium, Automation in Analytical Chemistry*, Vol. 1, Mediad, New York, 1967, 49.

87. Haas, M. L., Shenkman, L., Weissman, P., and Andres, R., A semiautomated technique for radioimmunoassay. Double antibody assay of insulin, *Diabetes*, 19, 127, 1970.

88. McGrew, B. E., DuCros, M. J. F., Stout, G. W., and Falcone, V. H., Automation of a flocculation test for syphilis, *Am. J. Clin. Pathol.*, 50, 52, 1968.

89. McGrew, B. E., Stout, G. W., and Falcone, V. H., Further studies of an automated flocculation test for syphilis, *Am. J. Med. Technol.*, 34, 2, 1968.

90. Stout, G. W. and Lewis, J. S., Automation of an indirect fluorescent antibody test for syphilis, *Ann. N.Y. Acad. Sci.*, 177, 453, 1971.

91. Khoury, A., Petrow, S., and Kasatiya, S. S., Automated seroagglutination test with *Brucella* and *Salmonella* suspensions, *Am. J. Clin. Pathol.*, 60, 467, 1973.

92. Miller, J. K., Nettleton, P. F., and Robertson, A. M., Evaluation of a two-channel automated system for the serodiagnosis of Brucellosis, *Vet. Rec.*, 92, 492, 1973.

93. Logan, L. C. and Cox, P. M., Evaluation of a quantitative automated micro-hemagglutination assay for antibodies to *Treponema pallidum*, *Am. J. Clin. Pathol.*, 53, 163, 1970.

94. Grunmeier, P. W., Gray, A., and Ferrari, A., Automated hemagglutination assays, *Ann. N.Y. Acad. Sci.*, 130, 809, 1965.

95. Hebeka, E. K., Brandon, F. B., and Molteni, J. C., Use of the autoanalyser for the assay of influenza-hemagglutinins, *Technicon Symposium, Automation in Analytical Chemistry*, Vol. 1, Mediad, New York, 1967, 273.

96. Morris, J. A., Jenkins, J. C., and Horswood, R. L., An automatic method for titration of influenza hemagglutinins, *Ann. N.Y. Acad. Sci.*, 130, 801, 1965.

97. Roumiantzeff, M., Emploi des méthodes cinétiques de fixation du complément pour la détermination des parents sérologiques des virus Aphteux, in Technicon Symposium, *Automation in Analytical Chemistry*, Vol. 2, Mediad, New York, 1968, 7.

98. Vargues, R., The use of the AutoAnalyzer for the automatic titration of antigenic preparations by means of complement-fixation, *Ann. N.Y. Acad. Sci.*, 130, 819, 1965.

99. Irwine, W. J., Automated determination of thyroid and gastric complement-fixation antibody; comparison with the fluorescent antibody and manual complement-fixation methods, *Clin. Exp. Immunol.*, 1, 341, 1966.

100. Sturgeon, P., Kwak, K. S., and Gitnik, G., Australian antigen detection by a rapid automated complement fixation method, *Vox Sang.*, 20, 533, 1971.

101. Kwak, K. S., Gitnik, G. L., and Sturgeon, P., Hepatitis associated antigen screening by automated complement fixation comparison with manual methods, *Am. J. Clin. Pathol.*, 59, 41, 1973.

102. Nydegger, U. E., Lambert, P. H., Celada, A., and Miescher, P. A., Complement dosage in continuous flow system and an application to quantitative complement fixation with hepatitis associated antigen, in *Automation in Microbiology and Immunology*, Héden, C. G. and Illéni, T., Eds., John Wiley & Sons, New York, 1975, 393.

103. Wagstaff, W., Firth, R., Booth, J. R., and Bowley, C. C., Large-scale screening by the automated Wassermann reaction, *J. Clin. Pathol.*, 22, 236, 1969.

104. Miller, J. K., Preliminary evaluation of an automated *Brucella* antibody screening system, *Res. Vet. Sci.*, 12, 199, 1971.

105. Miller, J. K., Diagnosis of Brucellosis, *Lab. Pract.*, 20, 173, 1971.

106. Joubert, L., Rowmiantzeff, M., and Valette, L., L'automation en sérologie spécifique: Application au déspistage de la Brucellos, *Rev. Med. Vet.*, 118, 307, 1967.

107. Valette, L., Application de la methode cinetiqué á la reaction de fixation du complement automatiseé dans Brucellose, *Rev. Hyg. Med. Soc.*, 16, 433, 1968.

108. Engelbrecht, E., The bioreactor, A robot technician, in *Automation in Microbiology and Immunology*, Héden, C. G., and Illéni, T., Eds., John Wiley & Sons, 1975, 409.

109. Sanderson, C. J., A multiple dropper for microtitration apparatus, in *Automation, Mechanization and Data Handling in Microbiology*, Society for Applied Bacteriology Tech. Ser. No. 4, Baillie, A. M. and Gilbert, R. J., Eds., Academic Press, New York, 1970, 191.

110. Mansberg, H. P. and Kusnetz, J., Quantitative fluorescence microscopy: Fluorescent Ab automatic scanning techniques, *J. Histochem. Cytochem.*, 14, 260, 1966.

111. Goldman, M., An improved microfluorimeter for measuring brightness of Ab reactions, *J. Histochem. Cytochem.*, 15, 38, 1967.

112. Hijmans, W. and Schaeffer, M., Eds. and Conference Chairmen, 5th International Conference on Immunofluorescence and Related Staining Techniques, *Ann. N.Y. Acad. Sci.*, 254, 1, 1975.

113. Ploem, J. S., in *Standardization in Immunofluorescence*, Holborou, E. J., Ed., Blackwell, Oxford, 1970, 63.

114. Holmberg, K. and Forsum, U., Identification of *Actinomyces, Arachnia, Bacterioneme, Rothia*, and *Propionibacterium* species by defined immunofluorescence, *Appl. Microbiol.*, 25, 834, 1973.

115. O'Neill, P. and Johnson, G. D., A semiautomatic procedure for routine screening by immunofluorescence techniques, *Ann. N.Y. Acad. Sci.*, 177, 446, 1971.

116. Binnings, G. F., Riley, M. J., Roberts, M. E., Barnes, R., and Pringle, T. C., Automated instrument for the fluorescent treponemal antibody-absorption test and other immunofluorescence tests, *Appl. Microbiol.*, 18, 861, 1969.

117. Hunter, E. F., Deacon, W. E., and Meyer, P. E., An improved FTA test for syphilis, the absorption procedure (FTA-ABS), *Public Health Rep.* 79, 410, 1964.

118. Deacon, W. E., Lucas, J. B., and Price, E. V., Fluorescent treponemal antibody-absorption (FTA-ABS) test for syphilis, *JAMA*, 198, 624, 1966.

119. Venereal Disease Research Laboratory, Provisional Technique for the Fluorescent Treponemal Antibody Absorption (FTA-ABS) Test, Venereal Disease Program, Communicable Disease Center, Atlanta, September 20, 1965.

120. Lewis, J. S., Duncan, W. P., and Stout, G. W., Automated fluorescent treponemal antibody test: instrument and evaluation, *Appl. Microbiol.*, 19, 898, 1970.

121. Markovits, A. and Burboek, P., A semiautomated system for mass immuno-fluorescent testing procedures, in *Automation in Microbiology and Immunology*, Héden, C. G. and Illéni, T., Eds., John Wiley & Sons, New York, 1975, 505.

122. Mishuck, E. and Roberts, M., Automation of fluorescent antibody techniques. Part II. The fast system, in *Automation in Microbiology and Immunology*, Héden, C. G. and Illéni, T., Eds., John Wiley & Sons, New York, 1975, 491.

123. Kaufman, G. I., Automation of fluorescent antibody tests: An overview of the U.S. situation, in *Automation in Microbiology and Immunology*, Héden, C. G. and Illéni, T., Eds., John Wiley & Sons, New York, 1975, 429.

124. Birry, A., Caloenescu, M., and Kasativa, S. S., Further evaluation of an automated fluorescent treponemal antibody (AFTA) test for syphilis, *Am. J. Clin. Pathol.*, 57, 391, 1972.

125. Coffey, E. M., Naritomi, L. S., Ulfeldt, M. V., Bradford, L. L., and Wood, R. W., Further evaluation of the automated fluorescent treponemal antibody test for syphilis, *Appl. Microbiol.*, 21, 820, 1971.

126. Kasatiya, S. A. and Birry, A., The use of incident light in immunofluorescence applied to syphilis serology, *Am. J. Chem. Pathol.*, 57, 395, 1972.

127. Bowman, R. L., Blume, P., and Vurex, G. G., Capillary tube scanner for mechanized microbiology, *Science*, 158, 78, 1967.

128. Blume, P., Johnson, J. W., and Matsen, J. M., Automated antibiotic susceptibility testing, in *Automation in Microbiology and Immunology*, Héden, C. G. and Illéni, T., Eds., John Wiley & Sons, New York, 1975, 243.

129. Wyatt, P. J., Automation of differential light scattering for antibiotic susceptibility testing, in *Automation in Microbiology and Immunology*, Héden, C. G. and Illéni, T., Eds., John Wiley & Sons, New York, 1975, 267.

130. Isenberg, H. D., Reichler, A., and Wiseman, D., Prototype of a fully automated device for determination of bacterial susceptibility in the clinical laboratory, *Appl. Microbiol.*, 22, 980, 1971.

131. Praglin, J., Curtiss, A. C., Longhenry, D. K., and McKie, J. E., Jr., Autobac 1 – A 3-hour, automated antimicrobial susceptibility system: I. System description, in *Automation in Microbiology and Immunology*, Héden, C. G. and Illéni, T., Eds., John Wiley & Sons, New York, 1975, 197.

132. McKie, J. E., Jr., Borovoy, R. J., Dooley, J. F., Evanega, G. R., Mendoza, G., Meyer, F., Moody, M., Packer, D. E., Praglin, J., and Smith, H., Autobac 1 – A 3-hour, automated antimicrobial susceptibility system: II. Microbiological studies, in *Automation in Microbiology and Immunology*, Héden, C. G. and Illéni, T., Eds., John Wiley & Sons, New York, 1975, 209.

133. Wretlind, B., Goertz, G., Illéni, T., Lundell, S., and Seger, B., Antibiotic sensitivity testing by zone-scanning of agar strips (a preliminary report), in *Automation in Microbiology and Immunology*, Héden, C. G. and Illéni T., Eds., John Wiley & Sons, New York, 1975, 293.

134. Martin, N. H., Data processing in clinical pathology, *J. Clin. Pathol.*, 21, 231, 1968.

135. Dickson, J. F., Automation of clinical laboratories, *Proc. Inst. Elect. Electron. Eng.*, 57, 1974, 1969.

136. Poletti, B. J., Zack, J. F., Jr., and Mueller, T. J., Computer control in the clinical laboratory, *Am. J. Clin. Pathol.*, 53, 731, 1970.

137. Anderson, N. G., The development of automated systems for clinical and research use, *Clin. Chem. Acta*, 25, 321, 1969.

138. Amsterdam, D. and Schneierson, S. S., Electronic data processing for the clinical microbiology laboratory, *Appl. Microbiol.*, 17, 93, 1969.

139. Lusted, L. B. and Coffin, R. W., Prime: an operative model for a hospital automated information system, *Proc. Inst. Elect. Electron. Eng.*, 57, 1961, 1969.

140. Whitby, J. L. and Blair, J. N., A computer-linked data processing system for routine hospital bacteriology, in *Automation, Mechanization and Data Handling in Microbiology*, Society for Applied Bacteriology Tech. Ser. No. 4, Baillie, A. M. and Gilbert, R. J., Eds., Academic Press, New York, 1970, 23.

141. Alexander, M. K., Connigale, J., Johnson, T., Poulter, I. R., and Wakefield, J., A data processing system for hospital bacteriology, *J. Clin. Pathol.*, 23, 77, 1970.

142. Baillie, A. and Gilbert, R. J., Eds., *Automation, Mechanization and Data Handling in Microbiology*, Society for Applied Bacteriology Tech. Ser. No. 4, Academic Press, New York, 1970.

143. Bengtsson, S., Bergqvist, F. -O., and Schneider, W., A data system for bacteriological routine and research, in *New Approaches to the Identification of Microorganisms* Héden, C. G. and Illéni, T., Eds., John Wiley & Sons, New York, 1975, 291.

INTRODUCTION

Bacteria are identified by conventional methods using morphological, biochemical, and serological tests. While the combinations of a small number of tests are easily matched against a table of results, problems can arise in the comparison of a large number of tests. It is especially difficult to evaluate individual tests in differentiating the steadily increasing number of described bacterial species. Beers and Lockhart[1] suggested the possibility of computer-assisted identification. One of the first schemes for computerized bacterial identification was an outgrowth of the NASA requirement[2,3] for microbial sampling of Apollo spacecraft. The system uses microbial categories and an identification scheme specifically designed for the program.[4] While other systems[5,6] deal with medically important taxa, the Apollo system was directed toward naturally occurring organisms. Probabilities for a given test result for a specific strain were not assigned in the Apollo system. Computers may be used for estimating the differentiating power of tests as well as for identification by estimating overall similarity with known organisms. Since a computer may be capable of storing descriptions of many species of bacteria, it can add or delete properties of new species from time to time as new information becomes available. Also, it can compare the properties of an unidentified organism with those in the computer store, giving the results in the form of a decision on the identification. The computer's speed and memory allow it to rapidly consider many possible species in the identification process and to furnish the laboratory with a list of probable diagnoses, including those species seen infrequently.

The use of a computer minimizes the possibility of error in identification due to infrequent occurrence of an organism, the occurrence of a more common organism with superficial resemblance to other organisms, or the inability to recognize the test pattern of an organism. The computer's sensitivity to subtle pattern variation exceeds that of even the skilled laboratory worker and enables it to make a diagnosis based on a smaller set of tests.

In addition to the usefulness of a computer in the identification of bacteria, clerical work in bacteriology can be substantially reduced by the introduction of computer techniques. Several computer systems for bacteriology have been reported, including "in-house" computer systems which produce reports as well as statistical information for epidemiology and control of infections.[7,8] The use of a computer bureau is valuable for batch processing of statistical data accumulated on punched cards or paper tape,[9,10] and a time-shared computer may be used to prepare reports and to store, process, and retrieve data in bacteriology.[11-13] However, it is essential to have a competent and patient programmer, a reliable computer, and a willingness on the part of the laboratory staff to adapt to the requirements of the computer. In a successfully instituted program, the benefits of the system would include a greatly increased capacity for issuing well-presented reports and also the compilation and analysis of statistics for epidemiological studies, laboratory quality control, and other purposes concerned with management. It is also possible that such use of computers in bacteriology may serve as an example to the physician in his research as well as in his daily work.

This chapter reviews possibilities of computer-assisted identification of bacteria in diagnostic and public health laboratories. A brief discussion of approaches and methods used in computer technology is given with a view toward their applications in diagnostic bacteriology and epidemiology and public health programs.[5,14-20]

APPROACHES TO NUMERICAL IDENTIFICATION OF BACTERIA

Three approaches to numerical identification of bacteria have been described. In the first approach, unidentified strains may be compared with previously constructed taxonomic groups using numerical clustering methods.[17,21,22] A second method is to mathematically construct keys using morphological, biochemical, and serological characteristics of bacteria.[19] A third approach, originally suggested by Beers and Lockhart,[1] is to estimate the probability of strains of each taxon giving positive results in each test. These probabilities are used to yield a statistical estimate of the reliability of the diagnosis. Dybowski and Franklin[16] and Lapage et al.[5,6] applied the conditional probability method to the identification of enterobacteria and to the identification of other Gram-negative rod-shaped bacteria.

Cluster Analysis

Concepts of numerical taxonomy are based on the ideas of Adamson, a French botanist of the 18th century. He thought that the classification of organisms should be based on their overall similarity, considering as many features as possible, with each feature being given equal weight. Sneath[23] first applied these ideas to the classification of bacteria. The group of organisms under study was subjected to a wide variety of tests, and the similarity of characteristics between pairs of organisms was expressed numerically. Usually, all test results were weighted equally. Those organisms with a high percentage of similarity were arranged together to form taxonomic groups or clusters. Thus, a means of classifying a group of organisms rather than identifying a single strain was demonstrated. Quadling and Colwell[21] and Gyllenberg[17] have described methods for identification by comparing unidentified strains and previously constructed taxonomic groups using numerical clustering methods.

Goodman[24] examined 103 strains from the family Enterobacteriaceae, 12 stock strains and 91 isolates; each strain was subjected to 51 tests (for motility, growth at different temperatures, fermentation of various substrates, etc.), and all were weighted equally. Data were punched onto IBM cards for programming on an IBM 360 computer. The program first computed the similarity between each pair of strains (103 strains = 5253 similarities to be computed), and then similar strains were placed into groups or clusters. Thirteen clusters were formed from the 103 strains. A taxonomic map was constructed as a means of displaying the results of the numerical analysis (Figure 6.1) according to the method of Carmichael and Sneath.[25] Each circle on the map represents a cluster, and the clusters are made up of organisms that have been grouped together by the computer on the basis of their overall similarity. The size of the circle relates to the degree of similarity; for example, the small circle for *Klebsiella* represents 18 closely related strains, and the larger circle for *Serratia* contains 7 strains with

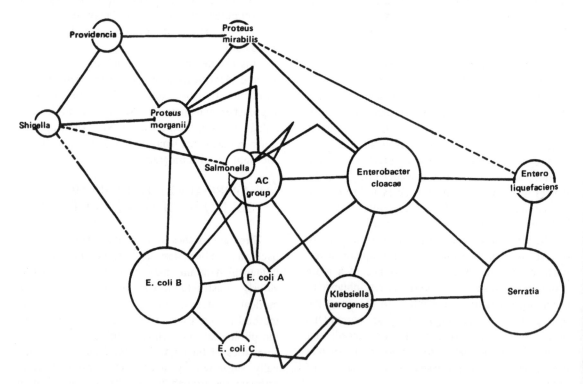

FIGURE 6.1. A taxometric map drawn from the results of cluster analysis of 103 strains of Enterobacteriaceae. (From Goodman, Y. E., *Can. J. Med. Technol.*, 34, 2, 1972. With permission.)

considerable variation within the group. The relationship between clusters is shown by the length of the line drawn between them. A broken line indicates that the clusters should be closer together than shown, and a bent line indicates that they should be further apart.

The clusters in the Goodman test correspond to presently recognized genera. Relationships between strains within a cluster and between clusters were displayed by computer analysis, which tends to confirm the grouping of Ewing and Edwards.[26] Based on the computer analysis, Goodman reduced the original 51 tests to 9 tests that provide a good separation of the groups. The computer may be used for bacterial identification by storing in the memory system tests and expected results for each species. Results of unknowns can then be compared with those for each species in the memory bank, and the percentage similarity to each organism can be printed out by a teleprinter.

Cluster analysis is of value in identifying bacteria at the species level, as shown by Juffs[22] using *Pseudomonas* species as an example. *Pseudomonas* comprise a significant portion of the psychrotrophic bacteria found as contaminants in raw and pasteurized milk supplies.[27-33] Many pseudomonads are proteolytic and/or lipolytic; identification is of importance to the dairy industry since it is believed that these organisms play a significant role in the deterioration of cold-stored milk.

The selection procedure used with initial plating of the sample was either penicillin agar modified from Lightbody[34] with incubation at 30°C for selection of colony types (i.e., translucent, low convex, butyrous), or bromcresol purple-chalk-milk agar (BCP agar) as described by Lightbody[35,36] for enumerating proteolytic colonies, with incubation at 7°C for 10 days.

Juffs[22] used primary characterization procedures to indicate whether isolates belonged to the genus *Pseudomonas* and to indicate characteristics of significance in dairy bacteriology, e.g., hydrolysis of casein and butterfat. Inocula for all diagnostic tests were grown on tryptone-glucose-yeast extract (TGE) slopes at 28°C for 18 hr. Gram stain was performed using the alcohol decolorization method. The oxidase test was that described by Kovacs[37] using a platinum loop. Dark purple color developing within 10 sec indicated a positive reaction. The oxidation-fermentation test was the single-tube modification by Park and Holding[38] of the test described by Hugh and Leifson.[39] The

procedure of Thornley[40] was followed for the arginine dihydrolase reaction. Isolates were evaluated for production of diffusible fluorescent pigments by streaking on medium B of King et al.,[41] with confirmation of fluorescence by examination at 350 nm. Isolates on medium A of King et al.[41] were checked for production of phenazine pigments. Red pigmentation was due to pyorubin; blue pigmentation was due to pyocyanin, which was confirmed by extraction with chloroform. For the casein hydrolysis test, isolates were streaked on BCP agar. A zone of casein hydrolysis was confirmed by flooding plates with trichloroacetic acid. To test for butterfat hydrolysis, isolates were streaked on medium described by Jones and Richards[42] with dyed butterfat prepared as described by the British Standards Institution.[43] Freshly prepared butter oil was used as the source of fat. Motility studies utilized the hanging drop technique of Sherman, and isolates were examined for flagella by electron microscopy and Bailey's flagella stain.[44] Colony pigmentation was observed on TGE agar. The isolates identified as *Pseudomonas* were Gram-negative rods giving a nonfermentative reaction in the oxidation-fermentation test and were motile with polar flagella.

Pseudomonas isolates were then identified to species level using some of the aforementioned tests (i.e., production of fluorescent and phenazine pigments, oxidase reaction, and arginine dihydrolase reaction) and other tests drawn from the list of characters considered by Stanier et al.[45] to be of value for the differentiation of species. Secondary tests also included utilization of creatine, sorbitol, L-leucine, and lactose, as well as lipolysis with incubation at 28°C.

Isolates identified as *Pseudomonas* were subjected to a numerical taxonomic analysis[46] using the complete-linkage cluster analysis B of Szabo.[47] The complete similarity matrix is calculated first; clustering begins around the most highly related pair(s) of organisms found in the similarity matrix. The clustering level is successively lowered by steps of equal magnitude (in this case 2%). An organism is admitted to a cluster when its coefficient of similarity with all members of the cluster is greater than or equal to the level set for cluster formation; two clusters can fuse only when the similarity coefficients between all organisms in the two clusters are greater than or equal to the set clustering level. In the single linkage method, linkage between an organism and any member of a cluster at a given similarity level provides admission to that cluster, and two clusters can fuse with linkage between any pair of organisms, one from each cluster.[46,48]

The similarity coefficients were calculated from the following relationship:

$$\text{Similarity coefficient} = \frac{m \cdot 100\%}{n - nc}$$

where

m = sum of + and − matches between two strains;

n = total number of characters;

nc = total number of characters for which either or both strains scored a no consequence reaction.

Twenty-nine characters were included in the analysis. Data cards were entered in random order, and the S-matrix print-out program of Szabo[47] was used to obtain the similarity matrix print-out. Data were entered for only one isolate from any group of isolates clustering at the 100% level. A dendrogram was constructed from the complete-linkage cluster analysis. Species identified in this manner included *Pseudomonas fluorescens, P. aeruginosa, P. putida, P. maltophilia, P. pseudoalcaligenes, P. cepacia,* and *P. alcaligenes.*

More recently, based on pigmentation, carbohydrate utilization, flagellation, and growth temperature requirements (Table 6.1), a speciation scheme for eight pseudomonads was presented by Johns and Tischer.[49] The pseudomonads included *aeruginosa, fluorescens, putida, alcaligenes, maltophilia, stutzeri, pseudomallei,* and *mallei* species.

Construction of Keys

Generally, classification is concerned with the arrangement of elements into groups on the basis of common characteristics. Identification collates an element's characteristics one-to-one with characteristics of elements in established classifications. When an unknown element's characteristics match in one-to-one correspondence with those characteristics of an element in an established classification (to the exclusion of all other known possibilities), an identification has been made. Bacteria are suited to the study of identification by numerical methods because their morphological, biochemical, antigenic, virulent and other characteristics may often be recorded on a three-valued basis: the character is absent, variable, or present. Higher valued logic, i.e., n-valued logic, can also be used in the schemes of identification of bacteria. Maccacaro[50] suggested that keys could be made by evaluation of the "information content" of each test. Hill and Silvestri[51] constructed a probability key for the diagnosis of Actinomycetales, and the mathematic basis for the key was described by Möeller.[52] Gyllenberg[53] used an estimation of differentiating power to choose tests, and this method was applied by Rypka et al.[18] and further developed by Rypka and Babb.[19] Niemelä et al.[54] also described a method for test selection based on information theory. In all of these methods, a mathematical model is used once to determine the "best" tests, and strains are then identified by comparing their test results with those expected for each taxon by means of conventional methods. Although the advantage of a reduction in the number of tests required for identification can be expected, these methods have the same disadvantage as conventional identification methods, that is, the difficulty of identifying strains aberrant in one or more characters.

Rypka et al. described[18] two-valued characters of bacteria (e.g., absence [= 0] or presence [= 1] of flagella) which were selected from an organism vs. character matrix. A minimal character set (Tmin) optimally separated all of the organisms (G) in the group (Tmin = $\log_2 G$). They constructed a table with all possible combinations (2^{Tmin}) of two-valued answers for the number of characters in the minimal set, and traced on the table the results of tests for characters of the bacteria in order to determine the identity of the isolate. Additional characters may be studied if the results of the tests for characters in the minimal set indicate more than one organism as a possible answer. Using this method, Rypka et al.[18] reported that the shortest route to an identification was based on an optimal choice of characters for separation of the bacteria. The method also stresses the importance of using exhaustive identification schemes for greater accuracy.

Rypka and Babb[19] described a semiautomated method of constructing and using inclusive (exhaustive) schemes for bacterial identification. Using Gram-positive cocci (staphylococci and streptococci isolated from the noses, throats, rectums, and tissues of humans and dogs) as a model system, they chose 32 characters for use in

TABLE 6.1

Diagnostic Characteristics of *Pseudomonas* sp.

% of Strains Positive

Organism	Pigment		Carbohydrates							Mot	Flagellation			Growth at	
	Flu	Pyo	Glu	Malt	Xyl	Tre	Gel	Arg	Lys		Mul	Mon	Atr	0–5°C	41–42 °C
P. aeruginosa	89.9	75.5	99.9	0	99.3	44.5	91.6	99.3	0	93.7	0	98	4.4	0.6	99.2
P. fluorescens	92.1	0	99.5	56	100	96.4	92.1	96.8	0	99.6	97	15.4	0	97.5	0
P. putida	99.3	0	100	8	99.2	0	0	100	0	100	98.3	28.8	0	81.8	10.8
P. alcaligenes	0	0	0	0	11	0	29.5	22	0	100	0	100	0	0	93.8
P. maltophilia	0	0	70	100	31	67.7	100	8.2	75.9	100	98.1	2.4	0	0	47.7
P. stutzeri	0	0	99	97.7	100	3.2	11.9	4.1	0	96.8	0	100	0	0	86.9
P. pseudomallei	0	0	100	100	100	100	98.7	100	0	100	100	0	0	2.8	98.7
P. mallei	0	0	91	91	4.1	87.5	64.2	97.5	0	0	0	0	100	0	44.1

Note: Flu, fluorescence; Pyo, pyocyanin; Glu, glucose (O/F); Malt, maltose (O/F); Xyl, xylose (O/F); Tre, trehalose; Gel, gelatin; Arg, arginine dihydrolase; Lys, lysine decarboxylase; Mot, motile; Mul, multitrichous; Mon, monotrichous; Atr, atrichous.

From Johns, P. A. and Tischer, R. G., *Am. J. Med. Technol.*, 39, 495, 1973. With permission. Copyright by the American Society for Medical Technology.

character selection and a minimal set of 8 characters based on Tmin = Log_2G, where T_{min} = minimal character set separating all organisms (G) in a group. Tests were performed to determine the absence or presence of the first eight characters, and the results were traced, in order, over a prepared table using the following key:

    ■ = character present = 1
    ⊟ = character absent = 0
    □ = not tested for
    ◪ = character variable

If an answer was found with a subgroup containing more than one organism, a subgroup table was used; tests were performed for characters indicated, results were traced, and the answer was determined for the isolate. If the size of the minimal character set was increased, the number of organisms in the subgroups was reduced and fewer characters were required to separate the bacteria. The procedures used in this identification scheme were programmed in ALGOL 60 for use with the Burroughs 5500 computer. This method may be applied to the identification of a pathogenic organism from a patient, for example, as a guide to antibiotic therapy.

Probabilistic Method

This method, suggested by Beers and Lockhart,[1] involves estimation of the probability of strains of each taxon giving positive results in each test and the use of these probabilities to yield a statistical estimate of the reliability of the diagnosis. Payne[55] adopted this method working with Dybowski and Franklin,[16] who gave a detailed account of a conditional probability method and its application to the identification of enterobacteria. Lapage et al.[5] described the use of a similar probability method in the identification of Gram-negative rod-shaped bacteria. A method was also described for the selection of tests, with the highest discriminating power between the possible taxa suggested as likely identifications by the probability computations. Sneath[56] discussed the theoretical effect that test differences would have on identification. Dito[57] described an economical system for the storage of strain data. The results found on unknown strains can be matched with those stored for previously identified strains. Unknown strains may thus be identified or the occurrence of a new pattern of results may be

determined by this system. Operation of the system on an electronic desk calculator has also been described by Dito.[57] A similar approach was used by Fey[58] and Baer and Washington.[59] Lapage et al.[5] successfully identified 70 to 80% of 279 freshly isolated strains examined in both a limited number of tests and in all the tests included in their survey. General aspects and perspectives of probabilistic identification have been discussed by Lapage et al.,[5] and the methods incorporated in the computer program in a trial of computer-aided identification of bacteria were described by Wilcox et al.[60]

Utilization of a computer in bacterial identification permits the amassing of characteristic properties of species and the updating of such information. The reaction of an unidentified microorganism may be compared with those in the computer memory system, yielding a decision on the identification or a listing of most similar species and effective discriminating tests to differentiate these species. With such a system, the identification method is programmed in three stages.[61] The first stage is the calculation of an identification score for each taxon in the matrix. The second stage entails consideration of these scores to determine if a definite identification is justified. If no identification is justified, stage three selection is made for further tests capable of differentiating the most likely taxa. The first detailed bacterial identification by a probabilistic method was presented by Dybowski and Franklin.[16]

Laboratory methods of bacterial identification fall into two categories.[62] The sequential method bases selection of each succeeding test on the results of previous tests until the lowest branch of a sequence is reached. The simultaneous or pattern recognition method tests the organism for a large number of characteristics at one time.

To utilize the speed and accuracy of the simultaneous method while overcoming the inherent complexity, Friedman et al.[62] adopted a Baysean model for bacterial identification; this is based on the Bayes theorem:

$$Pn = \frac{PP_n \underset{\text{all tests}}{\Pi} P_{nij}}{PP_n \underset{\text{all tests}}{\Pi} P_{nij} + (1 - PP_n) \underset{\text{all tests}}{\Pi} q_{nij}}$$

where

Pn   = relative probability of the unknown isolate being species n;

PP_n = prior probability of species n (i.e., given a random isolate, the probability of this unknown being species n);

P_{nij} = the probability that if the unknown were species n, it would have finding j for test i;

q_{nij} = the probability that if the unknown were not species n, it would have finding j for test i.

A detailed discussion of Bayes' theorem and formula derivation is given by Fisher[63] and Pratt et al.[64]

To properly use Bayes' theorem, a support of quantitative data is needed. Therefore, Friedman et al.[62] limited their identification scheme to dextrose-fermenting Gram-negative rods, the majority of which were Enterobacteriaceae. From data collected over a 2-year period, it was determined that 34 species represented 99.6% of all such isolates and 38 tests accounted for 99.4% of all procedures used in identifying these organisms. The 34 species and 38 tests were established as the boundaries of the data base. A computer program was constructed to analyze the data to determine the percentage of times each test result was found to be positive for each of the species. In this way probabilities were obtained for all entries in the data base.

In the study model it was assumed that all tests had only two possible findings, i.e., P_{nij}, j = positive finding or j = negative finding. The probability of a negative finding for test i for species n was calculated using the formula:

$$j = negative \quad P_{nij} = 1 - P_{nij} \quad J = positive$$

Once all P_{nij}'s were known, the necessary q_{nij}'s were calculated by Friedman et al.[62] using the corresponding P_{nij} and PP_n according to the formula

$$q_n \times {}_{ij} = \frac{\Sigma P_{nij} \times PP_n}{\Sigma PP_n}$$

with all species in n except the species n^x. Once these were entered into the storage band, the computer (presented with test results) would retrieve the needed probabilities from disk storage and (utilizing the Bayes' theorem) compute a score from zero to one, giving the "relative" probability

that each species in the data base was the identity of the unknown isolate. The species with the highest score (for each possible identification in the data base) was selected as the "most likely identification" (MLI). Additional output included "test results that are against this diagnosis," a lower limit score (LLS) based on the proximity of the MLI score to the highest possible score (1.0), and "additional tests that would assist in differentiating the unknown."

The program described was written in FORTRAN computer language to run on a Control Corporation 3200 computer. It requires less than 5000 words of core storage and less than 5 sec for complete analysis of a set of test results. The model was 100% accurate in identifying species included in its data base and more than 99% accurate in identifying a random selection of dextrose-fermenting Gram-negative rods submitted for laboratory identification.

A similar program was described by Friedman and MacLowry[65] utilizing the Baysean mathematical model to identify bacteria solely on the basis of antibiotic sensitivities. The antibiotic sensitivities of the unknown organism (isolated from a clinical specimen) were entered into the computer via a single punch card. The program then selected the necessary probabilities from disk memory and, using the Baysean formula, calculated a relative probability score (ranging from 0 to 1) for each species in the identification scheme. The computer then selected the identification that received the highest score and designated the "most likely identification" (MLI) (Table 6.2). Results of all antibiotic sensitivity tests by serial dilution techniques[66] over a 3-year period for over 13,000 clinical isolates became the data base. Two distinct data bases were developed, a urine source and a nonurine source. When nonurine isolates were subgrouped, the program's accuracy improved (Table 6.3).

The results indicated feasibility of identification of bacteria based on antibiotic sensitivities; however, the system fails to meet the requirements for accuracy currently used in most clinical laboratories. Several modifications permit an increase in accuracy.

1. As the data base is increased, accuracy of the model should improve.

2. A requirement of Bayes' theorem is that all factors used in calculating probability should be

TABLE 6.2

Most Likely Identification of Organisms from a Wound by
Computer Matching Based on Antibiotic Sensitivity Test[a]

Identification	Relative probability
Bacillus sp.	0.001
Bacteroides sp.	0.000
Bacteroides fragilis	0.000
Citrobacter	0.051
Enterobacter sp.	0.340
Enterobacter cloacae	0.289
Enterobacter cloacae, atypical	0.056
Enterobacter aerogenes	0.287
Enterobacter hafniae	0.004
Escherichia coli	0.001
Haemophilus sp.	0.000
Haemophilus influenzae	0.000
Haemophilus parahaemolyticus	0.000
Haemophilus parainfluenzae	0.000
Herellea vaginicola	0.006
Klebsiella sp.	0.993
Proteus vulgaris	0.005
Proteus mirabilis	0.000
Proteus morganii	0.000
Pseudomonas sp.	0.000
Pseudomonas aeruginosa	0.000
Pseudomonas sp. *mucoid*	0.000
Pseudomonas aeruginosa mucoid	0.000
Pseudomonas maltophilia	0.000
Salmonella sp.	0.005
Serratia sp.	0.000
Staphylococcus aureus	0.000
Staphylococcus epidermidis	0.000
Streptococcus beta hemolytic	0.000
Streptococcus viridans	0.000
Streptococcus faecalis	0.000

[a]Most likely identification, *Klebsiella* sp. (Identification
acceptable by "selection criteria" as described by
Friedman and MacLowry.[65])

From Friedman, R. and MacLowry, J., *Appl. Microbiol.*,
26, 314, 1973. With permission.

mutually independent. Use of antibiotics with
nonoverlapping spectra or calculation of the inter-
dependence of the antibiotics used should improve
accuracy.

3. Increased use of subgrouping.

4. Use of more strict selection criteria.

5. Incorporation of a method to continually
update the data base to compensate for changes in
antibiotic sensitivity resulting from genetic variation should enhance the program's ability to
correctly identify isolates.

Bascomb et al.[14] described the identification of

1079 reference strains of bacteria by a probabilistic method based on Bayes' theorem used in a
computer. Gram-negative, aerobic, rod-shaped
bacteria which grew on nutrient agar at 37 or 22°C
were chosen as a model, and a matrix was
constructed which gave the probability of a strain
of any given taxon yielding a positive result in
each of the chosen tests. The individual strains
were tested and identified on the basis of the
constructed matrix. Identification was based on 48
tests.

The rate of success in the identification varied
between groups of taxa. The fermentative taxa
gave the highest rate of identification (90.8%); in
54% of these taxa, 100% of the strains were
identified. By contrast, nonfermentative taxa gave
an identification rate of 82.1%; in 29% of these
taxa, 100% of the strains were identified. It was
not clear what factors operate in successful identification. For example, the number of biopatterns
given by strains in a taxon showed no correlation
with success in identification in that taxon. Bascomb et al.[14] attributed the difference in identification rates for fermenters and nonfermenters to
the lack of discriminating tests for nonfermenters.
There appears to be no method *a priori* in
a biological system of determining whether a probabilistic matrix will operate successfully or
whether an empirical trial is needed.

The identification method of Willcox et al.[60] is
also based on Bayes' theorem. It allows for
dependent tests and missing data in the probability
matrix. The Willcox method suggests a definite
identification only if the Baysean probability of
one of the taxa exceeds a threshold level; if not, a
separate procedure for selecting the best tests is
used to continue the identification. In the Willcox
method, although the tests were selected as though
for a diagnostic table, identification was carried
out by a probability calculation, not by the logical
matching process used in such tables. A consideration of the identification calculation suggests that
with an identification level of 0.999, if two tests
are required to separate each taxon pair, there is a
reasonable chance that the discrimination assumed
by the test selection model has been achieved. The
test selection procedure is similar to methods for
selecting tests for diagnostic tables previously
described by Gyllenberg[53] and Rypka et al.[18] The
theory of the identification method adopted is an
extended trial of computer-aided identification of
Gram-negative rods as described by Lapage et al.[6]

TABLE 6.3

Effect of Subgrouping on the Identification of Bacteria

Species	No. of isolates in nonurine data base	No. of isolates in subgroup	Correctly identified by computer[a](%)	
			Without subgrouping	With subgrouping
Enterobacter cloacae	246	26	35	42
Escherichia coli	862	152	65	73
Klebsiella sp.	943	82	67	78
Pseudomonas aeruginosa	1398	87	91	95
Staphylococcus aureus	2004	162	82	85
Staphylococcus epidermidis	415	45	48	61

[a]Most likely identification (same as laboratory identification).

From Friedman, R. and MacLowry, J., *Appl. Microbiol.*, 26, 314, 1973. With permission.

APPLICATION OF COMPUTER IDENTIFICATION PROGRAM

Diagnostic Bacteriology Laboratory

In a routine diagnostic bacteriology laboratory, an organism is first isolated from a clinical specimen and then its biochemical activity is investigated. Most laboratories utilize a standard profile of tests (such as catalase, oxidase production, the mode of carbohydrate degradation, etc.) from which a characteristic and diagnostic pattern can usually be obtained. While the combinations of a small number of tests are easily matched against a reference table of results, problems can arise in the comparison of a large number of tests. Experimental errors in conventional bacteriological tests are usually appreciable, even under conditions of careful standardization. Taylor et al.,[67] analyzing data of Gottlieb,[68] reported that reproducibility of tests routinely used for streptomycetes varied among laboratories from 63% to 97%. Sneath and Johnson[69] discussed methods of estimating experimental errors in microbiological tests. They calculated p values (i.e., the probability that a result is erroneous) from an analysis of variance and for the purpose of including or excluding chosen factors that affect test reproducibility. It was suggested that when p is over 10%, the error

of similarity values becomes unacceptably large. Taylor et al.[67] found that the tests commonly used for streptomycetes must be of limited value for taxonomy (p value of almost 20%) because the expected similarity between identical strains would be only 69%, while the standard error based on the 37 tests considered would be about 9%. Lapage et al.[5] obtained an average of 6.9% discrepancies for replicate comparisons between laboratories for a conventional set of tests used to identify Gram-negative rods. This corresponds to p of 3.9% if one assumes neither series is an accurate standard.

Washington et al.[70] examined the agreement between microtests and conventional tests in Enterobacteriaceae; there was a minimum of 4.2% discrepancy after 24 hr of incubation and only a little less at 48 hr. Sneath and Johnson[69] suggested that within one laboratory p can usually be kept below 5%, but the discrepancies between laboratories may often be much larger. They further suggested that it is generally better to use many tests, even if the tests are not as reproducible as desired, rather than to use only a few extremely reproducible tests.

In order to recognize similarities and differences when making multiple comparisons simultaneously by the many tests used for bacterial identification schemes, Spencer and Hyde[71] suggested the use of a minicomputer for small diagnostic bacteriology laboratories. Their approach utilized a computer program in which the results from a standard battery of tests were coded so that (0) represented a negative result of the absence of a particular characteristic; (1) represented a positive result or the fact that acid and gas were produced together in a fermentation reaction; and (2) represented a doubtful or late result or the production of acid alone. In this scheme, the physical and biochemical data used to identify microorganisms and assembled as "character tables" were converted into a binary number which was subsequently processed by a program according to the manufacturer's instructions (Olivetti P101 computer). The basic algorithm of the program converted the binary number into decimal form, that is, using the base 10. Two groups of organisms (the first group included 15 species representing the Enterobacteriaceae family and the second group included species of miscellaneous organisms designated as "nonfermentative Gram-negative bacteria") were characterized using this approach. It was considered that this approach offered both versatility and potential to small diagnostic laboratories for identification of bacteria on a routine basis.

Computer-based programs for the identification of bacteria have focused primarily on the Gram-negative rods.[5,16,18,55,61,72] Taxonomic difficulties are also associated with the coryneform bacteria.[73-79] The coryneform group is characterized mainly on a morphological basis and includes the genera *Corynebacterium, Arthrobacter, Brevibacterium, Microbacterium, Cellulomonas, Listeria, Erysipelothrix, Mycobacterium,* and certain species of *Nocardia.*

Studies of the taxonomy of coryneform bacteria by numerical methods have been performed by Da Silva and Holt,[80] Chatelain and Second,[81] Harrington,[82] Mullakhanbhai and Bhat,[83] Splittstoesser et al.,[84] Davis et al.,[85] Davis and Newton,[86] Masuo and Nakagawa,[87] and Skyring and Quadling.[88] Nonnumerical studies have been performed by Keddie et al.[89] Robinson,[90-93] Abe et al.,[94] Komagata et al.,[95] and Yamada and Komagata.[96,97] DNA base composition studies of coryneform bacteria have been reported by Bouisset et al.,[98] Abe et al.,[94] Skyring and Quadling,[88] and Yamada and Komagata.[97,99] Bousfield[100] combined a numerical analysis of a range of properties of coryneform bacteria with the determination of DNA base composition of representative strains. A total of 158 strains (including the aforementioned strains plus new isolates from sea water, fish, soil, and plant surfaces) were subjected to numerical analysis. As a result of the studies, Bousfield has proposed considerable reorganization of most coryneform genera and has made suggestions for the reclassification of species and particular strains.[100]

The use of numerical techniques in the study of bacterial taxonomy has been reviewed by Sneath.[101] Payne[55] first attempted identification of medically important bacteria with a computer. Lapage et al.[5] described a method of computer identification of medically important bacteria (Figure 2). In this system, 50 tests were chosen on the basis of ease of reading, reproducibility, and adaptability for objective interpretation. Most morphological and cultural tests were excluded on these bases. Groups and species of Gram-negative bacteria of medical importance were selected for this study. Programs were written in Algol, and computations were performed using an ICL Atlas computer. Matrices were constructed, and each entry was an estimate of the probability of a positive result occurring in strains of the taxon tested; values ranged from 0.99 for 100% positive results to 0.01 for 100% negative results.[16] The program consisted of three stages: (1) calculation of an "identification score" for each taxon, (2) consideration of scores to determine if the identification was absolute, and (3) selection of further tests to continue the identification. Probabilities of a strain belonging to a taxon were determined for each taxon by multiplying individual test probabilities. For test selection, all probabilities were converted to + and - signs (≥ 0.85 = +, ≤ 0.15 = -). Probabilities 0.85 to 0.15 were not used in evaluating tests. Selection and performance of the initial set of tests were accomplished by the laboratory providing strains for identification. The computer laboratory prepared a punched card from the initial set of data. Cards were sent to the computer center where program and matrix were loaded into the computer, and punched cards were fed in, followed by identification and/or test selection. The rate of successful computer identification was 81.4% when all 50 tests were used and 77.4% when 30 tests were used.

Using a combination of identification logic by Dybowski and Franklin[16] and Lapage et al.[5] for computer-assisted identification of micro-

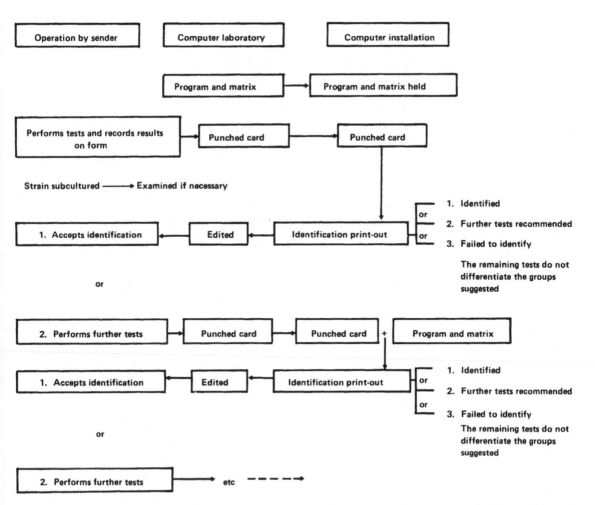

FIGURE 6.2. Flow chart of computer identification of bacteria. (From Lapage, S. P., Bascomb, S., Willcox, W. R., and Curtis, M. A., in *Automation, Mechanization and Data Handling in Microbiology,* Society for Applied Bacteriology Tech. Ser. No. 4, Baillie, A. M. and Gilbert, R. J., Eds., Academic Press, London, 1970. With permission.)

organsisms, Gullenberg and Niemelä[20] reported four possible identification states:

1. Identification of unknown based on both absolute and normalized probabilities for the group identified exceeding predefined limits.

2. An intermediate state indicated by an absolute probability that exceeds the limit of group membership, but a normalized probability below a predefined limit.

3. Neighborhood, in which normalized probability may or may not exceed the given limit, but absolute probability remains below the limit for group membership while exceeding a secondary neighborhood unit.

4. Unknown, classed as an outlier when the probability does exceed either membership or neighborhood limits.

Identification is possible only in relation to a predefined set of reference elements which is constructed as a matrix in which each group is defined by the relative frequencies of the different characters. This is referred to as the identification matrix, and all data in the matrix are treated as equal, regardless of the identification principle applied.

Alternative identification procedures have been reported.

1. Identification by the principle of probabilities in which characters are considered qualitative, constituting mutually exclusive alternatives, and equal weight is given all alternatives.

2. Identification based on Euclidean distances between unknown and group centroids

which are considered as points in an m-dimensional space (m is the number of characters); therefore, characteristics should be quantitative.

3. Identification by correlation coefficients which entails evaluation of similarity between the unknown and the group based on correlation coefficient as tested by the t-test. When needed, the normalization of the characters is performed by transforming the variance mean to zero and the variance to one.

Epidemiology and Public Health Laboratory

The application of computer technology in epidemiology and public health is valuable for the purpose of infection control, detection of trends suggesting preventable cross infection, determination of antibiotic susceptibility patterns, surveillance of nosocomial infections in hospitals, and provision of the potential for rapid analysis of all recorded data required at any time by the epidemiological situation. Ryan and Paplanus[102] described a computerized nosocomial infection surveillance system which combines clinical, epidemiological, and laboratory data. Following an infectious disease episode, cards are punched in the English language on the typewriter-like keyboard of a keypunch. The cards are computer analyzed in a matter of hours or held to be batch processed at intervals required by the infection control program. The computer system utilizes a Control Data Corporation 6400 computer, as described by Paplanus et al.,[103] for storage and analysis of diagnoses and other descriptive information used in surgical pathology and cytology. The system is suited to epidemiological investigation as it allows sorting of data files for one or a combination of words used to describe each case. Information is entered in categories such as description of the patient, his name, hospital number, age, location in the hospital, and time of hospital stay. Another data set describes the infection, its timing and anatomic location, and the infecting organism, together with appropriate antimicrobial susceptibility patterns and typing.[104] The final and largest category of the data entry system includes all of the factors which might predispose the patient to develop an infection in the hospital setting (e.g., catheters, surgical procedures, underlying disease, immunosuppressive therapy, antibiotic therapy, the use of complex equipment such as respirators and artificial kidneys, and many other factors).

The system was used to generate monthly infection reports. Data were rapidly analyzed by the computer, providing assistance in the control of nosocomial outbreaks, the identification and definition of problems which were previously unrecognized, the determination of pharmacy- or manufacturer-contaminated medications, and the determination and prevention of urinary infection via indwelling catheters. The application of the computer system was not only useful for the generation of the monthly infection report, but was also useful in the investigation of nosocomial urinary and respiratory infections.

A bacteriology laboratory-based information system can be used for hospital epidemiology because, in addition to bacteriological results, a report contains patient identifying data, hospital location, name of attending physician, and clinical information, all of epidemiological significance. However, an ideal computer system for epidemiological investigations must include integrated direct input of all recorded administrative, clinical, laboratory, pharmacy, and housekeeping activities in the hospital and must be programmed in a manner appropriate for the definition and solution of infection control problems. Data appropriate to the problem or investigation at hand can then be selected, sorted, and analyzed from this broad data base. Alternatively, information can be gathered and synthesized by the hospital epidemiologist aided by clinical, laboratory, and epidemiological data already being collected and analyzed by hand. Greenhalgh[105] and Porter[106] described a computer service which processes information sheets filled out by the epidemiologist and returns an infection report 1 to 2 weeks after submission of the sheets.

Meers[107] reported a computer-aided study of a bacteriological examination to compare the incidence of infection and the distribution of infecting organisms in patients of either sex at various ages. Detailed information on every patient within a 6-month period who had urine sent to a laboratory for bacteriological examination and the results of each examination were entered on computer cards. A total of 15,606 cards were completed containing coded information recording the sex and age of each patient, the origin of the request, the presence or absence in the urine of an excess of protein or cells, the culture result, and the name of any significant organism isolated together with its sensitivity to various antimicrobial drugs. Organisms leading to positive culture reports were divided into 12 types (Table 6.4) which were separately coded for entry on computer cards according to the criteria described in the table. This information was interrelated in a computer, and in some cases the resulting numerical details were expressed as rates so as to eliminate the effect of uneven sex and age distribution. The sensitivity pattern of each type of significant organism isolated was established according to its source. It was found that hospital and general practice experience of urinary tract infections differed widely with regard to the age and sex distribution of patients, the causative organisms concerned, and sensitivity to antimicrobial drugs.

This epidemiological study supported the earlier findings of Loudon and Greenhalgh,[108] Mond et al.,[109] and Brooks and Mauder,[110] which indicated that it is unreliable to use a test for an excess of either protein or cells as the sole criterion in diagnosing a urinary tract infection. When both tests were considered together, the rate of error was diminished but still remained un-

acceptably high. The distribution of the groups of pathogens involved in urinary tract infection varied according to the age, sex, and source of the patients yielding them. Also, the sensitivity of strains from hospital and general practice was found to differ widely.

From the information presented in the foregoing discussion, it should be clear that the benefits of a computer system in epidemiology and public health include a greatly increased capacity for issuing well-presented reports and an increased capacity for compiling and analyzing statistics for epidemiological studies, laboratory quality control, and other purposes concerned with patient management.

COMPUTERIZED DATA PROCESSING, STORAGE, AND RETRIEVAL SYSTEM FOR BACTERIOLOGY

In recent years, bacteriology laboratories in large hospitals and industries have begun to utilize computer facilities for processing test results. These microbiological data are useful in many

TABLE 6.4

Criteria Used in a Computer-aided Study of the Bacteriological Examination of Urine[a]

Gram-negative bacilli

Escherichia coli	Characteristic colonial morphology, lactose positive, indole positive
Klebsiella spp.	Characteristic colonial morphology, lactose positive, indole negative
Proteus spp.	Characteristic colonial morphology, lactose negative, strong urease production
Paracolons	Variable colonial morphology, lactose negative, negative or weak urease production
Coliforms	Variable colonial morphology, lactose positive
Pseudomonas spp.	Characteristic colonial morphology and pigment (oxidase positive)

Gram-positive cocci

Fecal streptococci	Characteristic colonial morphology (catalase negative)
Staphylococcus albus	Characteristic colonial morphology, coagulase negative, novobiocin sensitive (catalase positive)
Micrococcus spp.	Characteristic colonial morphology, coagulase negative, novobiocin resistant (catalase positive)
Staphylococcus aureus	Characteristic colonial morphology, coagulase positive (catalase positive)

Others

Candida spp.	Characteristic morphology
Other organisms	

[a]Criteria in parentheses were only used in cases of doubt.

From Meers, P. D., J. Hyg., 72, 229, 1974. With permission.

ways other than those involved in direct patient care. They are useful for Infections Committee reviews of possible epidemic strains of organisms and for following the development of antibiotic-resistant strains. The utilization of computer facilities for data reviews is also possible, provided that the capabilities for long-term storage and retrieval are incorporated into the reporting system.

Ellner[111] suggested that a computer system for a clinical microbiology laboratory should provide the following:

1. A method for entering data when the specimen is received and when results become available.

2. A method for rapidly retrieving data in the laboratory for telephone inquiries.

3. Hard copy (printed) reports of results suitable for inclusion in the patient's chart.

4. An alert warning to the medical staff of potentially life-threatening or epidemiological situations.

5. Management assistance for the laboratory director and supervisor.

6. Information to the accounting department regarding laboratory charges.

7. Storage of data for statistical retrieval.

Computerized data processing in a bacteriology laboratory includes the input or the entry of data into the system, the processing of data within the system, and the output or the retrieval of data from the system. Since chemistry and hematology tests in many hospitals are invariably ordered with microbiological tests, the biographical information for patients may already be stored in the computer. However, patients' names and the hospital numbers should be included in a data entry system used for accessioning specimens. At the time of specimen registration, the computer should display the patient's age, sex, and location. The computer may assign a laboratory number, and a cathode ray tube (CRT) may be used for data input and error checking. Data such as the patient's hospital number and type of test can be entered on the CRT by means of a keyboard. Various types of hardware that may be used for manual data entry are listed in Table 6.5. A mark-sense card reader would satisfy all of the above requirements. Using pencil strokes on preprinted cards, technicians could quickly indicate the results of the various tests (e.g., Gram stain, morphology, sugar fermentation), together with identification of the organism and its antibiogram. The complete card would then be dropped into the card reader to enter data. A medium-speed line printer near the mark-sense card reader would indicate errors, rejects, or unusual isolates that have not been subjected to error checking. The card can also serve as a permanent record.

Several types of punched cards have been used for handling routine hospital bacteriological records, including key sort notched cards,[112] peripherally punched cards,[113] and punched cards for use with the computer.[114] The use of punched cards serves to supplement and store laboratory records. Gabrieli[115] used punched cards to generate the immediate report. Whitby and Blair[116] described fourfold aims in the use of punched cards.

TABLE 6.5

Types of Hardware for Manual Data Entry

Machine	Display	Disadvantages
Keypunch	None	Batch process; depends on another person or requires special skill
Electric typewriter	None CRT	Slow; subject to error
Multiple key terminal	Medium-speed printer CRT	Subject to transcription error; several required —takes up space; expensive
Mark-sense card reader	Medium-speed printer	Cards must be treated with care — "do not fold, spindle, or mutilate"

From Ellner, P. D., in *Microbiology – 1975,* Schlessinger, D., Ed., American Society for Microbiology, Washington, D.C., 1975, 73. With permission.

1. To link up with the method of punched card patient documentation used in registration.

2. To provide work cards on which laboratory findings are recorded.

3. To provide reports on all routine specimens sent to the bacteriology department.

4. To organize the data presented in Number 2 into a form that can be analyzed.

The system utilizes the IBM 80 column punch cards, an 836 control unit, and an IBM 1440 computer.

On admission to the hospital, punch cards are used for recording patient's identification details, and summary data are recorded to the left on a special 6 REQ card which serves as the basic identification of the patient for subsequent use in recording data. Such special cards are adapted to accompany specimens for bacteriology. There are five routine bacteriological specimen types: (1) urine; (2) feces; (3) sputum; (4) swab, body fluid, and other; and (5) cerebrospinal fluid. The five different card layouts can be superimposed, and all contain the basic patient identification information to the left of the card (columns 3 to 28); the specific study results are accommodated on a total of 80 columns. All information is automatically duplicated onto the work card and report card from the request card. A day book list is printed by a typewriter wired to the 836 control unit.

All laboratory data receive a numerical code, e.g., specimen source, results, etc. Coded data are converted to descriptive phraseology by means of a program fed into an IBM 1440 computer. Print-outs and storage are on magnetic tape. The two-digit code is based on frequency of specimen types received. Approximately 12 min is required to print 100 reports once the data cards have been prepared.

Harvey et al.[11] have described a time-shared computer system for data processing in the bacteriology laboratory of a large teaching hospital complex (Figure 6.3). The system can produce bacteriology reports, generate accounting information, and collate statistics concerning infection in hospital and domicillary practice. Test results from numerically coded work sheets may be punched onto paper tape using a teletype terminal located in the laboratory. The terminal may also be used to communicate with a Univac® 1108 computer (Computer Sciences of Australia)

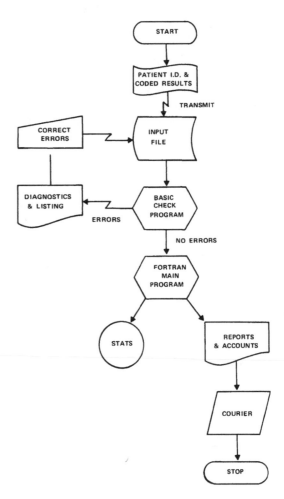

FIGURE 6.3. Flow chart of bacteriology data processing. (From Harvey, K. J., Were, M., Heys, W., and Smith, D. D., *Med. J. Aust.*, 2, 1076, 1972. With permission.)

through a telephone link. Conversational programs written in BASIC or FORTRAN can be used to check reports for errors which can be corrected before submitting the job for batch processing.[117] Programs written in FORTRAN V produce reports on a high-speed line printer, and these can be conveyed to the hospital by a daily courier service. Harvey et al.[11] report that the computerization of data processing produced reports rapidly with increased accuracy and quality and that the system was useful in the analysis of epidemiological statistics and streamlining of accounting procedures.

Many investigators have described various methods and systems for the storage and retrieval of clinical microbiology laboratory data.[10,11, 118-122] While some of these systems are designed for retrospective data entry, others are limited either in their capabilities for storage or retrieval of large amounts of data or in their utilization of adequate code dictionaries. Vermeulen et al.[13] reported a computerized clinical microbiology data storage and retrieval system which is highly flexible, maintains a continually updated cumulative data base, and is adaptable to most hospital or commercially available computer hardware (Figure 6.4).

Vermeulen et al.[123] also described a computerized microbiology reporting system used in the Baltimore Cancer Research Center (BCRC). Data are prepared in the proper format and are typed on punched paper tape by laboratory

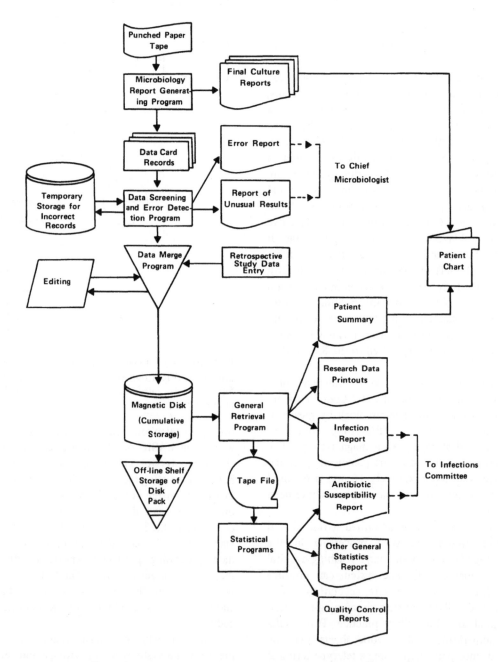

FIGURE 6.4. Flow chart diagram of data storage and retrieval system. (From Vermeulen, G. D., Gerster, J. W., Young, V. M., and Hsieh, R. K. C., *Am. J. Clin. Pathol.*, 61, 209, 1974. With permission.)

personnel after completion of all laboratory studies. Clinical source and result information is entered in the form of easily recognizable letter codes. Number or single letter codes are used only to designate the type of test performed (that is, routine bacteriologic culture, culture for acid-fast organisms, virus culture, etc.). Growth quantitations are entered as single numerical codes or are entered with a letter to express a range of numbers of colonies. Antibiotic disk susceptibility is also entered in the form of two series of ten digits each, with the following code: 0, the drug was not tested; 1, resistance; 2, intermediate susceptibility; 3, susceptibility to the drug. The order of the numbers indicates which antibiotic was tested. Bacterial serotyping results are recorded without coding. Virus culture results and serologic titers for viral and other infectious agents are recorded in a similar manner. The prepared paper tape then serves as the means of entering data for batch processing by the computer. A hard-copy (printed) final report for each sample submitted to the microbiology laboratory is printed by the computer. These reports contain no coded information and are placed in the patient's permanent medical records. A complete listing of data transmitted from the laboratory on each occasion is kept on file for future reference. Individual coded, fixed-format data records are produced by the computer simultaneously with the final hard-copy reports. These data records serve as the means of future data processing (storage and retrieval). The system also provides error detection and data screening capabilities by listing records containing errors and unusual or noteworthy results for the microbiologist.

In this system, all data are recorded in a cumulative microbiology data bank, and retrieval may be performed by multiple parameters (e.g., specific culture type, specified source and/or result codes, various combinations of genus and species codes, antibiotic susceptibility test results, specific susceptibility pattern, positive results, etc.). The different types of restrictions may be combined in any single retrieval operation. Thus, the retrieval system provides a means of rapidly reviewing and summarizing large volumes of microbiology data for clinical laboratory and research purposes.

Neblett[124] described a computerized national microbiological data retrieval and analysis system which is an antimicrobial susceptibility and nosocomial infection reporting service. Data on bacterial isolates from participating laboratories are reported on dual mark-sense forms daily or as completed, approximately by the 10th day of the following month. The data are tabulated ten ways (including trend reports and quarterly plus annual summaries) and are returned as meaningful reports to the laboratory's medical staff, infectious surveillance personnel, and interested administrators. The antibiotic sensitivity (Kirby-Bauer method) report includes both local hospital and national data expressed as the percentage sensitive to a particular drug, with the total number of an isolate given along with the number tested against the drug. New drugs are added to the sensitivity report periodically as they become available. Number and percentage of organisms are reported by department and body site. For each hospital department designated, the uppermost number tabulated opposite an organism is the number found during the previous month. Total isolates for the month are tabulated at the end of each departmental column, and the percentage opposite each name is its fraction of that department's total isolates. This provides clinical services with information about dominant organisms from their patients. The same type of tabulation is supplied for body site origin of isolates, providing laboratory and infectious disease personnel with information about dominant isolates from each source and its frequency of occurrence.

The BAC-DATA system reported by Neblett[124] is available to hospitals throughout the United States by subscription. Presently the approximate cost is $0.25 per patient medical bed.

Bengtsson et al.[125] developed a data system for bacteriological routine and research. This system uses punch cards for all patient data contained in the request form and uses optimal mark sheets for registration of the bacteriological diagnosis, antibiotic sensitivity pattern, phage type, etc. The data system (BACTLAB) delivers daily lists of all examinations performed and produces bimonthly lists containing all findings of interesting bacteria (such as *Staphylococcus aureus*) found in selected wards in the university hospital during the past month. The BACTLAB system is also used in the study of the causes of postoperative infections in a recently opened operating ward. In this research program, samples are taken from patients operated on in certain theaters and from the staff taking part in the operation. The bacterial content of the air in the theater is followed by means of settle plates. Data concerning the patient (e.g., current infections, antibiotic therapy, predisposing factors such as diabetes and malignancy) and his operation (e.g., type of operation and anethesia, occurrence of drains, vena cava infusion, tracheostoma), as well as information on any postoperative infection, are also collected.

Information flow for the data system used for epidemiological research is shown in Figure 6.5. The results are read in an IBM 130 computer equipped with a 1231 optical mark-page reader. The program language used is FORTRAN and 1130 Assembler. The information on the bacteriological examination is related to the data on the patient and his operation in order to produce a condensed version of the treatment period for future statistical analysis. The data (which are also stored on magnetic tape) can be used for a further analysis (using IBM 370/155 computers) of factors that may influence the postoperative infection rate; this may help to reduce the number of postoperative infections (presently occurring at a rate of approximately 5% of all operations).

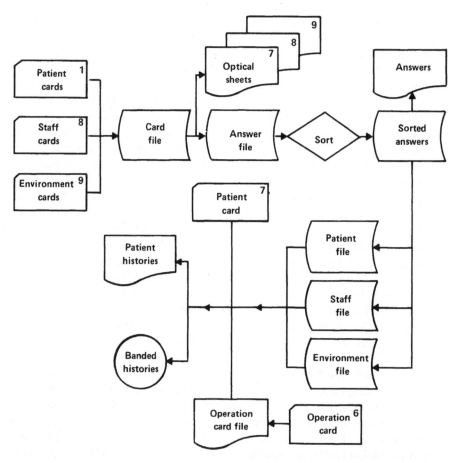

FIGURE 6.5.   Information flow for the data system used for epidemiological research. (From Bengtsson, S., Bergqvist, F.-O., and Schneider, W., in *New Approaches to the Identification of Microorganisms,* Hedén, C. G. and Illéni, T., Eds., John Wiley & Sons, New York, 1975. With permission.)

REFERENCES

1. Beers, R. J. and Lockhart, W. R., Experimental methods in computer taxonomy, *J. Gen. Microbiol.*, 28, 633, 1962.
2. Policy Directive, NPD 8020.7, NASA, Washington, D.C., 1963.
3. Policy Directive, NPD 8020.8A, NASA, Washington, D.C., 1967.
4. Dillon, R. T., Holdridge, D., Puleo, J. R., and Oxborrow, G. S., A computerized bacterial identification system as applied to planetary quarantine, *Space Life Sci.*, 3, 63, 1971.
5. Lapage, S. P., Bascomb, S., Willcox, W. R., and Curtis, M. A., Computer identification of bacteria, in *Automation, Mechanization and Data Handling in Microbiology*, Society for Applied Bacteriology Tech. Ser. No. 4, Baillie, A. M. and Gilbert, R. J., Eds., Academic Press, London, 1970, 1.
6. Lapage, S. P., Bascomb, S., Willcox, W. R., and Curtis, M. A., Identification of bacteria by computer: general aspects and perspectives, *J. Gen. Microbiol.*, 77, 273, 1973.
7. Whitby, J. L. and Blair, J., Data processing in hospital bacteriology: experience of 18 months trial, *J. Clin. Pathol.*, 25, 338, 1972.
8. Strumfjord, J. V., Spayberry, M. N., Biggs, H. G., and Notto, T. A., Electronic data processing system for clinical laboratories, *Am. J. Clin. Pathol.*, 47, 661, 1967.
9. Stirland, R. M., Hillier, V. F., and Steyger, M. G., Analysis of hospital bacteriological data, *J. Clin. Pathol.*, 3 (Suppl. 22), 82, 1969.
10. Alexander, M. K., Conningale, J., Johnston, T., Poulter, I. R., and Wakefield, J., A data processing system for hospital bacteriology, *J. Clin. Pathol.*, 23, 77, 1970.
11. Harvey, K. J., Were, M., Heys, W., and Smith, D. D., A time-shared computer system for data processing in bacteriology, *Med. J. Aust.*, 2, 1076, 1972.
12. Lindberg, D. A. B., Symposium on information science. VII. Electronic reporting, processing and retrieval of clinical laboratory data, *Bacteriol. Rev.*, 29, 554, 1965.
13. Vermeulen, G. D., Gerster, J. W., Young, V. M., and Hsieh, R. K. C., A computerized data storage and retrieval system for clinical microbiology, *J. Clin. Pathol.*, 61, 209, 1974.
14. Bascomb, S., Lapage, S. P., Curtis, M. A., and Willcox, W. R., Identification of bacteria by computer: identification of reference strains, *J. Gen. Microbiol.*, 77, 291, 1973.
15. Sneath, P. H. A., Computer taxonomy, in *Methods in Microbiology*, Vol. 7A, Norris, J. R. and Ribbons, D. W., Eds., Academic Press, London, 1972, 29.
16. Dybowski, W. and Franklin, D. A., Conditional probability and the identification of bacteria: a pilot study, *J. Gen. Microbiol.*, 54, 215, 1968.
17. Gyllenberg, H. G., A model for computer identification of micro-organisms, *J. Gen. Microbiol.*, 39, 401, 1965.
18. Rypka, E. W., Clapper, W. E., Bowen, I. G., and Babb, R., A model for the identification of bacteria, *J. Gen. Microbiol.*, 46, 407, 1967.
19. Rypka, E. W. and Babb, R., Automatic construction and use of an identification scheme, *Med. Res. Eng.*, 9, 9, 1970.
20. Gyllenberg, H. G. and Niemelä, T. K., Basic principles in computer-assisted identification of microorganisms, in *New Approaches to the Identification of Microorganisms*, Hedén, C. G. and Illéni, T., Eds., John Wiley & Sons, New York, 1975, 201.
21. Quadling, C. and Colwell, R. R., The use of numerical methods in characterizing unknown isolates, *Dev. Ind. Microbiol.*, 5, 151, 1964.
22. Juffs, H. S., Identification of *Pseudomonas* spp. isolated from milk produced in South Eastern Queensland, *J. Appl. Bacteriol.*, 36, 585, 1973.
23. Sneath, P. H. A., Some thoughts on bacterial classification, *J. Gen. Microbiol.*, 17, 184, 1957.
24. Goodman, Y. E., Bacterial identification: clues from computers and chemistry, *Can. J. Med. Technol.*, 34, 2, 1972.
25. Carmichael, J. W. and Sneath, P. H. A., Taxometric maps, *Syst. Zool.*, 18, 402, 1969.
26. Ewing, W. H. and Edwards, P. R., The principal divisions of *Enterobacteriaceae* and their differentiation, *Int. Bull. Bacteriol. Nomencl. Taxon.*, 10, 1, 1960.
27. Schultze, W. D. and Olson, J. C., Jr., Studies on psychrophilic bacteria. I. Distribution in stored commercial products, *J. Dairy Sci.*, 43, 346, 1960.
28. Witter, L. D., Psychrophilic bacteria – a review, *J. Dairy Sci.*, 44, 983, 1961.
29. Sandvik, O. and Fossum, K., Accumulation of bacterial proteinases during storage of milk due to the selection of psychrophilic bacteria, *Meieriposten*, 52, 639, 1963; quoted in *Dairy Sci. Abstr.*, No. 3519, 25, 1963.
30. Von Bockelmann, I., The composition of the thermoresistant flora in cold stored milk, *18th Int. Dairy Congr. Brief Commun.*, 1E, 105, 1970.
31. Von Bockelmann, I., Lipolytic and psychrotrophic bacteria in cold stored milk, *18th Int. Dairy Congr. Brief Commun.*, 1E, 106, 1970.
32. Kiuru, K., Edlund, E., Gyllenberg, H., and Antila, M., Psychrotrophic microorganisms in farm tank milk and their proteolytic activity, *18th Int. Dairy Congr. Brief Commun.*, 1E, 108, 1970.
33. Samagh, B. S. and Cunningham, J. D., Numerical taxonomy of the genus *Pseudomonas* from milk and milk products, *J. Dairy Sci.*, 55, 19, 1972.

34. Lightbody, L. G., Media containing penicillin for detecting post-pasteurization contamination, *Aust. J. Dairy Technol.*, 20, 24, 1965.

35. Lightbody, L. G., Inhibition of growth of butter cultures in naturally ripened cream after neutralization, *Queensl. J. Agric. Sci.*, 19, 249, 1962.

36. Lightbody, L. G. Ripening of raw farm cream in Queensland, *Queensl. J. Agric. Anim. Sci.*, 23, 57, 1966.

37. Kovacs, N., Identification of *Pseudomonas pyocyanea* by the oxidase reaction, *Nature*, 178, 703, 1956.

38. Park, R. W. A. and Holding, A. J., Identification of some common Gram-negative bacteria, *Lab. Pract.*, 15, 1124, 1966.

39. Hugh, R. and Leifson, E., The taxonomic significance of fermentative versus oxidative metabolism of carbohydrates by various Gram-negative bacteria, *J. Bacteriol.*, 66, 24, 1953.

40. Thornley, M. J., The differentiation of *Pseudomonas* from other Gram-negative bacteria on the basis of arginine metabolism, *J. Appl. Bacteriol.*, 23, 37, 1960.

41. King, E. O., Ward, M. K., and Raney, D. E., Two simple media for the demonstration of pyocyanin and fluorescein, *J. Lab. Clin. Med.*, 44, 301, 1954.

42. Jones, A. and Richards, T., Night Blue and Victoria Blue as indicators in lipolysis media, *Proc. Soc. Appl. Bacteriol.*, 15, 82, 1952.

43. British Standards Institution, Methods of Microbiological Examination for Dairy Purposes, British Standard 4285, 1968.

44. Conn, H., Ed., *Manual of Microbiological Methods: Society of American Bacteriologists Committee on Bacteriological Technique*, McGraw Hill, New York, 1957.

45. Stanier, R. Y., Palleroni, N. J., and Doudoroff, M., The aerobic pseudomonads: a taxonomic study, *J. Gen. Microbiol.*, 43, 159, 1966.

46. Sokal, R. R. and Sneath, P. H. A., *Principles of Numerical Taxonomy*, W. H. Freeman, San Francisco, 1963.

47. Szabo, E., Comparative Study of Numerical Taxonomic Clustering Methods, Ph.D. thesis, University of Queensland, 1970.

48. Sneath, P. H. A., The application of computers to taxonomy, *J. Gen. Microbiol.*, 17, 201, 1957.

49. Johns, P. A. and Tischer, R. G., Characterization of *Pseudomonas* species for identification in the clinical laboratory, *Am. J. Med. Technol.*, 39, 495, 1973.

50. Maccacaro, G. A., La misura della informazione contenuta nei criteri di classificazione, *Ann. Microbiol. Enzimol.*, 8, 231, 1958.

51. Hill, L. R. and Silvestri, L. G., Quantitative methods in the systematics of Actinomycetales. III. The taxonomic significance of physiological-biochemical characters and the construction of a diagnostic key, *G. Microbiol.*, 10, 1, 1962.

52. Möeller, F., Quantitative methods in the systematics of Actinomycetales. IV. The theory and application of a probabilistic identification key, *G. Microbiol.*, 10, 29, 1962.

53. Gyllenberg, H., A general method for deriving determination schemes for random collections of microbial isolates, *Ann. Acad. Sci. Fenn. Ser. A4*, 69, 1, 1963.

54. Niemelä, S. I., Hopkins, J. W., and Quadling, C., Selecting an economical binary test battery for a set of microbial cultures, *Can. J. Microbiol.*, 14, 271, 1968.

55. Payne, L. C., Towards medical automation, *World Med. Electron.*, 2, 6, 1963.

56. Sneath, P. H. A., Computers in bacteriology, *J. Clin. Pathol.*, 3 (Suppl. 22), 87, 1969.

57. Dito, W. R., Personal communication of desk-top computer applications in the clinical laboratory, presented as Application of Programmable Calculators to Microbiology, Commission of Continuing Education Workshops, American Society of Clinical Pathology, Spring and Fall Meetings, 1970 and 1971.

58. Fey, H., Differenzierungsscheme für gramnegative aerobe Stäbchen, *Schweiz. Z. Allg. Pathol. Bakteriol.*, 22, 641, 1959.

59. Baer, H. and Washington, L., Numerical diagnostic key for the identification of *Enterobacteriaceae*, *Appl. Microbiol.*, 23, 108, 1972.

60. Willcox, W. R., Lapage, S. P., Bascomb, S., and Curtis, M. A., Identification of bacteria by computer: theory and programming, *J. Gen. Microbiol.*, 77, 317, 1973.

61. Curtis, M. A., Lapage, S. P., Bascomb, S., and Willcox, W. R., Computer identification of Gram-negative rods, in *Automation, Mechanization and Data Handling in Microbiology*, Society for Applied Bacteriology Tech. Ser. No. 4, Baillie, A. M. and Gilbert, R. J., Eds., Academic Press, London, 1970, 96.

62. Friedman, R. B., Bruce, D., MacLowry, J., and Brenner, V., Computer-assisted identification of bacteria, *Am. J. Clin. Pathol.*, 60, 395, 1973.

63. Fisher, R. A., *Contributions to Mathematical Statistics*, John Wiley & Sons, New York, 1950.

64. Pratt, J. W., Raiffa, H., and Schlaifer, R., *Introduction to Statistical Decision Theory*, McGraw Hill, New York, 1965, 10.1.

65. Friedman, R. and MacLowry, J., Computer identification of bacteria on the basis of their antibiotic susceptibility patterns, *Appl. Microbiol.*, 26, 314, 1973.

66. MacLowry, J. D., Jaqua, M. J., and Selepak, S. T., Detailed methodology and implementation of a semi-automated serial dilution micro-technique for antimicrobial susceptibility testing, *Appl. Microbiol.*, 21, 46, 1970.

67. Taylor, G. R., Guthrie, R. K., and Shirling, E. B., Serological characteristics of *Streptomyces* species using cell wall immunizing antigens, *Can. J. Microbiol.*, 16, 107, 1970.

68. Gottlieb, D., An evaluation of criteria and procedures used in the description and characterization of the streptomycetes. A cooperative study, *Appl. Microbiol.*, 9, 55, 1961.

69. Sneath, P. H. A. and Johnson, R., The influence on numerical taxonomic similarities of errors in microbiological tests, *J. Gen. Microbiol.*, 72, 377, 1972.

70. Washington, J. A., Yu, P. K. W., and Martin, W. J., Evaluation of accuracy of multitest micromethod system for identification of Enterobacteriaceae, *Appl. Microbiol.*, 22, 267, 1971.

71. Spencer, F. and Hyde, T. A., An approach to microbiologic diagnosis and matrix problems for the small hospital laboratory using a small computer, *Am. J. Clin. Pathol.*, 60, 264, 1973.

72. Dominque, G. J., Dean, F., and Miller, J. R., A diagnostic scheme for identifying Enterobacteriaceae and miscellaneous gram negative bacilli, *J. Clin. Pathol.*, 51, 62, 1969.

73. Clark, F. E., The generic classification of the soil corynebacteria, *Int. Bull. Bacteriol. Nomencl. Taxon.*, 2, 45, 1952.

74. Jensen, H. L., The coryneform bacteria, *Annu. Rev. Microbiol.*, 6, 77, 1952.

75. Jensen, H. L., The genus *Nocardia* (or *Proactinomyces*) and its separation from other *Actinomycetales*, with some reflections on the phylogeny of the actinomycetes, in Proc. 6th Int. Congr. Microbiology, Vol. 1, Rome, 1953, 69.

76. Jensen, H. L., Some introductory remarks on the coryneform bacteria, *J. Appl. Bacteriol.*, 29, 13, 1966.

77. Gibson, T., The taxonomy of the genus *Corynebacterium*, Proc. 6th Int. Congr. Microbiology, Vol. 1, Rome, 1953, 16.

78. Bousfield, I. J., A Taxonomic Study of Some Aerobic Coryneform Bacteria, Ph.D. thesis, University of Aberdeen, 1969.

79. Veldkamp, H., Saprophytic coryneform bacteria, *Annu. Rev. Microbiol.*, 24, 209, 1970.

80. Da Silva, G. A. N. and Holt, J. G., Numerical taxonomy of certain coryneform bacteria, *J. Bacteriol.*, 90, 921, 1965.

81. Chatelain, R. and Second, L., Taxonomic numerique de quelques *Brevibacterium*, *Ann. Inst. Pasteur* (Paris), 3, 630, 1966.

82. Harrington, B. J., A numerical taxonomical study of some corynebacteria and related organisms, *J. Gen. Microbiol.*, 45, 31, 1966.

83. Mullakhanbhai, M. F. and Bhat, J. V., A numerical taxonomical study of *Arthrobacter*, *Curr. Sci.*, 36, 115, 1967.

84. Splitstoesser, D. F., Wexler, M., White, J., and Colwell, R. R., Numerical taxonomy of Gram-positive and catalase-positive rods isolated from frozen vegetables, *Appl. Microbiol.*, 15, 158, 1967.

85. Davis, G. H. G., Fomin, L., Wilson, E., and Newton, K. G., Numerical taxonomy of *Listeria*, streptococci, and possibly related bacteria, *J. Gen. Microbiol.*, 57, 333, 1969.

86. Davis, G. H. G. and Newton, K. G., Numerical taxonomy of some named coryneform bacteria, *J. Gen. Microbiol.*, 56, 195, 1969.

87. Masuo, E. and Nakagawa, T., Numerical classification of bacteria. II. Computer analysis of coryneform bacteria: comparison of group-formations obtained on two different methods of scoring data, *Agric. Biol. Chem.*, 33, 1124, 1969.

88. Skyring, G. W. and Quadling, C., Soil bacteria: principal component analysis and guanine-cytosine contents of some arthrobacter-coryneform soil isolates and of some named cultures, *Can. J. Microbiol.*, 16, 95, 1970.

89. Keddie, R. M., Leask, B. G. S., and Grainger, J. M., A comparison of coryneform bacteria from soil and herbage: cell wall composition and nutrition, *J. Appl. Bacteriol.*, 29, 17, 1966.

90. Robinson, K., A Study of Coryneform Bacteria with Particular Reference to the Genus Microbacterium, Ph.D. thesis, University of Aberdeen, 1966.

91. Robinson, K., Some observations on the taxonomy of the genus *Microbacterium*. I. Cultural and physiological reactions and heat resistance, *J. Appl. Bacteriol.*, 29, 607, 1966.

92. Robinson, K., Some observations on the taxonomy of the genus *Microbacterium*. II. Cell wall analysis, gel electrophoresis and serology, *J. Appl. Bacteriol.*, 29, 616, 1966.

93. Robinson, K., An examination of *Corynebacterium* species by gel electrophoresis, *J. Appl. Bacteriol.*, 29, 179, 1966.

94. Abe, S., Takayama, K., and Kinoshita, S., Taxonomical studies on glutamic acid-producing bacteria, *J. Gen. Appl. Microbiol.*, 13, 279, 1967.

95. Komagata, K., Yamada, K., and Ogawa, H., Taxonomic studies on coryneform bacteria. I. Division of bacterial cells, *J. Gen. Appl. Microbiol.*, 15, 243, 1969.

96. Yamada, K. and Komagata, K., Taxonomic studies on coryneform bacteria. II. Principal amino acids in the cell wall and their taxonomic significance, *J. Gen. Appl. Microbiol.*, 16, 103, 1970.

97. Yamada, K. and Komagata, K., Taxonomic studies on coryneform bacteria. III. DNA base composition of coryneform bacteria, *J. Gen. Appl. Microbiol.*, 16, 215, 1970.

98. Bouisset, L., Breuillard, J., and Michel, G., Etude de l'ADN chez les *Actinomycetales*, *Ann. Inst. Pasteur* (Paris), 104, 756, 1963.

99. Yamada, K. and Komagata, K., Taxonomic studies on coryneform bacteria, *Int. Conf. of Culture Collection Tokyo* (Abstr.), University of Tokyo Press, Tokyo, 1968, 23.

100. Bousfield, I. J., A Taxonomic Study of Some Aerobic Coryneform Bacteria, Ph.D. thesis, University of Aberdeen, 1969.
101. Sneath, P. H. A., New approaches to bacterial taxonomy: Use of computers, *Annu. Rev. Microbiol.*, 18, 335, 1964.
102. Ryan, K. J. and Paplanus, S. H., Computer applications in hospital epidemiology, in *Microbiology — 1975*, Schlessinger, D., Ed., American Society for Microbiology, Washington, D.C., 1975, 76.
103. Paplanus, S. H., Shepard, R. H., and Zargulis, J. E., A computer based system for autopsy diagnosis storage and retrieval without numerical coding, *Lab. Invest.*, 20, 139, 1969.
104. Bolano, C. R., New tool to help control infection: the computer, *Mod. Hosp.*, 121, 89, 1973.
105. Greenhalgh, P. J., An automated system for infection control, *Off. J. Assoc. Oper. Room Nurses*, 19, 77, 1974.
106. Porter, K. W., Computerized infection surveillance, *Hospitals*, 48, 91, 1974.
107. Meers, P. D., The bacteriological examination of urine: a computer-aided study, *J. Hyg.*, 72, 229, 1974.
108. Loudon, I. S. L. and Greenhalgh, G. P., Urinary tract infections in general practice, *Lancet*, 2, 1246, 1962.
109. Mond, N. C., Percival, A., Williams, J. D., and Brumfitt, W., Presentation, diagnosis, and treatment of urinary-tract infections in general practice, *Lancet*, 1, 514, 1965.
110. Brooks, D. and Maudar, A., Pathogenesis of the urethral syndrome in women and its diagnosis in general practice, *Lancet*, 2, 893, 1972.
111. Ellner, P. D., Considerations in computer data entry, retrieval, and error checking, in *Microbiology — 1975*, Schlessinger, D., Ed., American Society for Microbiology, Washington, D.C., 1975, 73.
112. Beaulieu, M., A method for compiling bacterial findings in a medium sized hospital, *Can. Med. Assoc. J.*, 90, 1108, 1964.
113. Schnierson, S. S. and Amsterdam, D., A manual punch system for recording, filing and analyzing antibiotic sensitivity test results, *Am. J. Clin. Pathol.*, 47, 818, 1967.
114. Stirland, R. M., Data processing in clinical pathology. Report of a working party of the Association of Clinical Pathologists, *J. Clin. Pathol.*, 21, 256, 1968.
115. Gabrieli, E. R., Pessin, V., Thorpe, J., and Palmer, R. C. C., Initial experience with and potential of data processing and computer techniques in a hospital clinical laboratory, *Am. J. Clin. Pathol.*, 47, 60, 1967.
116. Whitby, J. L. and Blair, J. N., A computer-linked data processing system for routine hospital bacteriology, in *Automation, Mechanization and Data Handling in Microbiology*, Society of Applied Bacteriology Tech. Ser. No. 4, Baillie, A. M. and Gilbert, R. J., Eds., Academic Press, London, 1970, 23.
117. Dixon, W. J., *Biomedical Computer Programs*, University of California Press, Berkeley, 1970.
118. Cheatle, E. L., Computer tabulates bacteriology reports, *Hospitals*, 38, 91, 1964.
119. Pribor, H. C. and Hdyt, R. S., Can you use a reference laboratory effectively?, *Lab. Manage.*, 10, 16, 1971.
120. Schneierson, S. S. and Amsterdam, D., A punch card system for identification of bacteria, *Am. J. Clin. Pathol.*, 42, 328, 1964.
121. Percival, A. K. and Nunn, D. E., A cumulative reporting system for a bacteriology department, *Med. J. Aust.*, 1, 1047, 1970.
122. Petralli, J., Russell, E., Kataoka, A., and Merigan, T. C., On-line computer quality control of antibiotic sensitivity testing, *N. Engl. J. Med.*, 283, 735, 1970.
123. Vermeulen, G. D., Schwab, S. V., Young, V. M., and Hsieh, R. K. C., A computerized system for clinical microbiology, *Am. J. Clin. Pathol.*, 57, 413, 1972.
124. Neblett, T. R., Experiences within a computerized national microbiological data retrieval and analysis system, in *New Approaches to the Identification of Microorganisms*, Hedén, C. G. and Illéni, T., Eds., John Wiley & Sons, New York, 1975, 275.
125. Bengtsson, S., Bergqvist, F.-O., and Schneider, W., A data system for bacteriological routine and research, in *New Approaches to the Identification of Microorganisms*, Hedén, C. G. and Illéni, T., Eds., John Wiley & Sons, 1975, 291.

INDEX

A

D

E

fluorescent spectrophotometry, 124
GC analysis, 163–165
glutamate tube test, 93
growth media, 90
in water, 66–67, 122
India ink test, 123
membrane filtration test, 94, 187
radioactive antibody technique, 123
radiometric detection, 187
ultraviolet spectroscopy, 88
Esculin, 100
Ethanol, 14
Ethylenediaminetetraacetic acid (EDTA), 19
Ethylhydrocupreine hydrochloride (Optochin) test, 21, 34
Etiologic agents, identification, 168–170
Eubacterium
capsular material, 28–34
differentiation, 28–34, 50
identification, 10
structure, 5
Eubacterium alactoliticum
fatty acid analysis, 163
Eubacterium ruminantium
butyl ester analysis, 163
Eucaryotes
differentiation, 1–2
Ewing-Johnson modification, 19
Ewing's classification, table, 105
Exotoxins, 24

F

FA test, see Fluorescent antibody (FA) test
Facultatively aerobic bacteria, see Aerobic and facultatively aerobic bacteria
Farrell's medium, 69
FAST, see Fluorescent antibody staining technique
Fecal coliforms, see Coliforms
Fecal streptococci
differentiation, 27
in water, 66
isolation, 100–101
Fermentation acid, table, 97
Fermentation of sugars, 18–19, 107–110
Fermentation tests, 18–19, 36
Ferric chloride test, 20
FID, see Flame ionization detector
Field ionization mass spectrometry (FI-MS), 152–153
Filamatic® filler, 180
Filament vial filler, 180
Filtration membranes, 92–94
FI-MS, see Field ionization mass spectrometry
Fingerprint (GC profile) analysis, 158–159, 161, 166–168, 192–193
figures, 167–171, 193
Firefly bioluminescent reaction, 98–99
Firefly tails, 98–99
FITC (fluorescein isothiocyanate), 119
Flagella, 15
Flame ionization detector (FID), 67, 144, 161, 163, 166
Flame photometers, 144

Flavobacterium
characteristics, 44
table, 47
differentiation, 44
table, 47
Flocculation tests, see Precipitation tests
Flow chart of computer identification of bacteria, 221
Fluorescein isothiocynate, 91–92
Fluorescein isothiocyanate staining, 119, 202
Fluorescence, 100
Fluorescent antibody (FA) test, 34–35, 38–39, 46, 113–115, 197
tables, 115, 117–118, 121
Fluorescent antibody staining technique (FAST), 9, 113–115
figures, 199, 200, 202
Fluorescent autoantibody technique
automated, 197–198
figures, 199–202
table, 198
Fluorescent dyes, 14
Fluorescent microscopy, 35, 90–91
Fluorescent spectrophotometry, 123–124
Fluorescent treponemal antibody (ABS) test, 114
Fluorescent treponemal antibody absorption (FTA-ABS) test, 63, 196
figure, 197
Fluorimetric measurement, 112
Fluoro-Kits®, 197
Flying spot scanner for fluorescent antibody scanning, figure, 201
Food bacteriology
enrichment serology, 130
fecal streptococci, 100
fluorescent testing, 114–115
methods of testing, 68–71
table, 71
microbial contamination, 92–94
miniaturized microbiological techniques, 88
radiochemical analysis, 99–100
Salmonella, 101, 114–115
table, 115
Food poisoning, 27, 50, 66, 71–72
Foot-operated diluter/dispenser, 88
Formalinized saline solution, 28
FORTRAN, 217, 225, 228
FORTRAN V, 225
Four-tube R-B system, 103
Fractionation, 85
Fragmentation patterns of chemical constituents, 98
Frambesia, 63
Francisella
differentiation, 42
Francisella (Pasteurella) tularensis
biochemical characteristics, 42–44
differentiation, 42
Frankfurter processing
salmonellae in, 71
Freeze dried colonies, preparation, 87
Fretted glass (Biemann-Watson) separator, 153
Friedlander's bacilli, 23
Frost "little" plate method, 92

staining characteristics, 14
Vibrio comma
  pathogenicity, 44
Vibrio fetus
  pathogenicity, 44
Vibrionaceae
  characteristics, table, 6
Violet red-bile agar, 102
Viridans streptococci, 27
Virulence studies in animals, 23–24
Viruses
  immunoelectrophoresis tests, 124
Vitatron recorder, 184
Vitatron spectrophotometer, 184
Voges-Proskauer test, 21, 106
VPI schema of fatty acid profiles, 95

W

Wasserman test, 56–63, 128, 195, 196
Water
  bacteria in, 66–67
  chlorinated, 93–94
  growth media, 90
  immunofluorescence (IF) tests, 122
  membrane filtration tests, 93
  unchlorinated, 93–94
Wheat germ, 92
White soft paraffin, 94
Wiel-Felix test, 65
Wilcox method (probability test), 218
Willis and Hobbs medium, 22
Whooping cough, 42
Wolf Trap®, 186, 202
Wound infections, 50

X

X Factors, 42
XE-60, 143
XF-1150, 143

Y

Yaws, 56, 63
Yeasts
  Bactec detection, 187
Yersinia
  characteristics, table, 6
  differentiation, 42
    table, 46
Yersinia enterocolitica
  Analytab test, table, 106
  API system test, 105
  differentiation, 42–44
Yersinia (Pasteurella) pestis
  airborne pathogenicity, 67
Yersinia pestis
  differentiation, 42–44
Yersinia pseudotuberculosis
  differentiation, 42–44

Z

Zeiss halogen immunofluorescent microscope, 197
Ziehl-Neelsen carbolfuchsin stain, 14
Ziehl-Neelsen technique, 17, 90, 91
Zymomonas
  polyacrylamide gel electrophoresis, 193

TITLES OF INTEREST

CRC PRESS HANDBOOKS AND MANUALS:

CRC HANDBOOK OF BIOCHEMISTRY AND MOLECULAR BIOLOGY, 3rd Edition
Edited by **Gerald D. Fasman, Ph.D.,** Brandeis University.

CRC HANDBOOK OF CHEMISTRY AND PHYSICS, 59th Edition
Edited by **Robert C. Weast, Ph.D.,** Consolidated Natural Gas Co., Inc.

CRC HANDBOOK SERIES IN CLINICAL LABORATORY SCIENCE
Editor-in-Chief, **David Seligson, Sc.D., M.D.,** Yale University School of Medicine.

CRC HANDBOOK OF MICROBIOLOGY
Edited by **Allen I. Laskin, Ph.D.,** Exxon Research and Engineering Co., and **Hubert Lechevalier, Ph.D.,** Rutgers University.

CRC PRESS "UNISCIENCE"^{TM} TITLES:

CHEMOPROPHYLAXIS AND VIRUS INFECTIONS OF THE RESPIRATORY TRACT
Edited by **J. S. Oxford, Ph.D.,** National Institute for Biological Standards and Control, London, England.

CHEMOTHERAPY OF INFECTIOUS DISEASE
By **Hans H. Gadebusch, Ph.D.,** Squibb Institute for Medical Research.

DEVELOPMENTS IN LYMPHOID CELL BIOLOGY
Edited by **A. Arthur Gottlieb, M.D.,** Rutgers University.

CRC CRITICAL REVIEWS^{TM} JOURNALS:

CRC CRITICAL REVIEWS^{TM} IN BIOCHEMISTRY
Edited by **Gerald D. Fasman, Ph.D.,** Brandeis University.

CRC CRITICAL REVIEWS^{TM} IN CLINICAL LABORATORY SCIENCES
Edited by **John Batsakis, M.D.,** University of Michigan Medical School, and **John Savory, Ph.D.,** University of Virginia Medical Center.

CRC CRITICAL REVIEWS^{TM} IN MICROBIOLOGY
Edited by **Henry D. Isenberg,** Long Island Jewish-Hillside Medical Center.

Direct all inquiries to CRC Press, Inc.

T - #0644 - 101024 - C0 - 254/178/15 - PB - 9781138560857 - Gloss Lamination